KEY THEORETICAL FRAMEWORKS

KEY THEORETICAL FRAMEWORKS

Teaching Technical Communication in the Twenty-First Century

EDITED BY
ANGELA M. HAAS
MICHELLE F. EBLE

UTAH STATE UNIVERSITY PRESS
Logan

© 2018 by University Press of Colorado

Published by Utah State University Press
An imprint of University Press of Colorado
245 Century Circle, Suite 202
Louisville, Colorado 80027

All rights reserved
Printed in the United States of America

 The University Press of Colorado is a proud member of the Association of University Presses.

The University Press of Colorado is a cooperative publishing enterprise supported, in part, by Adams State University, Colorado State University, Fort Lewis College, Metropolitan State University of Denver, Regis University, University of Colorado, University of Northern Colorado, Utah State University, and Western State Colorado University.

∞ This paper meets the requirements of the ANSI/NISO Z39.48-1992 (Permanence of Paper)

ISBN: 978-1-60732-757-8 (paperback)
ISBN: 978-1-60732-758-5 (ebook)
DOI: https://doi.org/10.7330/9781607327585

The University Press of Colorado gratefully acknowledges the generous support of Eastern Carolina University toward the publication of this book.

Library of Congress Cataloging-in-Publication Data

Names: Haas, Angela M., editor. | Eble, Michelle F., 1974– editor.
Title: Key theoretical frameworks: Teaching technical communication in the 21st century / [edited by] Angela M. Haas, Michelle F. Eble.
Description: Logan : Utah State University Press, [2018] | Includes bibliographical references and index.
Identifiers: LCCN 2017045690 | ISBN 9781607327578 (pbk.) | ISBN 9781607327585 (ebook)
Subjects: LCSH: Communication of technical information—Study and teaching (Higher) | Technical writing—Study and teaching (Higher) | Communication of technical information—Moral and ethical aspects. | Technical writing—Moral and ethical aspects.
Classification: LCC T10.5 .K15 2018 | DDC 607.1—dc23
LC record available at https://lccn.loc.gov/2017045690

CONTENTS

List of Illustrations   vii
Foreword by Gerald Savage   ix
Acknowledgments   xiii

Introduction: The Social Justice Turn
   Angela M. Haas and Michelle F. Eble   3

PART I: EMBODIED KNOWLEDGE AND RISKS

1. Apparent Feminism and Risk Communication: Hazard, Outrage, Environment, and Embodiment
   Erin A. Frost   23

2. Validating the Consequences of a Social Justice Pedagogy: Explicit Values in Course-Based Grading Contracts
   Cruz Medina and Kenneth Walker   46

3. The University Required Accommodations Statement: What "Accommodation" Teaches Technical Communication Students and Educators
   Barbi Smyser-Fauble   68

PART II: SPACE, (EM)PLACE, AND DIS(PLACE)MENT

4. Spatial Orientations: Cultivating Critical Spatial Perspectives in Technical Communication Pedagogy
   Elise Verzosa Hurley   93

5. Indigenous Contexts, New Questions: Integrating Human Rights Perspectives in Technical Communication
   Godwin Y. Agboka   114

6. An Environmental Justice Paradigm for Technical Communication
   Donnie Johnson Sackey   138

PART III: INTERFACING PUBLIC AND COMMUNITY RHETORICS WITH TECHNICAL COMMUNICATION DISCOURSES

7   Stayin' on Our Grind: What Hiphop Pedagogies Offer to Technical Writing

　　Marcos Del Hierro　　*163*

8   Black Feminist Epistemology as a Framework for Community-Based Teaching

　　Kristen R. Moore　　*185*

9   Advocacy Engagement, Medical Rhetoric, and Expediency: Teaching Technical Communication in the Age of Altruism

　　Marie E. Moeller　　*212*

PART IV: ACCOMMODATING DIFFERENT DISCOURSES OF DIVERSITY

10 Using Narratives to Foster Critical Thinking about Diversity and Social Justice

　　Natasha N. Jones and Rebecca Walton　　*241*

11 Race and the Workplace: Toward a Critically Conscious Pedagogy

　　Jessica Edwards　　*268*

12 Shifting Grounds as the New Status Quo: Examining Queer Theoretical Approaches to Diversity and Taxonomy in the Technical Communication Classroom

　　Matthew Cox　　*287*

Afterword: From Accommodation to Transformation

　　J. Blake Scott　　*304*

About the Authors　　*313*
Index　　*317*

ILLUSTRATIONS

FIGURES

5.1. Nigeria government's memo to Shell on military crackdown on Ogoni residents to enable Shell to resume oil operations   127
8.1. Sample Activity for Tenets 1 & 2   200
8.2. Activity for Enacting Tenets 3 & 4   203
8.3. Black Feminist Approach to the Classroom   204
9.1. "One promise, two sisters"   229
9.2. Risk Factors and Prevention   230
9.3. Being Female   231
9.4. Barriers to screening   232
9.5. Lesbians and women who partner with women   233
10.1. Capacities of Narrative Suited to Facilitating Social Justice Pedagogy   244

TABLES

2.1. Grading Contract (Medina)   54
2.2. Rate My Professor Comments   57
2.3. Grading Contract (Walker): Collaborative Service-Learning Redesign Project   61
8.1. Enacting Tenets 1 & 2   199
8.2. Enacting Tenets 3 & 4   202
10.1. Heuristic for Using Narrative to Explore Social Justice and Diversity in the Technical Communication Classroom   260

FOREWORD

Gerald Savage

Cultural studies and social justice are terms that the technical communication field has been slow to take up, although both terms were invoked by a few scholars at least as long ago as the 1990s. Regarding the perspective of social justice, most teachers, scholars, and practitioners might substitute "ethics" as the preferable term—preferable because it does not seem to imply the risky domains of politics, ideology, or social activism frequently associated with social justice—despite Aristotle's observation in *On Rhetoric* that "Ethical studies may fairly be called political" (Aristotle 1954, 26). Nevertheless, I acknowledge the contemporary sense of ethics as a quality of individual character, as James Porter (1993) observed over twenty years ago. He said that "most conventional discussions of ethics ... have tended to focus not on the question of the social good so much as whether or not the individual is doing the right thing—that is, the focus has been personal salvation rather than social good" (142). Implicit in this convention is the assumption that ethics is a personal responsibility—again, to quote Porter, "thus promoting a sense of ethics as individual heroism" (130). Porter attempts to move us beyond that traditional notion because, among other reasons, it can pit one person's ethics against another's, which then becomes a question of who has the most power in the situation. Moreover, expecting the individual to act alone on the basis of a personal ethical standard effectively disempowers most people by forcing them to make a choice that could well destroy them professionally, as we have so often seen in the case of whistle-blowers over the years. Porter's solution is to move ethical decision-making into a more complex dynamic of contextual, collaborative decision-making on the basis of law and policy as well as ethics. This view may often be not only less risky for the individual but also more effective in some cases. It seems just as likely, however, that such cases will be decided on the basis of law and policy alone, leaving out ethics, which would remain in the realm of the personal and therefore a choice of "liking it or lumping it." Moreover, it has never seemed to me very likely to resolve broader issues that fall outside of, if not directly in conflict with, existing laws and policies—that is, those issues involving problems of social justice.

My own preference is to leave ethics in the category to which it is consigned in contemporary mainstream culture and to turn to social justice with all of its connotations of politics and ideology. I understand social justice to involve collective work on a large scale. For the most part, it can only be enacted on that scale. Social justice works to transform the social and cultural structures that have permitted injustice to exist, that have in fact made injustice invisible, or worse, have denied its existence.

Social justice work must begin by assembling a community of thinkers and actors who agree on the need for change. The work of such communities can appropriately involve disruptive actions such as demonstrations, strikes, marches, or sit-ins. But it can also, often more effectively, involve showing how: teaching, campaigning, studying, witnessing, and materially transforming the conditions that perpetuate injustice. Perhaps some of the contributors to the present book have acted disruptively in support of social justice, but what they do here is teach and show how to enact social justice in the curriculum, the classroom, and in the work of technical communication.

Although we might have wanted such a book six or seven years ago, I doubt it could have been done then. Only two or three of the contributors were even enrolled in their doctoral programs at that time. Although some published resources on social justice and cultural studies specific to technical communication were available, most of us teaching social justice and cultural studies perspectives in technical communication programs had to rely upon materials primarily from outside of the field. Some of the most stimulating thought of scholars within our field was being voiced in our national conferences, where new ideas are first tried out.

The idea for the collection erupted from a session on social justice—one of several—at the 2013 conference of the Council for Programs in Technical and Scientific Communication in Cincinnati. The names of about ten new, activist technical communication scholars were brought up in the panel presentations. During discussion after the presentations, several people asked for those names again, and suddenly the idea of an edited collection was in the air. Three years later, this collective brainstorm has transitioned into a manuscript, and most of the emerging scholars whose names were mentioned are contributors. Angela Haas and Michelle Eble have done an extraordinary job in gathering together this group of scholars to produce such an unprecedented and important collection. I attribute this accomplishment to the vision and organizing abilities of the two editors and to the industry and passion of the contributors.

But there is more to do. This is both a daunting and an inspiring prospect. If we are to naturalize the idea of technical communication as a field committed to social justice, I believe we need to adopt a stance that may seem alien to those who have worked in this field for many years. As this book makes clear, there is pedagogical work of social justice to do, which will draw hitherto unseen and unheard people into our classrooms and our professions. It is work that both exposes and undoes the complex and often not readily perceived ways in which technologies perpetuate and reinforce social, economic, and environmental injustices. It is action that is already natural to the emerging scholars speaking in this book, and I believe they are showing us the direction in which awaits some of the most important and rewarding work of our field for decades to come.

References

Aristotle. 1954. *The Rhetoric and the Poetics of Aristotle*. Translated by Ingram Bywater, translator of Poetics; W. Rhys Roberts, translator of Rhetoric. New York: Modern Library.

Porter, James E. 1993. "The Role of Law, Policy, and Ethics in Corporate Composing: Toward a Practical Ethics for Professional Writing." In *Professional Communication: The Social Perspective*, ed. Nancy Roundy Blyler and Charlotte Thralls, 128–43. Newbury Park, CA: Sage.

ACKNOWLEDGMENTS

We'd like to begin by acknowledging the intellectual work of Flourice Richardson, Marcos Del Heirro, and Jerry Savage, who were on the "Programmatic Perspectives on and Projections for Social Justice Curricula & Pedagogy" with Angela at the 2013 conference of the Council for Programs in Technical and Scientific Communication (CPTSC) in Cincinnati—as well as everyone in the audience with Michelle, including Katrina Dunbar and several contributors to this collection. The conversations in that place and space inspired this collection and continue to inform our work in important ways. A debt of gratitude goes to Jerry Savage for encouraging us to develop this project in the first place and to Blake Scott for his unwavering support—and to both for their foundational scholarship. As the citation evidence overwhelmingly reveals, their work has cleared the path for social justice scholarship and pedagogy offered in this collection. Many thanks also go to Jeff Grabill, who listened with empathy and generously advised us on finding our manuscript a welcoming home. With that said, we are grateful to Michael Spooner, Kylie Haggen, and their team at Utah State UP who thankfully believed in the importance of this collection. Your responsive and supportive feedback has been crucial to the success of this project. Thanks also to Laura Furney, Managing Editor; Beth Svinarich, Marketing Manager, and Anya Hawke, Copyeditor, for their keen professional skills in producing this project and getting it into print. We are also grateful to Daniel Pratt with Utah State UP for initiating and finalizing the book cover design and to Heather Noel Turner for using her visual design and rhetoric skills to help us imagine what we wanted through multiple iterations. We also appreciate the constructive feedback from our reviewers. A special shout-out to Reviewer #2 who provided especially useful and usable suggestions.

Most of all, we are honored and thankful that our contributors trusted us with their brilliant, groundbreaking, paradigm-shifting scholarship, curricula, and pedagogy. We are indebted to their collective patience, collaborative spirits, and embodied knowledges. Working with and learning from our contributors has been one of the most important intellectual and collaborative experiences of our careers. Their work inspires us to do and be better, and we feel confident that the future of the discipline and profession is in good hands. We will forever be grateful to have worked with these fine scholars, teachers, and human beings.

ANGELA:
First, I am thankful to my life partner, Keith Brotheridge, and our kitties, Bran & Arya, for providing me with the physical and emotional space and support to do this scholar-teacher-activist gig in the ways in which I do them. Their patience and love through this project has been critical to its completion. Next, I am grateful to the amazing, smart, hilarious, fierce Michelle Eble. The range of feelings and fun we have shared throughout this project has been vast and varied. I cannot imagine experiencing the unexpected scholarly and mentorly processes of this project with anyone else. I look forward to future collaborative decolonial trouble-making. Further, I'm deeply appreciative for the students enrolled in my undergraduate and graduate technical communication and rhetoric courses over the years. Their work and embodied experiences have shaped my thinking about technical communication theories, methodologies, pedagogies, and practices in critical ways and remind me of the important work technical communicators can and should do outside of academia. Finally, I am so appreciative to my colleagues Julie Jung and Elise Verzosa Hurley for their steadfast support and co-mentoring over the years and specific to this project—and to my Michigan State University (MSU) cohort family for the love and laughs we have shared over the years that continues to sustain me, and especially to Qwo-Li Driskill, Jim Ridolfo, and Jill Chrobak for their nurturing friendship through and helpful advice on this project.

MICHELLE:
The exigency for a collection like this came to mind as I participated in the interviewing of potential candidates for a couple of tenure-track positions in technical communication at East Carolina University (ECU). I met many of the contributors in this collection and learned of their scholarship and teaching practices through this process. And, as stated above, CPTSC 2013 absolutely solidified the need for such a collection. Because edited collections are the "cultural capital" in our field, we wanted to feature this incredibly innovative scholarship in one place, learn from these scholars, and make apparent this very important shift

in theories and pedagogies in technical communication. My heartfelt thanks to the work of the scholars featured in this collection and their commitment to doing this important work, especially when it's risky and disruptive to the status quo, and I also want to thank the numerous students at ECU that I've learned from over the last fifteen years. My deep appreciation goes to my smart colleagues and friends at ECU—Erin Frost, Matthew Cox, Nikki Caswell, and especially Will Banks, who provided mentoring, support, and encouragement for this project and all my pursuits, successes, and disappointments, and I would not at all be the scholar-teacher or person I am today without our conversations, laughter, happy hours, and trouble-making. There isn't enough room to express my gratitude to Angela Haas for her presence in my life, her mentoring, her friendship, and her dedication to making the world a better place one student at a time. Thanks, Angela, for all of your steadfast and consistent support in this project and beyond. We're certainly trouble, and we certainly know how to have fun! And last, a huge shout out to my partner, Shane Ernst, and Mara for their support, patience, and excitement in absolutely everything I do.

KEY THEORETICAL FRAMEWORKS

Introduction
THE SOCIAL JUSTICE TURN

Angela M. Haas and Michelle F. Eble

The practices of technical communication in the twenty-first century have become so diverse that we cannot possibly trace all the current industries in which technical communicators work, much less our contemporary workspaces, job titles, roles, and responsibilities. Globalization—and the complex and culturally-rich material and information flows that come with it—has forever changed who we think of as technical communicators, the work that technical communicators do, and thus where and how we understand technical communication happens. Alumni from Illinois State University and East Carolina University's technical and professional communication programs alone come from diverse cultural backgrounds and end up working around the world for major Fortune 500 companies, design firms, local factories, governmental agencies, grassroots/nonprofit organizations, hospitals, universities, publishing houses, and media outlets in a wide range of positions, including social media manager, user experience specialist, project manager, learning designer, professor, grant writer, community relations liaison, legal writer, information architect, journalist, translator, editor, and documentation specialist—just to name a few. Although some technical communicators today still work in traditional engineering and technical manual writing contexts, globalization has certainly influenced the potential places where this work transpires and audiences impacted by this work as well.

Because globalization is continuously broadening our understanding of who we are as technical communication scholars, practitioners, and pedagogues, we must systematically interrogate the relationships between globalization and technical communication. Globalization affects three critical spheres of technical communication's influence—technological, scientific, and cultural—and in highly complex ways. For example, on one hand, globalization has allowed us access to an unprecedented wealth of diverse material goods and technical and scientific information

DOI: 10.7330/9781607327585.c000

from diverse places across the globe, a sense of connectedness in terms of networks and networked people, communities, economies, products, and media outlets, as well as new digital spaces in and geographical places at which to work. On the other hand, access to global networks, what counts as global, and movement within and across global places and spaces has always already been facilitated and impeded by actors and rhetorics that legitimize inequitable rules and conditions informed by ideological assumptions about: ownership of land, technological and scientific resources, material goods, and information; what is understood to be technical and scientific and for what reasons; what can be bought, sold, and traded and by whom and for what reasons; who can travel and for what reasons; who is seen, unseen, re-seen as part of local and global networks, how so, by whom, and for what reasons—and much more.

Thus, while technical communicators may appreciate the international, professional, and economic gains afforded to us by globalization, we must also interrogate how we may be complicit in, implicated by, and/or transgress the oppressive colonial and capitalistic influences and effects of globalization. As Carolyn Rude (2009) reminds us, we have the potential to *both* "function as agents of knowledge making, action, and change" for some *and* function as agents of oppression—albeit often unwittingly—for others (183). As public intellectuals, knowledge workers, and advocates for users, technical communicators have a responsibility to advocate for equity in local and global networks of scientific, technical, and professional communication. To do so, technical communicators must be able to ascertain how these networks are constructed, by whom, toward what ends—as well as the stakeholders, power dynamics, distributed agency (distributed by whom/what; who/what benefits, is underserved, and disenfranchised within the network; in what ways), and the direction(s) of the material and information flows and within the network(s). Needless to say, we have a complicated relationship with globalization; thus, we have an obligation to critically assess that complexity.

This edited collection offers *social justice frameworks* that foster curricular and pedagogical approaches to this complex rhetorical and advocacy work. Social justice approaches to technical communication are often informed by cultural theories and methodologies, but they also explicitly seek to redistribute and reassemble—or otherwise redress—power imbalances that systematically and systemically disenfranchise some stakeholders while privileging others. Using cultural and rhetorical theories to redress social injustices, social justice approaches essentially and ideally couple rhetoric with action to actually make social, institutional, and organizational change toward equity happen. This collection, then, contributes to both the cultural studies turn (Scott 2003; Scott and Longo 2006; Scott, Longo, and Wills 2007) and what we suggest is the *social justice turn* in technical communication studies—or a turn toward a collective disciplinary redressing of social injustice sponsored by rhetorics and practices that infringe upon, neglect, withhold, and/or abolish human, non-human animal, and environmental rights. Ultimately, this collection imagines socially-just futures for our discipline, programs, and professions inspired by the work of emerging and established scholars and practitioners. Contributors from twelve different universities provide theoretical and curricular frameworks that support instructors teaching current and future technical communication practitioners how to be socially-just technical communicators and global citizens and to solve complex technical and scientific communication problems within diverse workplaces, work spaces, and organizational cultures by skillfully, ethically, and rhetorically negotiating contextual power dynamics. Using our privilege and skills as nimble, flexible, liminal, rhetorical, and ethical technical communicators, we can intervene in global and local technical communication problems at the macro and micro levels in the face of asymmetrical power relations and limited agency—and teach current and future practitioners to do the same.

Technical communication scholarship, practice, conferences, and pedagogies have rich histories of adapting with cultural, technological, and scientific changes. Over the past thirty years alone, our scholarship and practice has dramatically transitioned—with help from the work of the humanist, social, feminist, cultural, critical, intercultural, international, and global turns, and now the social justice turn—from understanding technology as neutral and science as objective to understanding that technologies and sciences are culturally-rich and thus informed by ideological agendas and uses. To be clear, technologies and sciences are unequally prescribed, controlled, and delegated. They have been used to empower and oppress. Technical communicators construct knowledge informed by multiple subjectivities that we can never fully shed. These critical shifts—just to name a few—demonstrate that the discipline and profession of technical communication is deeply committed to revisiting and revising our relationships with communication, technology, science, and culture in responsible and reflexive ways that have had great impact on our practices and users.

Decades of global changes, emerging research, programmatic changes, and evolving professional identities in technical communication have led us to an amazing place in our organizational conversations.

The most recent annual conferences held by the Association of Teachers of Technical Writing (ATTW) and the Council for Programs in Technical and Scientific Communication (CPTSC) have evidenced that the humanist, social, feminist, critical, cultural, and global studies turns in technical communication studies continue to inform our organizational and local programmatic identities, as well as our curricular and pedagogical approaches to teaching technical communication. The last five national conferences of ATTW, for example, facilitated conversations about culturally-specific perspectives on networks and networking, international technical communication, the shaping of data (and ourselves in relation to it), and advocacy and civic engagement in our research, pedagogy, and practice. The last six conferences sponsored by CPTSC have called for presentations on program development and revision informed by: relationships with public- and private-industries; workplace communication and technologies of/by/about underrepresented communities; local and global trends and practices; curricular, programmatic, institutional, disciplinary, social, political, or economic contexts and connections; and programmatic research that examines curriculum design, hiring and promotion, recruitment and retention, and innovative pedagogy. Especially noteworthy, Miriam Williams's keynote at the 2012 CPTSC conference critiqued colorblind approaches to technical communication and offered ways to think about ethnicity, race, and power in relation to technical communication research, pedagogy, and practice—and the plenary session showcased social justice approaches to technical communication informed by the rhetorical, technological, and scientific expertise of underrepresented scholars and communities.

These exciting trends in our organizational conversations about technical communication programming, curricula, and pedagogy have yet to be reflected in the scholarly books available to us. Not only are we presently in short supply of book-length projects focused on theoretical and methodological approaches to teaching, but the texts we currently have do not fully theorize the implications of the cultural studies turn nor attempt to address the social justice turn in relation to technical communication curriculum design and pedagogy. Nonetheless, there is much to learn from the scholarship designed to help teachers of technical communication respond to emerging disciplinary issues. For example, *Innovative Approaches to Teaching Technical Communication*, edited by Tracy Bridgeford, Karla Saari Kitalong, and Dickie Selfe, seeks to energize technical communication pedagogy in dynamic ways and, thus, offers assignments, activities, and practices for doing so (Bridgeford, Kitalong, and Selfe 2004). James Dubinsky's (2004) collection, *Teaching Technical*

Communication: Critical Issues for the Classroom, provides resources for first-time technical communication instructors. Finally, although not primarily focused on teaching, *Central Works in Technical Communication,* edited by Johndan Johnson-Eilola and Stuart A. Selber, includes a section on pedagogy with essays that respond to technological development, cross-cultural collaboration, and gender issues (Johnson-Eilola and Selber 2004).

Although these texts certainly remain key to the discipline, they are now over a decade old (and some of the chapters therein were re-published from earlier publications). Accordingly, some of the issues they respond to are no longer emerging, and other critical issues have emerged since their release. Further, although a few of the chapters offered in these collections explicitly engage theoretical approaches to teaching, the majority discuss pedagogical practices without identifying the theories informing them. Building on this work, our collection consists only of original essays that seek to offer novice and veteran teachers fresh but tested curricular and pedagogical approaches for identifying new emerging issues and reassessing former emerging issues in relation to social justice and globalization. Moreover, our contributors make explicit how their recommendations are informed by specific theories and methodologies, as a social justice pedagogy understands that the curricular and pedagogical choices that we *all* make are always already influenced by theories about teaching, learning, and communicating about science and technology. Thus, all teaching is ideological and political, even if we pretend it is not.

Since these foundational texts from 2004 and the cultural studies turn in 2006, five books have been published that make evident the value of considering cultural contexts when teaching technical communication. *Resources in Technical Communication: Outcomes and Approaches,* edited by Cynthia Selfe (2007), provides assignment sequences organized around meeting a set of student learning outcomes that should be modified by teachers to make them more appropriate for their local institutional contexts. Barry Thatcher and Kirk St.Amant's edited collection, *Teaching Intercultural Rhetoric and Technical Communication: Theories, Curriculum, Pedagogies, and Practice,* introduces a variety of ways to incorporate intercultural communication contexts into our curriculum and pedagogy (Thatcher and St.Amant 2011). *Online Education 2.0: Evolving, Adapting, and Reinventing Online Technical Communication,* edited by Kelli Cargile Cook and Keith Grant-Davie, addresses how technical communication programs and pedagogy can respond to trends in online education—such as changing student demographics, emerging Web 2.0 technologies, and multimodal pedagogies—in relation to institutional and departmental culture (Cargile Cook and Grant-Davie 2013). Especially inspirational to this project, Han Yu and Gerald Savage's *Negotiating Cultural Encounters: Narrating Intercultural Engineering and Technical Communication* is a collection of real-world stories written by technical communicators who narrate the complicated rhetorical and cultural dynamics at play when working in multicultural teams, and each story is followed by a list of related publications and discussion questions (Yu and Savage 2013). Finally, and most recently, the contributors to *Solving Problems in Technical Communication*, edited by Johndan Johnson-Eilola and Stuart Selber, employ contemporary research in the discipline to solve contemporary real-world technical communication problems—including those associated with new media, but most notably ethics and intercultural communication—in order to bridge the academic-practitioner and theory-practice splits (Johnson-Eilola and Selber 2013).

Our social justice approach to teaching technical communication contributes to the important cultural work of these edited collections by demonstrating that all technical communication contexts are multi- and inter-cultural and influenced by institutions and systems of power—and distributed agency therein—and that social justice approaches to technical communication better position us in any context to better advocate for technological and scientific change in equitable ways within these contexts. Thus, we borrow from our disciplinary traditions of responding to emergent issues, considering (inter)cultural contexts, and solving problems *in* technical communication toward better understanding how injustice is not just a problem in technical communication but also one that we can solve *with* technical communication. Just as cultural theories help us to create more culturally responsive and responsible documents and technologies that are usable and useful for their users, we posit that interfacing cultural theories with social justice frameworks have the same benefits for the users of our curriculum and pedagogy—but also better positions them as agents for redressing workplace, public, civic, and environmental inequities.

This edited collection is the first of its kind in bridging the theoretical with the pedagogical as a means of articulating, using, and assessing social justice frameworks for designing and teaching undergraduate and graduate courses in technical communication. To do so, this collection capitalizes on the momentum gained from the cultural studies and social justice turns in the discipline to make apparent[1] how cultural theory informs classroom practices and how these practices can work in service of redressing technical, technological, and scientific injustices in and outside of the classroom. Moreover, we position social justice inquiry and action as integral to teaching, learning, and practicing ethical technical, scientific, and professional communication in the twenty-first century by highlighting the connections between and across social justice and: intercultural, international, and transnational technical communication; medical, scientific, disability, legal, environmental, and cultural rhetorics; risk communication; civic engagement; and much more. In addition to better representing diverse workplaces, practices, and practitioners, we hope that this collection will also inspire other programmatic initiatives (e.g., recruiting and supporting increased representation of, participation from, and mentoring of historically underrepresented and underserved populations, forming social justice committees and special interest groups, etc.).

We are honored that Gerald Savage and J. Blake Scott, key scholars in the social justice and cultural studies turns in technical communication studies, composed the foreword and afterward to this collection, as their work has cleared a path for us, the teacher-scholars contributing to this project, and future generations of technical communication teacher-scholars. Collectively, our contributors take up Savage and Scott's work and put it into conversation with—and thus contribute to and clear paths in additional areas of—technical communication scholarship, including but not limited to: intercultural and international communication (Barnum and Huilin 2006; Ding 2009; St. Germaine-Madison 2006; Sun 2006, 2012), race and ethnicity studies (Evia and Patriarca 2012; Haas 2012; Johnson, Pimentel, and Pimentel 2008; Williams 2006; Williams and Pimentel 2014), diversity and technical communication programming and curriculum design (Savage and Mattson 2011; Savage and Matveeva 2011), gender and feminist studies (E. Flynn 1997; J. Flynn 1997; Frost 2013; Koerber 2000), postcolonial and globalization studies (Agboka 2013; Bokor 2011; Jeyaraj 2004), disability rhetorics (Palmeri 2006; Smyser-Fauble 2012; Walters 2010; Wilson 2000), and environmental rhetorics and risk communication (Blythe, Grabill, and Riley 2008; Bowdon 2004; Evia and Patriarca 2012; Grabill and Simmons 1998; Sauer 2003; Simmons 2007; Simmons and Grabill 2007; Youngblood 2011).

All of the chapters in the collection do similar rhetorical and intellectual work for users of this project and the discipline. Each chapter: introduces a specific interface for social justice work in technical communication studies—oftentimes in conversation with a cultural theory or a combination of cultural theories—and detail its importance to the discipline and practices of technical communication; offers a case study that demonstrates how the theory/theories informed their curriculum design for and teaching of a specific technical communication course;

and provides broader implications for technical communication curricula, pedagogy, and practice beyond their specific course context. Moreover, every chapter explicitly demonstrates why other teacher-scholars should (and theorizes how to) adapt its specific social justice framework for other institutional and curricular contexts.

Collectively, the chapters in our collection evidence the following rhetorical values foundational to social justice approaches to technical communication in a globalized world:

- All technical communication has the potential to be global technical communication. Even if one works in/for a local organization, the technical communication of those outside the organization could shape the technical communication that transpires within, not to mention that stakeholders and/or users of that technical communication may come from diverse global locations.
- Social justice is both a local and global necessity. This means that contrary to rhetorics of national exceptionalism, the United States, "first-world," and Western countries could also benefit from social justice approaches to technical communication.
- International and intercultural communication happens outside of non-Western and non-US contexts (and without Western and "first-world" interlocutors). Moreover, these cases, their stakeholders, their technical communication—thus, cultural and rhetorical—work, and the power dynamics therein are worthy of our study.
- International technical communication happens within the United States. There are over five hundred sovereign indigenous nations independent from the United States but are located within United States national borders. And this international technical communication can and does happen independent from United States and other "first-world" involvement.
- International and domestic technical communication is all a matter of rhetorical perspective. A case study of Chinese technical communication, for example, is not international technical communication for Chinese technical communicators.
- Intercultural technical communication happens within and across national borders given ethnic and other cultural diversity.
- Although social justice begins at home, it's important to understand the relationships between local and global injustices. Certainly, we should consider our agency as technical communicators in light of the social injustices within our own communities rather than positioning ourselves as rhetorical missionaries for Others. But we should also study the patterns and trends across and between local and global stories of injustice so that we may better identify, analyze, and redress the ideologies, institutions, stakeholders, and rhetorics that sponsor them—and to more effectively form intercultural technical communication teams to do so.

- Social justice includes justice for the environment, as injustices against any living species (not just humans and non-human animals) should impact the social. Moreover, many non-Western epistemologies understand non-human actors as social beings.
- Social justice benefits everyone. Working to achieve or restore equity for one population or community does not require anyone with access to those rights to relinquish them—quite the opposite actually. For technical communication, specifically, equity means fair and just access to and representation in scientific and technical communication for *all* stakeholders.

Admittedly, due in part to the scope and emergent nature of this collection, our contributors evidence some of these foundations more so than others. For instance, most of our contributors discuss US-based contexts for their curricular and pedagogical case studies. Despite this limitation, our collection highlights the necessity of social justice in the United States, as well as the relationships between local and global injustices (e.g., racism, sexism, etc.). Further, we suggest that a US focus is appropriate for the present disciplinary moment. To explain, the focus of most current technical communication scholarship is US centric, and we assume that the majority of our readers are from the United States and/or are teaching or studying technical communication in the United States; thus, most of us likely have more agency to make change happen within US borders. Put simply, we should pitch in to clean up the mess of injustice in our own backyards before pointing at the messes of others. Moreover, while scholars of international technical communication always already understand their work in relation to globalization, few scholars of US-based technical communication do. The latter position is one of privilege, and we should no longer feel comfortable in this position. If we truly understand the complex nature of globalization and truly desire to teach current and future practitioners how to communicate for/with/about global audiences, then we need to understand "international" and "domestic" technical communication is all a matter of perspective. A case study of US-based technical communication is international to a technical communicator in Mexico, and thusly we must be accountable for better understanding the complexities and complicities of the United States on global inequities and global technical communication.

An affordance of our collection is its broad approach to redressing injustice vis-à-vis technical communication practice and pedagogy, an approach that understands animal (human and non-human) and environmental rights as integral to social justice. In chapter 4, Elise Verzosa Hurley brings spatial justice into this conversation, as well, which helps us to better interrogate the complex interdependencies of human, non-human animal, and environmental rights as they pertain to the spaces and places we inhabit, work, and communicate. Thus, the chapters interrogating issues of disability, gender, race, and sexual orientation all clearly contribute to the human rights movement (Sapp, Savage, and Mattson 2013) within the social justice turn—and the chapters that address environmental and risk communication contribute to the environmental justice movement. But we also see these categories as dynamic, as one can read the chapters written by Erin Frost, Elise Verzona Hurley, Godwin Agboka, and Donnie Johnson Sackey as tending to human, non-human animal, and environmental rights as well. Suffice it to say, our contributors help to reveal the interdependent relationships between the local and global on the macro-, meso-, micro-, and even—literally—the cellular levels.

Ultimately, our collection seeks to mobilize social justice rhetorics of technical communication that trouble institutional and geopolitical boundaries toward an understanding of interrelationships and interdependencies between local and global cultures, organizations, borders, publics, and citizenry and to foster more critical understandings of:

- our responsibilities to the cultures and communities within which, to whom, and about whom we communicate
- systems of and rhetorics from hegemonic power—and how and why they have historically shaped how we regard specific cultures and communities in relation to their technical and scientific expertise, or lack thereof
- the effects of globalization on local environmental, scientific, technological, cultural, and rhetorical practices
- the relationships between rhetorics, places, power, agency, networks, infrastructures, and institutions—and how space and place have real political and embodied effects on (in)justice and rights
- how bodies, embodiment, and risks affect teacher, student, practitioner, professional, and public identities—as well as knowledge production and lived realities
- tactics for challenging, resisting, and transgressing systems and rhetorics from/of power
- non-Western, non-patriarchal, or otherwise underrepresented traditions and histories of technical, technological, scientific, and rhetorical expertise
- the relationships between diversity, cultural studies, community literacies, public rhetorics, participatory action research, and social justice—and what these relationships can teach us about local and global technical, technological, and scientific knowledge work
- our roles as technical communication instructors, public intellectuals, and user advocates for diverse communities and

stakeholders—and how we can work toward pedagogical, social, technological, spatial, and environmental justice in these roles

Altogether these rhetorics offer a social justice framework of intellectual, professional, and rhetorical skills necessary for communicating in and representing diverse twenty-first-century technical communication workplaces, spaces, and practices, and communicating with diverse practitioners and users.

A strength of this collection is that we bring together emerging and established scholars, cutting-edge research, and critical theories gaining traction in the discipline in one place. The twelve chapters in this collection are organized into four sections: Embodied Knowledges and Risks; Space, (Em)Place, and Dis(Place)Ment; Interfacing Public and Community Rhetorics with Technical Communication Discourses; and Accommodating Diverse Discourses of Diversity. Part I: Embodied Knowledges and Risks situates *all* bodies as critical to ethical technical communication pedagogy and practice. The first chapter, "Apparent Feminism and Risk Communication: Hazard, Outrage, Environment, and Embodiment," written by Erin A. Frost (East Carolina University), argues that apparent feminism is critical to socially-just technical communication pedagogies and practices. To demonstrate this, Frost discusses how she employed this framework to design and teach an online, graduate-level risk communication course that focused on exposing the ways in which environmental, technical, and public risk attaches to particular bodies and, thus, affects lived realities. Cruz Medina (Santa Clara University) and Kenny Walker (University of Texas-San Antonio) posit that grading contracts can be used in ways that support social justice approaches to teaching technical communication in their chapter "Validating the Consequences of Social Justice Pedagogy: Explicit Values in Course-Based Grading Contracts"—as long as teachers and students consider the ways in which their identities and bodies impact their positionality, power dynamics, and agency in relation to the grading contracts. To wrap up Part I, Barbi Smyser-Fauble's (Illinois State University) chapter, "The University Required Accommodations Statement: What 'Accommodation' Teaches Technical Communication Students and Educators," uses a feminist disability studies framework to expose how the Americans with Disabilities Act (ADA) accommodation statements on course syllabi can often perpetuate the normalizing practices of an ableist culture. Specifically, this chapter examines how the ambiguous language often used in required ADA accommodations statements in course syllabi can problematically construct student bodies and identities, what is perceived as "accessible" or "reasonable accommodation," and who is perceived as worthy of accessibility and inclusion consideration.

Building upon Part I's insistence that we value diverse bodies and embodied knowledges, the chapters in Part II—Space, (Em)Place, and Dis(Place)Ment—demonstrate that space and place have real political and embodied effects on (in)justice and rights and thus call for pedagogical and professional practices that support spatial justice, human rights, and environmental justice. In "Spatial Orientations: Cultivating Critical Spatial Perspectives in Professional and Technical Communication Pedagogy," Elise Verzosa Hurley (Illinois State University) argues that we pay closer attention to rhetorics of space, place, and location toward more *critical spatial perspectives* in our pedagogical practice. Specifically, building intellectual alliances between critical cultural geography, rhetoric, and technical communication theories, Verzosa Hurley offers a case study of a professional and technical communication graduate-level course focused on spatial and visual rhetorics—and, in the process, helps our discipline to make the "spatial turn" and teachers and students to understand, imagine, and enact *spatial justice* practices in diverse professional workplaces. Concerned with geopolitical struggles over space, Godwin Agboka's (University of Houston-Downtown) chapter, "Indigenous Contexts, New Questions: Integrating Human Rights Perspectives in Technical Communication," exposes technical documents used during the Ogoni oil crisis in the Niger Delta of Nigeria—and related genocide and displacement of indigenous Nigerians, political unrests, gas flares, oil spills, and pollution of water—as a case study to demonstrate the relationships between technical communication, globalization, and human rights violations. Further, Agobka provides a pedagogical framework (informed by the UN statement on human rights) for discussing these relationships in technical communication curricula so that we prepare future and current technical communication practitioners to "aggressively interrogate" the ways in which technical communication may be complicit in human rights violations. Finally, in "An Environmental Justice Paradigm for Technical Communication," Donnie Johnson Sackey (University of Texas at Austin): examines feminist materialist, feminist political ecology, ecofeminist, and environmental justice perspectives on space, place, and agency—and in relation to the primacy they give to either people or the environment within their analyses; offers a pedagogical heuristic that leverages the affordances of these theories toward solving environmental problems in technical communication; and discusses how he used this heuristic in an environmental rhetoric course he designed. Ultimately, Sackey argues that "technical communicators should always be attuned to the emplaced conditions of social justice concerns."

The chapters in Part III: Interfacing Public and Community Rhetorics with Technical Communication Discourses demonstrate that community literacies are always already deeply emplaced, that everyday community literacy practices deeply inform technical communication, and, thus, that technical communication pedagogy should be concerned with diverse technical and professional communication practices within diverse communities and workplaces. In "Stayin' on Our Grind: What Hiphop Pedagogies Offer to Technical Writing," Marcos Del Hierro (University of New Hampshire) advocates for the inclusion of hiphop pedagogies in the teaching of technical writing. Using non-Western theories and social justice approaches to technical communication, Del Hierro offers a hiphop pedagogical framework as a way of decolonizing the field and designing our curricula to be more inclusive of knowledges that have been historically marginalized on the basis of ethnicity, race, and class and offers specific examples for engaging non-traditional technical writing practices, theories, and workplaces. Kristen Moore (University at Buffalo—SUNY) recommends another non-traditional approach to teaching technical communication in her chapter, "Black Feminist Epistemology as a Framework Community-based Teaching." Moore puts the four tenets of Patricia Hill Collins's Black Feminist Epistemology into conversation with community-based scholarship and participatory action research in the discipline to demonstrate its usefulness in her study of public and community engagement and then suggests how other technical communication teacher-scholars might inform their community-based research and teaching with this framework. While Del Hierro locates his work as informed by the technical literacies of a larger cultural group—the hiphop community—and Moore positions hers as informed by black feminist thought writ large and a local black, female community group in St. Louis, Missouri, the last author in this section situates her attention in relation to the "cultural place" of nonprofit advocacy websites. Specifically, in her chapter "Advocacy Engagement, Medical Rhetoric, and Expediency: Teaching Technical Communication in the Age of Altruism," Marie E. Moeller (University of Wisconsin-La Crosse) argues for the importance of analyzing medical rhetoric in scientific and technical communication pedagogy—even medical rhetoric communicated by nonprofit advocacy organizations—by evidencing how such rhetoric can have harmful effects on the very users it's supposed to help. She provides a model and suggestions for doing so by using a feminist disability studies framework for rhetorically engaging with the popular medical advocacy website of The Susan G. Komen Foundation.

The last section of the collection, Part IV: Accommodating Different Discourses of Diversity, offers a diverse set of theoretical and practical approaches aimed at challenging our current perceptions of teaching and learning about diversity in the classroom and the workplace. Natasha Jones (University of Central Florida) and Rebecca Walton's (Utah State University) chapter, "Using Narratives to Foster Critical Thinking about Diversity and Social Justice," provides a working definition of social justice research and pedagogy, proposes narrative as a "useful tool for fostering critical thinking about social justice in the technical communication classroom," and calls for more explicit social justice pedagogy work in technical communication scholarship. Further, Jones and Walton provide a heuristic for developing and examples of narrative-driven in-class exercises, assignments, and discussion guides. In "Race and the Workplace: Toward a Critically Conscious Pedagogy," Jessica Edwards (University of Delaware) argues that it is "necessary for students to consider the ethical and social responsibilities that undergird their language use, and moreover, that professional communication classrooms are a vital site for promoting students' cultural competence and attention to issues of race." Edwards includes student voices that evidence the necessity of critical race theory for tending to issues of race and racism in our pedagogies and suggests ways that we might study systemic structures of oppression and the rhetorical practices that support them so that technical communicators can be better positioned to subvert those structures and revise those rhetorics. In the final chapter in this section, Matthew Cox (East Carolina University) proposes queer theory and cultural rhetorics as an intersectional, "culturally-conscious" pedagogical framework for studying diversity in "Shifting Grounds as the New Status Quo: Examining Theoretical Approaches to Diversity and Taxonomy in the Technical Communication Classroom." Cox describes a graduate-level course he designed that used this framework and reports on the implications for and affordances of intersectional approaches to engaging with diverse cultural issues in the broader field of technical communication.

While the collection provides a robust and wide array of usable and useful support for instructors invested in teaching practitioners how to skillfully, ethically, and rhetorically negotiate contextual power dynamics when solving complex technical and scientific communication problems within a variety of contexts, we hope that this collection also helps to clear a path for future social justice inquiry, discussion, practice, and promise. A path built upon the understanding that <u>all bodies—human, non-human animals, and landbases—are critical to ethical technical communication pedagogy and practice</u>. As technical communication researchers and practitioners, we can use our <u>privilege</u> and rhetorical skills to help equip others with <u>new habits of mind and practice that attune them to responsible citizenship and advocacy, self-awareness and consciousness, and critical thinking</u>. To recognize how the material realities of our lives are continuously and differentially impacted by technical communication that takes form in a wide range of public rhetorics, including local, state, federal, and transnational legislation; corporate and organizational policies and practices; and scientific, medical, technical, and technological communication—just to name a few. In short, technical communication does important work in the world—and we have the position, agency, and obligation to identify and intervene in discourses that authorize injustice.

Note

1. To borrow from Frost's articulation of apparent feminism (see chapter 1).

References

Agboka, Godwin Y. 2013. "Participatory Localization: A Social Justice Approach to Navigating Unenfranchised/Disenfranchised Cultural Sites." *Technical Communication Quarterly* 22 (1): 28–49. https://doi.org/10.1080/10572252.2013.730966.

Barnum, C. A., and L. Huilin. 2006. "Chinese and American Technical Communication: A Cross-Cultural Comparison of Differences." *Technical Communication (Washington)* 53:143–66.

Blythe, Stuart, Jeffrey T. Grabill, and Kirk Riley. 2008. "Action Research and Wicked Environmental Problems: Exploring Appropriate Roles for Researchers in Professional Communication." *Technical Communication Quarterly* 22 (3): 272–298. https://doi.org/10.1177/1050651908315973.

Bokor, Michael Jarvis Kwadzo. 2011. "Connecting with the 'Other' in Technical Communication: World Englishes and Ethos Transformation of U.S. Native English-Speaking Students." *Technical Communication Quarterly* 20 (2): 208–237. https://doi.org/10.1080/10572252.2011.551503.

Bowdon, Melody. 2004. "Technical Communication and the Role of the Public Intellectual: A Community HIV-Prevention Case Study." *Technical Communication Quarterly* 13 (3): 325–340. https://doi.org/10.1207/s15427625tcq1303_6.

Bridgeford, Tracy, Karla Saari Kitalong, and Dickie Selfe, eds. 2004. *Innovative Approaches to Teaching Technical Communication*. Logan: Utah State University Press. https://doi.org/10.2307/j.ctt46nzds.

Cargile Cook, Kelli, and Keith Grant-Davie, eds. 2013. *Online Education 2.0: Evolving, Adapting, and Reinventing Online Technical Communication*. Amityville, NY: Baywood.

Ding, Huiling. 2009. "Rhetorics of Alternative Media in an Emerging Epidemic: SARS, Censorship, and Extra-Institutional Risk Communication." *Technical Communication Quarterly* 18 (4): 327–350. https://doi.org/10.1080/10572250903149548.

Dubinsky, James M. 2004. *Teaching Technical Communication: Critical Issues for the Classroom.* Boston: Bedford St. Martin's.

Evia, Carlos, and Ashley Patriarca. 2012. "Beyond Compliance: Participatory Translation of Safety Communication for Latino Construction Workers." *Journal of Business and Technical Communication* 26 (3): 340–367. https://doi.org/10.1177/1050651912439697.

Flynn, Elizabeth A. 1997. "Emergent Feminist Technical Communication." *Technical Communication Quarterly* 6 (3): 313–320. https://doi.org/10.1207/s15427625tcq0603_6.

Flynn, John F. 1997. "Toward a Feminist Historiography of Technical Communication." *Technical Communication Quarterly* 6 (3): 321–329. https://doi.org/10.1207/s15427625tcq0603_7.

Frost, Erin A. 2013. "Transcultural Risk Communication on Dauphin Island: An Analysis of Ironically Located Responses to the Deepwater Horizon Disaster." *Technical Communication Quarterly* 22 (1): 50–66. https://doi.org/10.1080/10572252.2013.726483.

Grabill, Jeffrey T., and W. Michele Simmons. 1998. "Toward a Critical Rhetoric of Risk Communication: Producing Citizens and the Role of Technical Communicators." *Technical Communication Quarterly* 7 (4): 415–441. https://doi.org/10.1080/10572259809364640.

Haas, Angela M. 2012. "Race, Rhetoric, and Technology: A Case Study of Decolonial Technical Communication Theory, Methodology, and Pedagogy." *Journal of Business and Technical Communication* 26 (3): 277–310. https://doi.org/10.1177/1050651912439539.

Jeyaraj, Joseph. 2004. "Liminality and Othering: The Issue of Rhetorical Authority in Technical Discourse." *Journal of Business and Technical Communication* 18 (1): 9–38. https://doi.org/10.1177/1050651903257958.

Johnson, Jennifer Ramirez, Octavio Pimentel, and Charise Pimentel. 2008. "Writing New Mexico White." *Journal of Business and Technical Communication* 22 (2): 211–236. https://doi.org/10.1177/1050651907311928.

Johnson-Eilola, Johndan, and Stuart A. Selber, eds. 2004. *Central Works in Technical Communication.* New York: Oxford University Press.

Johnson-Eilola, Johndan, and Stuart A. Selber, eds. 2013. *Solving Problems in Technical Communication.* Chicago: University of Chicago Press.

Koerber, Amy. 2000. "Toward a Feminist Rhetoric of Technology." *Journal of Business and Technical Communication* 14 (1): 58–73. https://doi.org/10.1177/105065190001400103.

Palmeri, Jason. 2006. "Disability Studies, Cultural Analysis, and the Critical Practice of Technical Communication Pedagogy." *Technical Communication Quarterly* 15 (1): 49–65. https://doi.org/10.1207/s15427625tcq1501_5.

Rude, Carolyn. 2009. "Mapping the Research Questions in Technical Communication." *Journal of Business and Technical Communication* 23 (2): 174–215. https://doi.org/10.1177/1050651908329562.

Sapp, D., G. Savage, and K. Mattson, eds. 2013. "Human Rights and Professional Communication." Special Issue of *Journal of Rhetoric, Professional Communication, and Globalization* 4 (1): 1–139.

Sauer, Beverly A. 2003. *The Rhetoric of Risk: Technical Documentation in Hazardous Environments.* Mahwah, NJ: L. Erlbaum Associates.

Savage, Gerald, and Kyle Mattson. 2011. "Perspectives on Diversity in Technical Communication Programs." *Programmatic Perspectives* 3:5–57.

Savage, Gerald, and Natalia Matveeva. 2011. "Seeking Inter-Racial Collaborations in Program Design: A Report on a Study of Technical and Scientific Communication Programs in Historically Black Colleges and Universities (HBCUs) and Tribal Colleges and Universities (TCUs) in the United States." *Programmatic Perspectives* 3:58–85.

Scott, Blake, and Bernadette Longo, eds. 2006. "Guest Editors' Introduction: Making the Cultural Turn." *Technical Communication Quarterly* 15 (1): 3–7. https.dx.doi.org/10.1207/s15427625tcq1501_2.

Scott, J. Blake. 2003. *Risky Rhetoric: AIDS and the Cultural Practices of HIV Testing.* Carbondale: Southern Illinois University Press.

Scott, J. Blake, Bernadette Longo, and Katherine V. Wills. 2007. *Critical Power Tools: Technical Communication and Cultural Studies.* Albany: State University of New York.

Selfe, Cynthia, ed. 2007. *Resources in Technical Communication: Outcomes and Approaches.* Amityville, NY: Baywood.

Simmons, W. Michele. 2007. *Participation and Power: A Rhetoric for Civic Discourse in Environmental Policy Decisions.* Albany: State University of New York Press.

Simmons, W. Michele, and Jeffrey T. Grabill. 2007. "Toward a Civic Rhetoric for Technologically and Scientifically Complex Places: Invention, Performance, and Participation." *College Composition and Communication* 58 (3): 419–48.

Smyser-Fauble, Barbi. 2012. "The New Scarlet Letter A: An Exploration of the Power of Online Informational Websites to Influence and Brand Those Impacted by Autism Spectrum Disorders." *Rhetoric, Professional Communication, and Globalization* 3:110–39.

St. Germaine-Madison, Nicole. 2006. "Instructions, Visuals, and the English-Speaking Bias of Technical Communication." *Technical Communication (Washington)* 15:182–90.

Sun, Huatong. 2006. "The Triumph of Users: Achieving Cultural Usability Goals with User Localization." *Technical Communication Quarterly* 15 (4): 457–481. https://doi.org/10.1207/s15427625tcq1504_3.

Sun, Huatong. 2012. *Cross-Cultural Technology Design: Creating Culture-Sensitive Technology for Local Users.* Oxford: Oxford University Press. https://doi.org/10.1093/acprof:oso/9780199744763.001.0001.

Thatcher, Barry, and Kirk St.Amant, eds. 2011. *Teaching Intercultural Rhetoric and Technical Communication: Theories, Curriculum, Pedagogies, and Practice.* Amityville: Baywood.

Walters, Shannon. 2010. "Toward an Accessible Pedagogy: Dis/ability, Multimodality, and Universal Design in the Technical Communication Classroom." *Technical Communication Quarterly* 19 (4): 427–454. https://doi.org/10.1080/10572252.2010.502090.

Williams, Miriam, and Octavio Pimentel, eds. 2014. *Communicating Race, Ethnicity, and Identity in Technical Communication.* Amityville, NY: Baywood.

Williams, Miriam F. 2006. "Tracing W.E.B. Dubois' 'Color Line' in Government Regulations." *Journal of Technical Writing and Communication* 36 (2): 141–165. https://doi.org/10.2190/67RN-UAWG-4NFF-5HL5.

Wilson, James C. 2000. "Making Disability Visible: How Disability Studies Might Transform the Medical and Science Writing Classroom." *Technical Communication Quarterly* 9 (2): 149–161. https://doi.org/10.1080/10572250009364691.

Youngblood, Susan. 2011. "Balancing the Rhetorical Tensions between Right to Know and Security in Risk Communication: Ambiguity and Avoidance." *Journal of Business and Technical Communication* 26 (1): 35–64. https://doi.org/10.1177/1050651911421123.

Yu, Han, and Gerald Savage. 2013. *Negotiating Cultural Encounters: Narrating Intercultural Engineering and Technical Communication.* Hoboken, NJ: Wiley. https://doi.org/10.1002/9781118504871.

PART I

Embodied Knowledge and Risks

1
APPARENT FEMINISM AND RISK COMMUNICATION
Hazard, Outrage, Environment, and Embodiment

Erin A. Frost

Many emerging theories informing technical communication scholarship involve a shift in focus that recognizes the importance of how different people—operating out of different embodied experiences—interface in a diversity of ways with technical documents, ideas, and conversations (Scott, Longo, and Wills 2007; Williams and Pimentel 2014). For example, cultural theories, feminist theories, critical race theories, queer theories, and more have recently been brought to bear on technical communication's traditional disciplinary spaces, and Haas (2012) has called for scholars to create "openings for further considering how to broach conversations about design and technology theory from and about scholars and users who are people of color—and from other underrepresented populations—in technical communication studies" (301). As the field recognizes the value of such theories, the importance of introducing and building upon them in pedagogical contexts becomes increasingly apparent. This chapter accounts for how apparent feminism—a methodological approach that seeks to make apparent the importance of feminist critiques and identities[1] in modern politics—was used in a graduate-level course on risk communication.[2] This course focused extensively on the ways risk attaches to particular bodies, and it asked students to use apparent feminism to consider how risk functions culturally and how technical and public understandings of risk communication messages affect lived realities.

This chapter situates risk communication in relationship to the field of technical communication; it explains what design and technology have to do with risk and why technical communicators should be paying attention. Further, it explores what critical approaches to risk communication pedagogy can teach instructors of technical communication. In service of this exploration, I offer a detailed account of my work with an apparent feminist pedagogical approach to risk communication and students' responses to this framing. As Moore notes, "Not only is our research better when we diversify our pedagogies but so too can the sites and workplace practices we study become more equitable" (296). Finally, this chapter offers implications for our discipline; it extrapolates ways the field can intervene in constructions of risk that do not take into account the ethical effects on those who speak from the margins. Simultaneously, I offer my own theoretically informed reflections and ideas about the risks and benefits associated with teaching a course like the one I describe. Ultimately, apparent feminism—one of the many theories operating under the umbrella of cultural and critical theories—is a necessary and important approach to technical communication pedagogy that can help the discipline move toward socially just pedagogies and practices.

RISK AND CULTURAL STUDIES IN TECHNICAL COMMUNICATION / HAZARD AND OUTRAGE

Risk communication is an important subset of technical communication, but it also is a robust discipline in its own right. Areas of study that exist within or intersect with risk communication include crisis communication, environmental communication, organizational communication, disaster communication, health communication, risk assessment, and public relations, among others. Many scholars[3] have done work demonstrating the importance of risk communication to the discipline of technical communication, often with an eye to "co-producing ethical, responsible, and responsive risk communication" (Haas and Frost 2017). In fact, Grabill and Simmons (1998) advocate for explicit citizen participation—which obviously values local cultures—in the construction of rhetorics of risk. They were, in part, taking up a call made by Plough and Krimsky (1990), who express the need for situating risk communication within a cultural framework. Plough and Krimsky assert that "if the technosphere begins to appreciate and respect the logic of local culture toward risk events and if local culture has access to a demystified science, points of intersection will be possible" (229–30).

In other words, this is an argument for moving risk and technical communication rhetorics toward considering culture and paying attention to embodiment. Embodiment takes into account a user's body as well as the ways in which they have used that body (or have been coerced into using that body, or have endured that body being used) over her or his lifetime.[4] A focus on embodiment means more than just paying attention to the normative ways that technical documents too often construct bodies and critiquing the common assumption of an objective or default body (usually meaning a white, straight, able, male body). It also means paying attention to the ways that these patterns, which have existed for generations, have influenced users' experiences. As N. Katherine Hayles (1999) explains, "embodiment differs from the concept of the body in that the body is always normative relative to some sort of criteria" (196); embodiment, meanwhile, is a far more complex phenomenon derived from the physical presence of the body itself as well as the many experiences associated with being in a particular kind of body that must navigate a variety of cultural contexts.

Thus, the aforementioned rhetorical move—the trend toward the production of socially just risk communication as one of technical communication's disciplinary agendas—is one that I seek to follow in this project, partially because risk communication is an area in need of critical attention. To explain, it may help to understand disciplinary definitions of risk. Risk is often articulated as (the perception of) the probability of a potential hazard. Risk communication consultant Peter Sandman (2014) offers a simple equation for thinking about risk: Risk = Hazard + Outrage. This standard definition, exemplified by Sandman's equation, makes perception/outrage explicit. In other words, our perceptions of a hazard affect our understanding of the risk it poses. For example, I might consider swimming in the ocean a risky idea because of the possibility of shark attack. Even though statistics tell me the actual hazard is extremely low, the public outrage (think *Jaws*) associated with such events is a significant enough value to deeply influence my perception of the risk. The hazard associated with driving a car is relatively quite large, but the public outrage associated with car accidents is much lower, and I therefore take that risk on a regular basis. Thus, it is easy to see that public perception—which is influenced by many things, including embodied experiences—is a major factor in designing risk messages. However, public perception is too often invoked as a firm, easily discerned value; minority or marginalized publics who may have different perceptions of risk than the often-assumed "standard" public based on historical inequity are not often taken into account.

Thus, cultural studies and risk communication should be linked—especially when it comes to how we think about design and technology, concepts which are too often considered technical and objective rather than culturally situated (Haas 2012; Haas and Eble, this collection; Jones and Walton, this collection; Savage, this collection; Williams 1991). While this connection is rarely explicitly made in risk communication textbooks and other introductory materials, several risk communication journals—*Risk Analysis; Risk, Health and Society;* the *Journal of Risk Research*—have recently published work that could be considered complementary to cultural studies approaches (Hamilton 2014; Stengel 2014; Zhou et al. 2014). These examples notwithstanding, much more work remains to be done in expanding such approaches and making them explicit within the discipline of risk communication. Scott (2003) and Sauer (2003) offered models for culturally informed studies of risk with their attention to populations exposed to HIV/AIDS and populations whose socioeconomic class requires they live and work in risk, respectively. In addition, scholars who may not identify or appear to be risk communicators have already been doing this important work for some time. Much of it has focused on environmental hazard—note Sackey's (this collection) contention that there is not a "line of demarcation between social justice and environmental justice" (138)—which is a common theme in my own research as well. Andrea Smith (2005), for example, has made a powerful argument about the relative risks experienced by native women; "American Indian women are twice as likely to be victimized by violent crime as women or men of any other ethnic group" (28), and "[p]erpetrators of sexual violence can usually commit crimes against Native women with impunity" (31) because "U.S. attorneys decline to prosecute about 75 percent of all cases involving any crime in Indian country" (32). Winona LaDuke (1999, 2005) has demonstrated how corporate and national apathy toward sacred places and traditions puts indigenous bodies in hazard, as when superfund sites are located near the water supplies of indigenous populations. Several authors (Woods 2010) have uncovered situations where class and ethnicity correlated inequities put particular bodies at greater risk of hazard to health and livelihood in post-Katrina New Orleans.

The scholars I cite above provide precisely the sort of exposure that technical communication scholars, practitioners, and students need in order to think about the interrelatedness of risk situations and the deeply unjust ways that bodies are exposed to risk. Critical and cultural approaches to risk communication pedagogy make the risks that critical theories address more apparent. Because of the connections between cultural theories and risk, culturally informed risk communication pedagogies are a particularly productive lens through which to think about technical communication pedagogies writ large.

Building on this foundation, apparent feminism provides a vital approach to risk and technical communication courses. Apparent feminism—like other cultural theories and cultural theory-based pedagogies—is a response to recognized risks. It is a response to risks faced by those who inhabit bodies that are not traditionally present in the academy. It is a response to risks faced by those who inhabit bodies too often devalued by hegemonic political forces. It is a response to risks not already recognized by those with significant cultural power. It is a response to risks too often reified in supposedly objective, neutral technical communication. While apparent feminism is just one of many important cultural theories with pedagogical applications, it is an approach with a wide variety of applicable benefits, as described below. Apparent feminism can help teach instructors of technical communication how to recognize and mediate a variety of risks on behalf of themselves, their students, and the future audiences those students design and write for.

DESIGNING AN APPARENT FEMINIST PEDAGOGY OF RISK + TEACHING RESEARCH

Apparent feminism is a methodology I have developed in response to my fear that "technical communication students take far too few courses that use feminisms and other critical approaches to explicitly question rhetorics of objectivity, neutrality, efficiency, and truth" (Frost 2014, 111). Apparent feminism functions at the nexus of social, ethical, political, and practical technical communication domains (Hart-Davidson 2001; Johnson 1998; Miller 1989) in its goal of making apparent the urgent—and too often covered-up—need for feminist critique of contemporary politics, education, and technical communication. Apparent feminism emphasizes the importance of being explicit about feminist identity in response to socially unjust situations, invites participation from allies who do not identify as feminist but do complementary work, and seeks to make apparent the ways in which efficient work depends firmly upon the inclusion of diverse audiences. This approach, although not easy in and of itself, does transfer smoothly to the technical communication classroom. I have written elsewhere (Frost 2013a; Frost 2014) about the exigence and importance of apparent feminist pedagogies, and this chapter builds my project of demonstrating through pedagogical action what apparent feminist pedagogies look like and what benefits they offer to their users (both instructors and students) and the field of technical communication.

My focus in this chapter is on how I used an apparent feminist pedagogy in a graduate, distance education course on risk communication by modeling my own research, confronting the fallacy of pedagogical objectivity, encouraging students to engage risks to their own online identities, and asking students to think about disciplinary trends regarding the embodied distribution of risk. I have employed various iterations of apparent feminist pedagogies in a variety of undergraduate and face-to-face courses. However, this was only the second time I'd employed apparent feminist pedagogy in a graduate, distance education course; I'd also used it in a methods course in the fall semester. This is important because several students in English 7765: Risk Communication were also in that fall methods class and thus were familiar with my apparent feminist approach. This is also important information because my experience in the fall methods class helped me to fine-tune the ways I designed the course. For example, in the fall I began the class with a video introduction. Based on the response to that—which was largely that my embodiment (in particular, my age) was surprising/distracting—I did not utilize an introductory video in Risk Communication.

Rather, I designed English 7765: Risk Communication to focus on embodiment, and the applied meta-lesson I asked students to engage with throughout the semester was their own online embodiment. I asked students to undertake a project that operationalized the cyberfeminist finding that the Internet is not a democratizing medium where people pretend bodies—and categories of bodies—don't exist (Fernández, Wilding, and Wright 2002; Hayles 1999; Turkle 1995). Thus, I asked them to take risks. Cox (this collection) tells us, "We live in a culture where happiness and smoothness and efficiency as success are so expected that they seem to keep us from taking risks. But risks may be the very thing that our students need most to truly have transformational experiences" (292). I wanted students to take risks and to do so critically. Thus, students engaged in extended conversations on a public website where they mediated their online identities (through the avatars and names they used, among other methods) in ways that they understood affected how others perceived their embodiment—and thus their rhetorical power and relationship to various risks.

I also explicitly discussed my apparent feminist approach to research and teaching with students, and I framed the critical texts we'd be reading throughout the semester with the following language in my syllabus course description:

> This course will bring together current understandings of risk communication—its theories, methodologies, and ideologies—with the gendered realities that both support and contradict particular understandings of risk. Beginning with popular risk communication staples such as *Readings in Risk* and *The Peter M. Sandman Risk Communication Website*, this course

will then move into interrogating constructions of risk that are situated historiographically and culturally, including Beverly A. Sauer's *The Rhetoric of Risk,* J. Blake Scott's *Risky Rhetoric,* and articles that highlight the processes of risk construction (Bowdon; Grabill and Simmons). Further, participants in this study will work to understand how constructions of risk that hegemonic forces frame as neutral are anything but for indigenous populations (LaDuke; Smith; Wildcat) and other marginalized peoples (Woods). Finally, participants will theorize ways to intervene in constructions of risk that do not take into account the ethical effects on those who speak from the margins.

In addition to reading these critical texts and mediating their own perceived embodiment—and thus defining the risks they were willing to take—I also made my own research agenda apparent to students. Students knew from the beginning of the course that I would be studying them and their interactions. They also were familiar with the story of how I came to be interested in risk communication, what my research history in that area is, and where I'm headed next with my work. Because I was at a place in my work where I felt "stuck"—something I'll elaborate on below—I used that as a teaching opportunity and explicitly asked students what they thought of the options available to a researcher in my situation. Modeling research agendas for graduate technical communication students is not only good practice for the purpose of helping them create and manage their own research agendas, but it is also important for them as they analyze and react to the ethos of their professor.

To extrapolate on this feature of the course, I talked to students about my dissertation project on apparent feminism (Frost 2013a) and how it was influencing my thinking about an ongoing project related to the Deepwater Horizon Disaster. (I go into some detail here about this project, because this detail is necessary in order to contextualize students' responses to this tactic.) As part of this ongoing research project, I've been paying attention to how public documents situate risk in relationship to the Deepwater Horizon Disaster. Some of the artifacts I've found—like particular publicly archived images—have circulated widely (Wikipedia 2010[5]). Others, like accident reports with corporate or government authors (*Deepwater Horizon Accident Investigation Report* 2010; *Deepwater Horizon Marine Casualty Investigation Report* 2011; *Macondo: The Gulf Oil Disaster, Chief Counsel's Report* 2011; *On Scene Coordinator Report* 2011), are public documents that have enjoyed far more limited amounts of what Ridolfo and DeVoss (2009) term *rhetorical velocity.* What I have very rarely found are documents directly concerned with the effects of the oil spill on human bodies. I am interested in why the two divergent narratives of oil spill recovery focus on economics or ecology but don't often recognize human bodies as significant actors in those systems. Even more specifically, I take up Katsi Cook and Winona LaDuke's concerns about the treatment of female bodies because, as Cook says, "women are the first environment" (LaDuke 1999, 18). My argument is that people interested in environmental risks and rhetorics certainly ought to be paying more attention to embodied risks.

I shared with students that I had returned to Mobile County in November 2013 with the intention of studying how the Mobile County Health Department talks to citizens about the effects of the spill. At this point in the semester, I was modeling the development of a research question, and I began with: "What data sets inform(ed) the action plans of health care facilities with responsibilities to Gulf Coast citizens in the aftermath of the Deepwater Horizon Oil Spill?" As it turned out, I was able to find almost no evidence that such action plans—or any conversations about health and oil—were happening at all. When I talked with locals, the consistent answer I received was that there is lots of information available about the economic effects of the spill, and some information on ecological effects, but not so much about health effects and very little about how any of these things might be connected. I mentioned this frustration during a conversation with a scientist/curator at a museum-type location in Mobile County, and he nodded knowingly before leading me from the Deepwater Horizon display we'd been looking at around the corner to the wall listing the facility's sponsors—where BP and Chevron logos were prominently displayed.

The graduate students in English 7756 took this story to heart, imagining how thinking about stakeholders might impact their potential research projects; they also provided me feedback on potential research questions for the future. It was clear to all of us that the research I was seeking to do had to be shifted because I was asking the wrong question. The data sets that were relevant to most public agencies in Mobile County were data sets having to do with money; they were the data sets that showed how the oil affected the fishing and shipbuilding enterprises that are the backbone of the Mobile area economy. Graduate students in ENGL 7765 have come up with some new ways of thinking about this kind of research that I'm excited to put into practice. As they considered a wide variety of sources on risk communication, critically defined risk, and thought about my positionality, the biases of their instructor, and how this shapes their learning experience, they also began to develop theories about my research agenda—theories that were later able to help them shape their own work. Students have suggested that the lack of prevalence of visible embodied problems as a result of the oil has decreased the need for a risk response because the potential hazard seems low, and that the lack of focus on embodiment might be an issue of not making connections between health issues and the disaster. Several students also chose to take up projects that deal directly with embodiment, and three focused particularly on gender. More details on students' work throughout the course is included in the following section.

In sum, designing an apparent feminist pedagogy of risk for an online graduate course involved modeling my own research as a way of teaching research practices, focusing students' attention on the fallacy of pedagogical objectivity, creating situations in which students had to mediate risks to their own online identities, and encouraging students to think about disciplinary trends and the ways that interdisciplinary inquiry can help us to move scholarly conversations toward recognizing unequally distributed risks and supporting social justice.

SPRING 2014: HAZARD, OUTRAGE, ENVIRONMENT, EMBODIMENT

In this section, I report on a number of the conversations and learning encounters that occurred in English 7765: Risk Communication during the Spring 2014 semester at East Carolina University (ECU). To better contextualize this information, I first briefly discuss the institutional space that influenced this course and the users of my apparent feminist risk communication pedagogical design.

Institutional Context

This course was taught through East Carolina University, a fast-growing public doctoral/research university in Greenville, North Carolina. ECU is located in a primarily rural area and serves an ethnically diverse student population. Technical and Professional Communication (TPC) courses are taught in the Department of English. As a result of hazards—particularly flooding—introduced by Hurricane Floyd in 1999, the department began a push to offer online courses (Michelle Eble, personal communication, May 15, 2014). Today, students can earn a graduate certificate in TPC (which requires five courses), an MA in English (with a variety of concentrations, one of which is TPC), or a PhD in Rhetoric, Writing, and Professional Communication. The certificate is offered only online, the MA can be earned entirely online, and the PhD is a face-to-face program; both online and face-to-face courses are available for MA and PhD students.

English 7765: Special Topics in Technical and Professional Communication first appeared in the catalog in 2003–2004. Historically, the course has been used by faculty members wanting to expose students to cutting-edge topics that do not yet have their own courses. Some TPC courses run as special topics courses initially and are later developed into stand-alone courses that become a permanent part of the curriculum, which seems a likely future for Risk Communication. Prior to my offering this course as risk communication in Spring 2014, the course had been offered a number of times with this topic and has always been popular with students.

Users

I conceptualize users for this course in three groups: (1) the professor, (2) un-enrolled participants, and (3) enrolled students. As a user, I engaged in many of the same ways students did, in addition to my professorial duties. I participated in public conversations with purposeful decreasing frequency over the semester in order to move students into the role of experts. I also spent considerable energy working within the course's private Blackboard space and in private email and phone conversations with students. Ultimately, I gleaned many of the same benefits as students. That is, I was able to develop my research and my understanding of the nature of risk communication as a result of this course.

Un-enrolled participants included a faculty observer (Michelle Eble) who was writing a teaching evaluation for me, as well as anyone who accessed the class's public website (including observers of and participants in the conversations on our public website). For example, in the first week of the course, students spent time reading *The Peter M. Sandman Risk Communication Website*, and one student, Paul, extended Sandman's arguments for full-disclosure by noting that full-disclosure encourages self-reflection. Sandman himself replied to this comment: "I've been lurking on this page (how could I help it!), not intending to comment. But I just had to say that your point that full disclosure helps protect the communicator from self-disclosure is superb. I wish I'd included it in the column!"[6] From this time on, students were very aware that the scholars whose work they were discussing were among their potential audiences and that this affected the professional risks they were willing to take.

The students who enrolled in the course were the main user group; my focus was on promoting benefits (as articulated above) for the students participating in the course. At the end of the first week of class, fourteen students were enrolled in the course. Of those fourteen students, eleven ultimately completed the course. One student dropped in Week 3 as a result of financial aid issues. One student dropped in Week 9 and one student dropped in Week 10 due to the heavy workload. Of the eleven students who completed the course, one was an undeclared student-at-large, five were TPC certificate students, and five were pursuing an MA in English. Of the MA students, four were pursuing concentrations in TPC, and one was pursuing a concentration in multi and transnational literatures (and enrolled in the course based on the reading list).

Contextual Reflections

In addition to learning about the pedagogical theoretical framework described above, students did much of their work (which included weekly participation in discussions, a requirement to lead discussion, a research proposal, an annotated bibliography, a digital presentation, and a term paper) in a public space (riskcomm7765.wordpress.com) that required them to constantly engage and mediate risks to their own professional reputations as they worked toward course objectives. Those course objectives included:

- define the field of professional communication and its intersections with risk communication
- research the connections between methods and methodologies in risk communication
- articulate the importance of cultural studies to work in risk and technical communication
- learn about publications (such as proceedings, peer reviewed journals, and books) especially relevant to the risk communication
- gain an understanding of what disciplines aside from technical communication are concerned with risk communication
- review strategies for evaluating both print and digital publications/presentations in risk communication
- acquire an understanding of research strategies (including digital research strategies) you can use to find secondary research about risk communication

The requirement to engage in a public, digital space—which made it possible for scholars whose work students read to respond to discussions—served to underscore the importance of public intellectualism (Bowdon 2004) and engaging graduate students in scholarly conversations, as well as requiring a meta-level conversation about risk at every stage of the course. My focus on the digital as space aligns with Verzosa

Hurley's (this collection) arguments about the importance of "how attention to rhetorics of space, place, and location may help to cultivate critical spatial perspectives that can enrich our teaching of technical communication" (94). In addition, this meta-conversation created an opportunity for me, the instructor, a first-year assistant professor, to talk to students about risks associated with technical communication curriculum design and teaching.

To this end—because getting students to think critically about the design of a course itself is productive and important to apparent feminist pedagogies—I assigned my own work (Frost 2013b) in Week 2. I made my goals for doing this apparent to students by telling them, "I assigned my own work because I think it's important for students to know where their instructors are coming from and how they see the world. Even if this article did not teach you much about risk communication, it should definitely have taught you a bit about how I think and thus how the class will be framed . . . which then puts you in a place to take more control of your own learning." Further, my prompt for students' responses that week included the following: "Based on what you know now about my work in risk communication, what conclusions might you draw about the framing of this course? What might this mean you need to do on your own to get the maximum benefit?" Students were, understandably, careful about answering this question. This was, after all, only the second week of the course and they couldn't have had much of a sense of the risks they'd run by critiquing the professor's work. However, several students did engage productively with the question. Katrina, for example, said that "after reading Dr. Frost's article, I could tell that culture does play a huge role in risk communication. This shows that there will be a personalized solution for each situation, being that every culture is not the same, and will have to be approached differently." Katrina's response was fairly representative in that most students could identify, as Lacey does here, "Dr. Frost's belief that communication created by the few, is not effective when it is separated from society and the culture at-large." Likewise, Heather identified the course's focus on social justice work and further articulated a question she would keep in mind throughout the course: "what is the best solution for this specific group of people?" Pierre—a student who had previously taken coursework with me—even explicitly connected at that early juncture that the course's orientation was "a reflective impression of 'apparent feminism.'"

For me, the biggest surprise of the class came in students' resistance to a foundational reading. This certainly seemed to be a case where "student resistance is a site to begin an inquiry into the ways in which students have internalized the dominant cultural narratives" (Medina and Walker, this collection, 52). Students were generally on board with the course's focus on the importance of culture, but several students struggled with Grabill and Simmons's (1998) groundbreaking article on risk as socially constructed. Students read this work alongside my own work (Frost 2013b) in Week 2, so this resistance/struggle came in direct relationship with their own acceptance of culture as a part of risk, as described in the paragraph above. For example, Paul engaged in a lengthy critique of the article, and one of his complaints was that the article was "too 'theoretical' and lacking any sustained concrete examples or case studies to illustrate what the authors really mean." Because Paul was not the only student who was critical of the article but had done the most detailed and careful job articulating his concerns with it, I chose to engage him in direct conversation. I pushed him to suspend his disbelief, and I tried to provide some context for the importance of the article to disciplinary conversations. I also tried to gently correct what appeared to me at the time to be a misunderstanding by saying, "it's important to note that Grabill and Simmons don't claim their new approach to be a fix-all; rather, they claim that it might be a better-but-not-perfect perspective." In hindsight, I believe many of the students who were critical of Grabill and Simmons had likely read the syllabus and my introductory materials carefully and were expecting largely contextual readings rather than theoretical ones. Thus, the Grabill and Simmons piece produced a significant amount of cognitive dissonance given the explicit managing I had required of students' expectations for the course.

One of the most successful aspects of the course was students' work on the connections between risk and embodiment. In thinking about Sandman's (2014) formulaic approach to risk (Risk = Hazard + Outrage), I tried to get students to think about whose bodies are invested with the privileges to determine hazard, and whose outrage is most commonly taken up.[7] John, an MA student whose thesis focuses on teaching and learning cultural competence in TPC, focused on Sandman's health-related posts and demonstrated the cultural complexity of dealing with risk in health situations by asking a series of questions:

> So, as an example, how would a technical and professional communicator work with public health officials to communicate to a cultural group who may have no facility with English, or access to television, radio, or the Internet? Will their employers care about providing such information at all? And, as Sandman states, as the threat levels of such possible pandemics ebb and flow, how do you keep your audience interested in making

periodic checks for updates. When will vaccines be available? Where will they be available? What are the possible risks? Are there special risks for children or women? How do you competently communicate the special risks to women in a culturally sensitive way?

John's response indicates that even at this early point in the class, he was considering and balancing the effects of cultural experiences on risk messages, the rhetorical effects of articulating too much risk too often, and methods for reaching particularly embodied audiences in ethical ways. Meanwhile, Paul, who works as a medical writer, spent significant time in his introduction explaining the various audiences he writes for when developing labels for medications:

> This labeling has to accurately communicate the risks of using the product to both doctors and patients and provide them with a sufficient amount of information to make an informed decision about using the product. Before the marketing approval, there also has to be risk-related communication to doctors involved in the clinical trials, to the patients who volunteer to take part, and to the regulatory authorities that watch over the process.

Ultimately, peers asked Paul additional questions that led to him providing further materials on the process of writing medical labels and the various cultures that qualify as stakeholders in that process. This provided early impetus for students to think about transcultural risk communication, a theme that continued to develop throughout the course.

Another theme of the class involved users' interest in gender-based inquiry. Ultimately, three students ended up doing final projects with gender studies perspectives as a major component of their theoretical approach. However, I was unsurprised to have two students who clearly and purposefully self-identified as *not* feminist to me in private emails during the semester. Many of my teaching strategies for demonstrating connections between risk and embodiment have to do with women. This makes logical sense to me, as feminists have long critiqued the tendency of hegemonic discourses to make unapparent the value of embodied epistemologies, and many feminist works have argued that experiential knowledge is both important and overlooked (Belenky et al. 1986; Hesse-Biber 2007; Ramazanoğlu and Holland 2009). As an example— one that I often use with students—the technomedical takeover of medicine was all about bodies and consolidating power with certain bodies; women dominated the field of healing until enterprising men realized they could use technical procedures like licensure and documentation to criminalize and marginalize experiential/embodied knowledges and practices (like midwifery, for example) in order to elevate their own prestige (Ehrenreich and English 1973; Seigel 2013). I argue to students that there is no such thing as an unbiased source since all people have lived, embodied realities and experiences that affect the way we think. The two students who identified as not feminist were assuring me that they understand this and agree, but that they do not have to be feminists to do so.

The feedback I have received on the class leads me to believe that apparent feminism was a productive but challenging vehicle for students to think through gendered inequities and their own self-identification. My apparent feminist acceptance and encouragement of non-feminist perspectives on enculturated and embodied risk appear to have been a productive—if not always comfortable—learning experience for most of the students who were enrolled in this course. Teaching evaluation results were quite positive, with many students noting that they especially liked the way the course brought together theory and application; one student stated, "I honestly feel that I could assist an organization or governmental entity in preparing and analyzing risk communication products for almost any given event or problem." Others appreciated the accessibility and scope of the course. Prevalent critiques were the cost of books and the heavy reading load. When asked to rate how demanding the course was compared with other courses they've taken at this institution, the eleven students in this class produced a mean 4.27 response on a 5-point scale where 5 indicates "very demanding." Two students expressed discomfort with the amount of work done on the public website. One student suggested that "[I]f students would alternate responses so that they do not respond to the same individuals each week, classroom discussions might be more engaging." Several students noted that my apparent feminist pedagogy facilitated a semester-long collaboration between classmates, which they deeply valued.

Student work throughout the course scaffolded toward the final project.[8] Students wrote a research proposal before or during Week 7 that required them to articulate a research area and a succinct research question, create a research plan, and identify potential venues for their proposed work. They also completed a detailed annotated bibliography in Week 11. These assignments were turned in via Blackboard discussion boards so that all students could benefit from observing others' work and so they could share sources. During the final two weeks of the course, students completed an article-length research project, as well as a digital final project presentation of 15–20 minutes in length to disseminate their findings. The final project presentations ranged from an interactive website to direct audiovisual recordings of conference papers and a website with audiovisual and captioning components. The final projects themselves focused on a variety of topics related to risk and embodiment:

- a detailed study on texting and driving among youth where the student/author, a high-school teacher, surveyed students to see how they defined risk and led them in multiple primary research projects (Heather)
- an interrogation of the coal ash spill and Duke Energy's risk practices (Mary)
- an exploration of the FDA's guideline document, titled "Presenting Risk Information in Prescription Drug and Medical Device Promotion," for evaluating the ethics and appropriateness of advertisements for prescription drugs and medical devices (Paul)
- an in-depth study of how risk communicators establish trust in a community, focusing on HIV/AIDS activism (Katrina)
- a localized study of risk-taking behaviors and perceptions by youth (precipitated by teen pregnancy rates) in two North Carolina counties (John)
- a theoretical exploration of the relationships between risk communication theories and public intellectualism (Pierre)
- a study of risks communicated and created by depictions of women's bodies in mass media (Lacey)

Finally, it bears a careful mention that one student was deeply concerned about the public nature of the course conversations. This student—who I've worked with previously and have an ongoing intellectual relationship with—shared these concerns with me during Week 4. As a solution, I made private discussion boards for each remaining week available in the course's Blackboard space and provided links to material in the public discussion space, as well as reminding the class that they should remember also to engage with the private boards from that point on. During Week 4, this student used the private Week 4 board to begin an extended and productive conversation that I and one other student participated in. However, this student never again used the private boards. Instead, she began thinking about other ways to mediate privacy risks in the public space. Over the course of the semester, she used her full name, an abbreviated version, and a set of initials; she also significantly limited her participation in the public space because of privacy concerns. While this was certainly a challenge for both her and me to navigate throughout the course, it also demonstrated that my goal of getting students to carefully consider the risks associated with intellectual debate in public spaces was successful.

IMPLICATIONS AND RISKS

In my own research on the aftereffects of the Deepwater Horizon disaster, I found that public health/embodiment sometimes runs counter to existing cultural and environmental practices. I was surprised, as I modeled for students, that my research question was unanswerable because people were, for the most part, not talking about health concerns related to the oil spill. However, this error on my part led to an important realization for teaching English 7765: risk communicators must emphasize the difficulty of navigating moral conduct and cultural relativity, and we must recognize the interconnectedness of economic and embodied concerns. For example, I must tread a careful path between prescribing increased attention to health concerns and respecting the economic benefits oil companies provide to coastal cultures like those in Mobile County. In short, risk communicators—including risk communication students—need to use a variety of epistemological perspectives, including apparent feminist perspectives, to think about how we define risk, and we need to re-consider how our perspectives influence our perceptions of the connections between risk, outrage, hazard, environment, economics, and embodiment.

Through this graduate-level risk communication course where I used an apparent feminist pedagogy to ask students to reconceptualize how risk attaches to particular bodies, I found that students were able to articulate and analyze a variety of embodied and enculturated risk situations that I was aware of but hadn't identified. These students helped me to understand and extrapolate some of the ways technical communicators can intervene in constructions of risk that do not take into account the ethical effects on those who speak from the margins. For example, students chose a variety of media for the final project presentations precisely because they were conceptualizing different ways of getting their messages out into the world. At least two students in this course will continue working on their final projects as they pursue publication, with my support. If published, their work will provide much-needed case studies demonstrating the ways that risk affects diverse populations and what can be done to mediate those effects.

The work completed by students in this class was specific to risk communication, but such apparent feminist pedagogies are applicable to any technical communication course. Because apparent feminism makes a point of interrogating texts often considered objective or neutral—as technical communication in many forms often is—any technical communication course is a productive site for apparent feminist inquiry. To enact an apparent feminist pedagogy, instructors can:

- Work to make their own self-identifications and positionalities understandable to students.
- Encourage students to think about how the instructor's situatedness affects the design and function of the course.
- Model behavior that is socially just and culturally aware.
- Ask students to think about how trends in disciplines and in teaching (at the university level, in English departments, in technical communication classes—whatever context is useful in a particular setting) affect the way that students and the instructors approach the class.
- Ensure that they are assigning a range of material for students to read. This means paying attention to the genre of materials (from discipline-based textbooks to bestsellers that widely influence public perception) and to the sex, gender, ethnicity, theoretical orientation, and goals of the authors of course texts.[9]
- Lead students to consider the situatedness of the authors of class texts and other technical documents and stimulate their interest in thinking about the authors of apparently authorless documents that purport to be neutral or objective (e.g., legislation, manuals, reference works).
- Require interaction with gender-based and race-based issues.
- Facilitate collaboration between students.
- Support alternative forms of knowledge, learning, and production.
- Create space for and value complex reflection on students' own identities, embodiments, cultural situatedness—and attendant risks.

By encouraging apparent feminist inquiry through this kind of critical and responsive apparent feminist pedagogy, apparent feminism can offer the field of technical communication new ways to understand, critique, revise, and produce technical documentation. The use of apparent feminist pedagogies also supports the development of a new generation of critical technical communicators (including instructors) who understand the importance of social justice to seemingly neutral, objective, or rational documents and spaces.

While I provide this chapter as a model for other instructors of technical communication to take up, I also must note that there are risks involved in teaching a course like this. Focusing a seemingly technical course on cultural issues would likely invite student resistance in many contexts. Although I did not experience much of this kind of resistance, instructors who are embodied in different ways than I am, who have a different level of job security, who work in more traditional departments, and/or who work at less diverse institutions than I might run different sorts and levels of risks. As Del Hierro (this collection) notes, teacher-scholars "followed through the tracks set by the academy, and we should not lose sight of how our bodies have been disciplined by the academy, whether we resisted or not" (177). Further, instructors who lack the departmental support that I had for this project might feel differently about the potential hazard—to their livelihood, professional reputation, and future job prospects—associated with teaching such a course. While I firmly advocate for critical and apparent feminist approaches to teaching risk and technical communication, I would be remiss not to emphasize that the embodied, lived nature of risk also makes such a project one that bears unequal risks for potential instructors.

In sum, this chapter:

- supports attention to social justice, cultural theory, and intellectual diversity in technical communication's practitioners, scholars, users, products, theories, pedagogical approaches, and stories
- analyzes the risks that technical communication teacher-scholars and students run in doing such work
- offers one approach for navigating this challenging but worthwhile intellectual space
- suggests that risk communication courses are a particularly productive space to do this work because this sort of pedagogy is a response to embodied risks

Ultimately, apparent feminism—one of many productive cultural theories emerging in technical communication—is a necessary, useful, and usable approach to technical communication teaching that can help the discipline move toward socially just pedagogies and practices.

Notes

1. For more on modern, intersectional feminist approaches to technical communication, readers need look no further than this collection. Chapters by Moeller and Smyser-Fauble utilize feminist disability studies, Moore offers an informed look at how womanism can complement (and complicate) feminisms, and Edwards explains how "Feminism and legal indeterminacy, together, make for a rich theoretical tool that has structural racism issues at the center and encourages interrogation of other forms of oppression in the process" (273).
2. This study has been approved by the East Carolina University and Medical Center Institutional Review Board, protocol number 13–002453.
3. See, for examples, the following: Blythe, Grabill, and Riley 2008; Bowdon 2004; Ding 2009; Dombrowski 1994; Evia and Patriarca 2012; Grabill 2007; Lindeman 2013; Ross 1994; Sauer 1996, 2003; Scott 2003, 2004; Simmons 2007; Simmons and Grabill 2007; Stratman et al. 1995; Youngblood 2012.
4. Some might make the argument that events before one's own lifetime also affect embodiment; I agree with this, but I retain the language above to emphasize that it is an individual user's *perception* of such events that will affect his or her embodied reality.
5. I initially accessed this photo, which is publicly archived on Wikipedia, via Alexander Street Press's *Engineering Case Studies Online* archive.

6. Please note that quotations from students, the instructor (me), and un-enrolled observers throughout this article are taken from the class's public website at https://riskcomm7765.wordpress.com, unless otherwise noted. Further, in an effort to "be ever vigilant in [my] reflective practice, thinking through possible harms to participants but also selecting carefully among the research methods that *are* and *are not* possible in digital environments" (Eble and Banks 2007, 40), I have done my best to provide a measure of privacy protection for the students in this class. I quote only from those students who consented to participate in the study, and I utilize pseudonyms of my own making in this chapter for all student quotations. However, the public website where much of the work of this course took place is still publicly available and, as such, ultimate agency over whether they are identifiable lies with students and the choices they made (and continue to make) about risk, embodiment, public intellectualism, and self-identification.
7. See Haas and Frost 2017 for a chapter in which we adopt an apparent decolonial feminist response to common patterns in risk communication, and we note that risks as we perceive them are most often attached to "the bodies of those with less access to mainstream or traditional power—Indians who remain on reservation land, women whose bodies accrue toxic chemicals, those without the economic means to flee an impending 'natural' disaster" (Haas and Frost 2017).
8. All required assignments were directly aimed at supporting the final project, except weekly conversations and a discussion leader assignment, which were intended to facilitate disciplinary thinking and exploration. For many students, much of this work still tied directly to the area they ultimately chose as their final project focus.
9. One of the critiques in my teaching evaluation was that I assigned a textbook that a student considered central to the field at the end of the course. The student said she/he would have liked to read this text first so that she/he could have read the other texts through that one. I had purposely encouraged the opposite—reading the textbook through the less traditional lenses we'd discussed earlier—but I did not make this point well enough and/or the student simply disagreed.

References

Belenky, Mary Field, Blythe Mcvicker Clinchy, Nancy Rule Goldberger, and Jill Mattuck Tarule. 1986. *Women's Ways of Knowing: The Development of Self, Voice, and Mind.* New York: Basic Books.

Blythe, Stuart, Jeffrey T. Grabill, and Kirk Riley. 2008. "Action Research and Wicked Environmental Problems: Exploring Appropriate Roles for Researchers in Professional Communication." *Journal of Business and Technical Communication* 22 (3): 272–298. https://doi.org/10.1177/1050651908315973.

Bowdon, Melody. 2004. "Technical Communication and the Role of the Public Intellectual: A Community WHV-Prevention Case Study." *Technical Communication Quarterly* 13 (3): 325–340. https://doi.org/10.1207/s15427625tcq1303_6.

Deepwater Horizon Accident Investigation Report. 2010. British Petroleum, 2010. Engineering Case Studies Online.

Deepwater Horizon Marine Casualty Investigation Report. 2011. Marshall Islands: Republic of the Marshall Islands Office of the Maritime Administrator, 2011. Engineering Case Studies Online.

Ding, Huiling. 2009. "Rhetorics of Alternative Media in an Emerging Epidemic: SARS, Censorship, and Extra-Institutional Risk Communication." *Technical Communication Quarterly* 18 (4): 327–350. https://doi.org/10.1080/10572250903149548.

Dombrowski, Paul M. 1994. "Challenger Through the Eyes of Feyerabend." *Journal of Technical Writing and Communication* 24 (1): 7–18. https://doi.org/10.2190/WXNY-RQ8C-J9L8-0QEQ.

Eble, Michelle, and Will Banks. 2007. "Digital Spaces, Online Environments, and Human Participant Research: Interfacing with Institutional Review Boards." In *Digital Writing Research: Technologies, Methodologies, and Ethical Issues*, ed. Dànielle DeVoss and Heidi McKee, 27–47. Cresskill, NJ: Hampton Press.

Ehrenreich, Barbara, and Deirdre English. 1973. *Witches, Midwives, and Nurses: A History of Women Healers*. Old Westbury, NY: Feminist.

Evia, Carlos, and Ashley Patriarca. 2012. "Beyond Compliance: Participatory Translation of Safety Communication for Latino Construction Workers." *Journal of Business and Technical Communication* 26 (3): 340–367. https://doi.org/10.1177/1050651912439697.

Fernández, Maria, Faith Wilding, and Michelle M. Wright. 2002. *Domain Errors!: Cyberfeminist Practices*. Brooklyn, NY: Autonomedia.

Frost, Erin A. 2013a. "Theorizing an Apparent Feminism in Technical Communication." PhD diss., Illinois State University.

Frost, Erin A. 2013b. "Transcultural Risk Communication on Dauphin Island: An Analysis of Ironically Located Responses to the Deepwater Horizon Disaster." *Technical Communication Quarterly* 22 (1): 50–66. https://doi.org/10.1080/10572252.2013.726483.

Frost, Erin A. 2014. "Apparent Feminist Pedagogies: Interrogating Technical Rhetorics at Illinois State University." *Programmatic Perspectives* 6 (1): 110–31.

Grabill, Jeffrey T. 2007. *Writing Community Change: Designing Technologies for Citizen Action*. New York: Hampton Press.

Grabill, Jeffrey T., and W. Michele Simmons. 1998. "Toward a Critical Rhetoric of Risk Communication: Producing Citizens and the Role of Technical Communicators." *Technical Communication Quarterly* 7 (4): 415–441. https://doi.org/10.1080/10572259809364640.

Haas, Angela M. 2012. "Race, Rhetoric, and Technology: A Case Study of Decolonial Technical Communication Theory, Methodology, and Pedagogy." *Journal of Business and Technical Communication* 26 (3): 277–310. https://doi.org/10.1177/1050651912439539.

Haas, Angela M., and Erin A. Frost. 2017. "Toward an Apparent Decolonial Feminist Rhetoric of Risk." In *Topic-Driven Environmental Rhetoric*, ed. Derek Ross, 168–86. New York: Routledge.

Hamilton, Myra. 2014. "The 'New Social Contract' and the Individualisation of Risk in Policy." *Journal of Risk Research* 17 (4): 453–467. https://doi.org/10.1080/13669877.2012.726250.

Hart-Davidson, William. 2001. "On Writing, Technical Communication, and Information Technology: The Core Competencies of Technical Communication." *Technical Communication (Washington)* 48 (2): 145–55.

Hayles, N. Katherine. 1999. *How We Became Posthuman: Virtual Bodies in Cybernetics, Literature, and Informatics*. Chicago: University of Chicago Press. https://doi.org/10.7208/chicago/9780226321394.001.0001.

Hesse-Biber, Sharlene Nagy. 2007. *Feminist Research Practice: A Primer*. Thousand Oaks: SAGE Publications. https://doi.org/10.4135/9781412984270.

Johnson, Robert R. 1998. *User-Centered Technology: A Rhetorical Theory for Computers and Other Mundane Artifacts*. Albany, NY: SUNY Press.

LaDuke, Winona. 1999. *All Our Relations: Native Struggles for Land and Life*. Cambridge, MA: South End.

LaDuke, Winona. 2005. *Recovering the Sacred: The Power of Naming and Claiming*. Cambridge, MA: South End.

Lindeman, Neil. 2013. "Subjectivized Knowledge and Grassroots Advocacy: An Analysis of an Environmental Controversy in Northern California." *Journal of Business and Technical Communication* 27 (1): 62–90. https://doi.org/10.1177/1050651912448871.

Macondo: The Gulf Oil Disaster, Chief Counsel's Report. 2011. National Commission on the BP Deepwater Horizon Oil Spill and Offshore Drilling, 2011. Engineering Case Studies Online.

Miller, Carolyn R. 1989. "What's Practical About Technical Writing?" In *Technical Writing: Theory and Practice*, ed. Bertie E. Fearing and W. Keats Sparrow, 14–24. New York: MLA.

On Scene Coordinator Report, Deepwater Horizon Oil Spill: Submitted to the National Response Team, September 2011. 2011. United States Coast Guard, 2011. Engineering Case Studies Online.

Plough, Alonzo, and Sheldon Krimsky. 1990. "The Emergence of Risk Communication Studies: Social and Political Context." In *Readings in Risk*, ed. Theodore S. Glickman and Michael Gough, 223–30. Washington, DC: Resources for the Future.

Ramazanoğlu, Caroline, and Janet Holland. 2009. *Feminist Methodology: Challenges and Choices*. London: Sage.

Ridolfo, Jim, and Dànielle Nicole DeVoss. 2009. "Composing for Recomposition: Rhetorical Velocity and Delivery." *Kairos* 13 (2). http://kairos.technorhetoric.net/13.2/topoi/ridolfo_devoss/.

Ross, Susan Mallon. 1994. "Technical Communicative Action: Exploring How Alternative Worldviews Affect Environmental Remediation Efforts." *Technical Communication Quarterly* 3 (3): 325–342. https://doi.org/10.1080/10572259409364575.

Sandman, Peter M. 2014. *The Peter M. Sandman Risk Communication Website*. http://www.psandman.com.

Sauer, Beverly A. 1996. "Communicating Risk in a Cross-Cultural Context: A Cross-Cultural Comparison of Rhetorical and Social Understandings in U.S. and British Mine Safety Training Programs." *Journal of Business and Technical Communication* 10 (3): 306–329. https://doi.org/10.1177/1050651996010003002.

Sauer, Beverly A. 2003. *The Rhetoric of Risk: Technical Documentation in Hazardous Environments*. Mahwah, NJ: L. Erlbaum Associates.

Scott, J. Blake. 2003. *Risky Rhetoric: AIDS and the Cultural Practices of HIV Testing*. Carbondale: Southern Illinois University Press.

Scott, J. Blake. 2004. "Tracking Rapid HIV Testing through the Cultural Circuit: Implications for Technical Communication." *Journal of Business and Technical Communication* 18 (2): 198–219. https://doi.org/10.1177/1050651903260836.

Scott, J. Blake, Bernadette Longo, and Katherine V. Wills. 2007. *Critical Power Tools: Technical Communication and Cultural Studies*. Albany: State University of New York.

Seigel, Marika. 2013. *The Rhetoric of Pregnancy*. Chicago: University of Chicago Press. https://doi.org/10.7208/chicago/9780226072074.001.0001.

Simmons, W. Michele. 2007. *Participation and Power: A Rhetoric for Civic Discourse in Environmental Policy Decisions*. Albany: State University of New York Press.

Simmons, W. Michele, and Jeffrey T. Grabill. 2007. "Toward a Civic Rhetoric for Technologically and Scientifically Complex Places: Invention, Performance, and Participation." *College Composition and Communication* 58 (3): 419–48.

Smith, Andrea. 2005. *Conquest: Sexual Violence and American Indian Genocide*. Cambridge, MA: South End.

Stengel, Camille. 2014. "The Risk of Being 'Too Honest': Drug Use, Stigma and Pregnancy." *Health Risk & Society* 16 (1): 36–50. https://doi.org/10.1080/13698575.2013.868408.

Stratman, James F., Caroyn Boykin, Marti C. Holmes, M. Jane Laufer, and Marion Breen. 1995. "Risk Communication, Metacommunication, and Rhetorical Stases in the Aspen–EPA Superfund Controversy." *Journal of Business and Technical Communication* 9 (1): 5–41. https://doi.org/10.1177/1050651995009001002.

Turkle, Sherry. 1995. *Life on the Screen: Identity in the Age of the Internet*. New York: Simon & Schuster.

Wikipedia. 2010. "File:Deepwater Horizon oil spill - May 24, 2010 - with locator.jpg." June 18, 2010. https://en.wikipedia.org/wiki/File:Deepwater_Horizon_oil_spill_-_May_24,_2010_-_with_locator.jpg.

Wildcat, Daniel R. 2009. *Red Alert! Saving the Planet with Indigenous Knowledge.* Golden, CO: Fulcrum.

Williams, Miriam F., and Octavio Pimentel. 2014. *Communicating Race, Ethnicity, and Identity in Technical Communication.* Amityville, NY: Baywood.

Williams, Patricia J. 1991. *The Alchemy of Race and Rights: Diary of a Law Professor.* Cambridge, MA: Harvard University Press.

Woods, Clyde. 2010. *In the Wake of Hurricane Katrina: New Paradigms and Social Visions.* Baltimore: Johns Hopkins University Press.

Youngblood, Susan. 2012. "Balancing the Rhetorical Tension Between Right-To-Know and Security in Risk Communication: Ambiguity and Avoidance." *Journal of Business and Technical Communication* 26 (1): 35–64. https://doi.org/10.1177/1050651911421123.

Zhou, Yang, Ning Li, Wenxiang Wu, Jidong Wu, and Peijun Shi. 2014. "Local Spatial and Temporal Factors Influencing Population and Societal Vulnerability to Natural Disasters." *Risk Analysis* 34 (4): 614–639. https://doi.org/10.1111/risa.12193.

2
VALIDATING THE CONSEQUENCES OF A SOCIAL JUSTICE PEDAGOGY
Explicit Values in Course-Based Grading Contracts

Cruz Medina and Kenneth Walker

In 2012, in Tucson, Arizona, conservative Superintendent of Education Tom Horne used House Bill (HB) 2281 to outlaw Tucson High School's Mexican American Studies (TUSD/MAS) program. Despite demonstrated increases in graduation rates and state test scores (Cabrera, Milem, and Marx 2012), the social justice program was dismantled and books from the curriculum were banned, including Paulo Freire's (1970) *Pedagogy of the Oppressed*. As teacher-scholars concerned with critical consciousness[1] and the application of social justice theories to the classroom, we found these events highly disturbing and demonstrative of what Angela Haas and Michelle Eble refer to in the Introduction of this collection as "the mess of injustice in our own backyards" (11). In the TUSD/MAS program, we saw how a model of social justice pedagogy at a programmatic level can have a positive impact on underrepresented student populations. For us, this model provoked questions about implementing social justice practices into our own technical communication assignments, courses, and program. However, TUSD/MAS was also a cautionary tale: even the most successful social justice pedagogies, curricula, and programs can come under perennial critique by those who feel threatened by teaching critical engagement with the unequal distribution of privilege.

The story of the TUSD/MAS program tells us that resistance to social justice pedagogies should be sites of scholarly inquiry as much as they are sites of political struggle. In technical communication, social justice represents a set of theories, methods, and practices that illuminate and respond to social and institutional inequality in courses, vertical curricula, and degree programs. In seeking our own pedagogical praxis based in critical consciousness, we recognized a need for this framework to interrogate institutional power relations throughout these sites, but particularly at the

DOI: 10.7330/9781607327585.c002

overlapping sites of student-teacher interactions, instructor evaluations, and course assessment. As Marcos Del Hierro explains in chapter 7 of this collection, classroom cultures and course-based assessments need to be interrogated because "[s]tudents with power and privilege dominate classroom discussions, expect to make the highest grades, and feel no obligation to interrogate their power and privilege" (175). In order to avoid re-inscribing systems of exclusion and oppression, evaluation and assessment should both work to critique the exercise of privilege and be inclusive of non-white students with varying levels of privilege.

Given these concerns, grading contracts seemed an intriguing place to start enacting social justice course-based assessments because of their purported ability to respond well to culturally diverse student populations and open potentials for student agency. Critical pedagogues like Ira Shor (2009) and Jerry Farber (1990), for example, integrated grading contracts into their curricular practices because it allows them to partially de-center power and enter into more authentic dialogues with students about course material and the social implications of the curriculum. Despite the general scarcity of the use of grading contracts in technical communication pedagogy (Wolvin and Wolvin 1975), grading contracts open up certain social justice affordances that contribute to what Haas and Eble call the "turn toward a collective disciplinary redressing of social injustice" (3). But we also saw a need to frame grading contracts critically, so that they might carefully attend to unequal distributions of power and access and perhaps obviate the perennial critique of social justice efficacy that allows educators to open up course-based assessments to a negotiation beyond the instructor's perception of success.

To meet this need, we offer consequential validity as a broadly applicable framework and grading contracts as a broadly applicable tool for integrating social justice values into the course-based assessment designs of technical communication (TC) pedagogy. Consequential validity is an inquiry framework that uses explicit values to interrogate the potential and current effects of our pedagogy for all students, but in this case, particularly for systemically marginalized students. At first glance, an inquiry into the intersections between classroom assessment and social justice may seem suspect. After all, the current culture of assessment is complicit in fostering inequality through its bias for normative student subjectivities, discourses, competencies, and performances that historically have served to oppress non-normative students of all kinds. Grading contracts have this potential as well. But it is the deliberative processes encouraged by the framework of consequential validity that we believe has ability to travel, mobilize, and build powerful frameworks for course-based assessments, especially for students of color (Gallagher 2012; Inoue 2009; 2012). Social justice "[advocates] for those in our society who are economically, socially, politically, and/or culturally underresourced" (Agboka 2013, 28), and consequential validity at the site of grading contracts provides a needed contribution for critical teacher-scholars in technical communication who seek ways to assert more socially just evaluations of their interactions with students.

In practice what this means is that grading contracts can potentially serve as a site to facilitate a conversation about the values students and teachers should be held to *and* how we might use the teacher/student dynamic to faithfully represent these values throughout the course of a semester. Much like Frost's apparent feminist pedagogy (see chapter 1), explicitly foregrounding social justice through consequential validity is another way to question the rhetorics of objectivity and neutrality common in technical communication classrooms. With this framework, instructors of technical communication can use grading contracts to ask questions like: how is the student constructed in this grading system? What kinds of agencies, competencies, and performances are valued, and do these concepts align with the values of social justice to advocate for the socially, politically, and/or culturally under-resourced? In other words, designing grading contracts with the frameworks of consequential validity makes explicit the values that are generally implicit in other methods of grading, and by foregrounding the values of social justice, grading contracts have the potential to destabilize the exercise of privilege in the technical communication classroom.

In what follows, we outline the affordances of associating technical communication course assignments and student/teacher/institutional power dynamics within the framework of consequential validity for social justice pedagogy at the site of grading contracts. Next, we provide two models for developing and using consequential validity as a framework for transforming grading contracts into de-centered negotiations of privilege in technical communication curricula. The first model examines student responses to grading contracts and course readings by a scholar of color to highlight the ways in which students can resist these assessment-based social justice tools. The second model shows how community-based projects with grading contracts can expose students to under-resourced organizations that deepen and complicate student understandings of social justice issues beyond the classroom. Finally, through a personal reflection on student/teacher rhetorical situations, we speak to the limits of consequential validity in grading contracts and outline a few avenues for future research.

SOCIAL JUSTICE PEDAGOGY, CONSEQUENTIAL VALIDITY, AND EXPLICIT VALUES IN GRADING CONTRACTS

In the past fifteen years, technical communication has undergone a "cultural turn" in terms of the scholarship developed since Bernadette Longo's (1998) call for cultural studies inquiry (Scott and Longo 2006; Scott, Longo, and Wills 2006). Scott (2004) advocated for integrating critical practices into curriculum and pedagogy in ways that parallel the efforts of critical pedagogues negotiating the praxis of social justice education. Our own integration of grading contracts stems from critical reflection on pedagogical and curricular efforts to both highlight and effect change with regard to the unfair playing field for nonwhite students who might enter our classrooms with less preparation, cultural capital, or institutionally authorized knowledge (Yosso 2006). Thus, our own approaches to technical communication pedagogy begin by asking what kinds of student performances do we value, and what are the potential consequences on systemically under-privileged students?

A part of the cultural turn was the advocacy for radically contextualized knowledge production that demystified notions of a universal audience and acknowledged those voices who do not echo the bourgeois white, male voice privileged by both the academy and industry (Herndl 2004, 3–8). The social justice turn in technical communication has extended this advocacy in part by acknowledging that technical communication has been shown to contribute to the erasure of people of color (Johnson, Pimentel, and Pimentel 2008). To reconcile omissions in the field, Haas (2012) posited a critical race approach sensitive to the representational and relational dynamics of cultural histories and material bodies. Agboka (2013) recently argued that social justice can be accomplished in technical communication through participatory localization that considers the user of texts in under-resourced cultures, communities, and other contexts (28–29). Yet, in conducting this work, it is important to highlight the racial component of cross-cultural communication that illuminates the unequal balance of power relations between document composer and audience. Turning our attention to the power relations between students and teachers, particularly through the institutionally sanctioned mode of grading, should account for unearned privileges such as race and class. In advocating for social justice at the site of grading, we attend to the construction of students by curricular tools and assessments by asking what student performances, literacies, and competencies can we value that might advocate for under-privileged students?

Scholars of technical communication recognized fairly early on the value of engaging assessment on their own terms (Allen 1993; Beard, Rymer, and Williams 1989; Coppola 1999). A recent special issue of *Technical Communication Quarterly* shows how the assessment of multimodal practices can successfully shape the conversation about effectiveness of teaching with new media (Ball 2012; Barton and Heiman 2012; Manion and Selfe 2012; Morain and Swarts 2012). Han Yu (2008, 2012) has made similar agentive claims for assessment and the increasingly overlapping areas of workplace writing and intercultural competence. And while cultural, racial, and social justice theories have had a broad influence on technical communication scholarship and pedagogy (Haas 2012; Scott, Longo, and Wills 2006; Williams and Pimentel 2012), the field needs more scholarship that examines how these bodies of literature might also shape practices in course-based assessments. The time seems right, then, to begin to inquire into the ways in which social justice theories might have an influence on our course-based assessments that may also afford interrogations into programmatic and institutional relations of power.

One line for this inquiry might begin with Gallagher's (2012) suggestion to make assessment a critical rhetorical practice by rearticulating outcomes with consequences. In Gallagher's view, outcomes reproduce institutional and ideological logics that divert attention away from important contextual variables like resources, working conditions, and the race and class inflected notion of student preparation. Outcomes can privilege efficient measurements for institutional purposes, often at the expense of critical inquiry for pedagogical purposes (46). To counter these practices, Gallagher suggests that inquiring into consequences and consequential validity can foster a sense of potentiality in our assessments that attends to both the intended and unintended results of our interactions with students. This position, he suggests, can also help negotiate the inherent tension between programmatic coherence on the one hand and singularity and potentiality on the other (56).

Traditionally understood, validity is assessment's evaluation of truth. Validity asks us to inquire: did our tools capture what we set out for them to capture? In this way validity defines the degree to which theory and evidence adequately and appropriately support the kinds of inferences and actions that assessments warrant (Messick 1989, 5–11). But as Inoue (2009) notes, validity inquiries do not represent universal theories, values, or rationales that warrant acontextual decisions; rather, validity theories are embedded with the values and expectations of a particular group: assessments do not give us "Truth" but rather "the best one can hope for is that assessment faithfully represents one's values" (109). This means validity is deeply rhetorical and hegemonic (109). So if our notions of validity are fundamentally about a representation of one's values, and as critical teacher-scholars we acknowledge a set of values rooted in social justice, then our assessments will have to somehow acknowledge the uneven distribution of knowledge, power, and access to resources found among student populations.

Because social justice asks that pedagogues endeavor to transform education so that it is liberatory rather than oppressive, curriculum, technology, and assessment all offer opportunities to address inequality, particularly when they are coherently integrated. Grading contracts represent agreements about classroom assessment that, when used effectively, put into action well-known commonplaces about motivation in student writing: students should be self-directed; students should have a sense of improvement; and students should write often with a clearly defined purpose. Contracts allow students to choose their own grade up-front, thereby agreeing to produce a corresponding amount of work, which the instructor grades based on meeting requirements such as page limits rather than its quality (albeit with the assumption that quality is a function of quantity). While grading contracts date decades back (Poppen and Thompson 1971; Taylor 1971; Yarber 1974), they have rarely been discussed with regard to countering institutional and social inequality. In writing studies broadly, scholarship has been dedicated to the adoption of grading contracts as a tool for dismantling, or at the very least de-centering, the hierarchical and intercultural relationships between students and professors within the rhetorical situation of the classroom (Farber 1990; Inoue 2009, 2012; Moreno-Lopez 2005; Shor 2009; Spidell and Thelin 2006). Less critically, Danielewicz and Elbow propose the use of grading contracts to reduce the time-consuming grading process, while improving learning and teaching. Approaching grading contracts from a technical perspective, Danielewicz and Elbow provide a useful definition of contract:

> the term "contract" aptly describes the type of written document that spells out as explicitly as possible the rights and obligations of all the parties—a document that tries to eliminate ambiguity rather than relying on "good faith" and "what's implicitly understood." (Danielewicz and Elbow 2009, 247)

Their definition elides the glaring contradiction between the legally-binding corporate connotations of "contract," and the humanitarian ethos of a social justice approach. Yet, they push back against the dehumanizing effect of contracts by explaining that they allow "us to present ourselves and our teaching authority more openly, humanly, and directly than most syllabi do" (253). The negative legal and capitalistic connotations of a contract also concern scholars such as Farber, Shor, Spidell, and Thelin who recognize contracts as agreements that reconstitute the asymmetrical relationship between student and teacher, and between individual students and the class as a whole. Whatever the terminology, however, the effectiveness of grading contracts should be evaluated up against the consequences for students, and in our case in particular, the intended and unintended consequences for instructors and students of color.

Contract grading has been shown to illuminate the privilege of students who resist this mode of assessment because of disrupted social power. For example, Spidell and Thelin (2006) equate student resistance to grading contracts as a form of elitism marked by "adherence to the status quo and little or no tolerance for those viewed as subservient or undeserving of the chance to better themselves. . . ." (44). Grading contracts can be challenging for both students and educators because institutionalized inequality is supported by systems of power that anesthetize students to their potential to transform their relationship with education and privilege. Yet Inoue's (2012) findings suggest ways in which grading contracts could undermine the expectations of privileged students accustomed to benefiting from institutionalized systems of power that uphold inequality (78–93). So rather than viewing student resistance to grading contracts negatively, critical pedagogy asserts that student resistance is a site to begin an inquiry into the ways in which students have internalized the dominant cultural narratives of grades, technologies, and instructors and, more particularly in our case, instructors of color. In applying concepts of consequential validity to the assessment site of grading contracts, our hope is that they work to both disrupt traditional exercises of privilege and advocate for the marginalized.

CONSEQUENTIAL VALIDITY INQUIRY MODELS FOR GRADING CONTRACTS IN TECHNICAL COMMUNICATION PEDAGOGY

Consequential validity as a framework is not about mainstreaming shared values, which is problematic, especially for systemically lesser privileged. Instead, it is about making course values explicit. In this case, consequential validity can make social justice values explicit in the course grading system so as teachers we can explicitly value equitable labor and processes, not privilege. Here our goal is to provide two models for developing and using grading contracts in technical communication pedagogy. Both models use consequential validity inquiries by attending to the ways in which student agencies, competencies, and performances are valued, and by aligning these performances with the values of social justice. In other words, despite the differences in these models, they both use grading contracts as course-level evaluation systems that open up conversations about explicit values and refocus our attention on potentiality of student-teacher interactions rather than the limited measures of what is observable or not at any one given time. But explicit values also open the space for student resistance—a key site for the interrogation of power relations.

MODEL 1: CONSEQUENTIAL VALIDITY AND GRADING CONTRACTS IN A DIGITAL WRITING COURSE

Medina's experience began in Jerry Farber's *Teaching Literature* graduate seminar in 2003, where Farber employed grading contracts as outlined in his anthologized "Learning How to Teach: A Progress Report" (Corbett, Myers, and Tate 1999). Medina willingly fulfilled the requirements for an "A" that included more presentations and facilitations to the class, for which Farber handwrote feedback. Responding to Medina's presentation titled "I am a bad teacher" based on the year he taught third grade in Puntarenas, Costa Rica, Farber explained in his note that he had the "it" for teaching that could not be taught. Although Medina entered his class with decidedly less pedigree and experience than the other pre-teachers, Farber's feedback gave the confidence to Medina to further his professional development and eventual academic career.

As a Latino, Medina proved he did have "it" for teaching: awards, remarkable student reviews, and repeatedly teaching in a summer bridge program for underrepresented student populations. While pursuing his dissertation on Latin@ (Latino/a) student writing, Medina taught two digital writing courses wherein he used grading contracts to make his social justice pedagogical values more explicit, to teach students to recognize the difference between *deserving* and *earning* a grade, and to level the privileged access often associated with digital technologies. The levels of competencies with digital writing vary broadly, so consequential validity asks that educators attend to these varying competencies while creating space for student agency where less prepared students feel as confident about their ability to earn their chosen grade as more privileged students. Less prepared students excel alongside more prepared students because the course values the performance of the assignments rather than the hegemonic standards of the

Table 2.1. Grading Contract (Medina)

In order to earn an A, you must satisfactorily complete the following:	• Present Public Argument at English Department Event • Volunteer with a one page write-up about the event/organization • Upload Public Argument slideshow/video to YouTube and get 20 comments • Present your research/public argument to the class with power point presentation • Blog that documents your research • 20 Tweets about class assignments, campus or community resources • 7–8 page Research Paper (6 Academic, 2 Popular Sources) • Research Proposal and Annotated Bibliography (6 Academic, 2 Popular Sources) • Read your research paper on webcam, upload to YouTube and send me the link • 4–5 page Rhetorical Analysis Paper, Reflective Essay, email textbook author, online discussion posts and the online library tutorials
In order to earn a B, you must satisfactorily complete the following:	• Upload Public Argument slideshow/video to YouTube and get 10 comments (or) write a one page rhetorical analysis of a publication and query letter that you plan to submit either a magazine article/short story with proof of submission • Present your research/public argument to the class with power point presentation • At least 10 Tweets about class assignments or resources • 4 page Rhetorical Analysis Paper, 6–7 page Research Paper (4 Academic, 2 Popular Sources) • Research Proposal and Annotated Bibliography (4 Academic, 2 Popular Sources) • Reflective Essay, email textbook author, online discussion posts and the online library tutorials
In order to earn a C, you must satisfactorily complete the following:	• 3–4 page Rhetorical Analysis Paper, 6 page Research Paper (2 Academic, 4 Popular Sources) • Research Proposal with Annotated Bibliography (2 Academic, 4 Popular Sources) • Reflective Essay • At least 5 productive Tweets, e-mail WPL author, online discussion posts and library tutorials

In order to earn a D grade, any of the above C requirements will not have been completed, or unsatisfactorily completed, not accomplishing the goals of the assignments or meeting the level of college writing.

F grades will be earned by those students who fail to satisfactorily complete more than one of the assignments in the C requirement.[2]

institution that dictate quality in technology and writing. Access to technology continues to affect non-white populations at higher rates (Banks 2006; Monroe 2004), so courses that de-emphasize subjective concerns of quality in turn emphasize the agency of students previously limited by resources of time and technology rather than effort.

Medina designed the grading contract to attend to social justice pedagogy that emphasizes student agency and under-resourced students; however, the semester with the research assignment contract was the first that he experienced numerous formal complaints, lower than normal teacher-course evaluations, in addition to the resistance experienced across institutional and non-institutional student evaluations as he also experienced with the semester-long contract. The student agency as resistance directly commented on the use of the grading contract, and below we offer a sample of this student resistance from both inter and extra institutional sources such as teacher-course evaluations, written student responses, and comments on the popular website *Rate My Professor* (*RMP*) (Ritter 2008). Through an examination of student responses to grading contracts, we hope to demonstrate how they might be used as a site for consequential inquiry of a social justice pedagogy that attends to student agencies, competencies, and performances. If grading contracts are sites for blurring lines of authority, flattening hierarchies, encouraging experimentation, and rewarding excellence, then they have potential to frame student-teacher interactions as aligning with the values of social justice. These events inspired a necessary pause for reflective interrogation of this pedagogical practice that was designed to address student agency and unsettle cultural capital and privilege in the classroom.

STUDENT REVIEWS

The following table includes *RMP* posts for Medina that evaluate grading contracts from the perspective of students.

In responses on TCEs and *RMP*, students negotiated the ambiguity of how the course was assessed with their perception of the workload. A student who responded positively in the written TCEs still commented, "I think he grades fairly. The only con to the class was how much work was required in order to get an A in the class." The theme of workload resurfaced within a relatively positive evaluation of instructor effectiveness: "He grades you for effectiveness and does not pick you appart [*sic*]. He is very good because he grades you for all the effort you give and recognizes it," although the same student noted "some of the technology oriented stuff was tedious and unnecessary." A student majoring in education responded positively with: "Experienced a new way of learning and grading styles." Even within positive feedback, there are misgivings about the workload: "It was good. I felt that there was a little too much unnecessary work required but overall it was a good class and I learned a lot." In a voice similar to what we found on *RMP*, a student responds that the class overall was "Okay, too much b.s. work." Many of these responses on *RMP* and TCEs reflect the unsettling of rigid hegemonic beliefs that students hold about education and which social justice pedagogy can elicit; however, "[t]his unsettling state may have produced the student confusion and resentment" noted in the evaluations (Spidell and Thelin 2006, 54). Because students can feel unsettled by the requirements of grading contracts, to integrate consequential validity means to pose the requirements of projects as problems that they can grapple with and negotiate as a part of the process of understanding the goals and consequences of individual components of a project.

Clearly, students have been enculturated to view grades as a power exercise and so student resistance to grading contracts comes as no surprise—why would students ever think that an evaluation system termed "contract" and "grading" would ever be a site for them to exercise their agency? But here the dominant cultural narratives around grades stand in relation to dominant cultural narratives of technological mastery, and the dominant raced and gendered narratives often associated with instructors of color—the "cool," "hip," and "nice and funny guy" who uses "pop cultural references." So the bad reviews Medina received are a space to begin interrogating dominant views of race as they stand in relation to dominant views of grades as an exercise of power and of technology as a tool to master. As a visibly raced instructor, Medina is described by microaggressions, or discursive exchanges that belittle people of color, that weaken his credibility by positioning him as inferior (Yosso 2006). Many attributes describing Medina's personality are used to set up a critique for the explicit values he asserts in his class—particularly the readings from authors of color and the thorough integration of technology in all assignments. In technical communication, Angela Haas argues that writers of color provide a necessary voice "to consider more deeply how race affects the ways in which technologies and documents are designed and used, how national and political values can inspire users to transform the work of technologies beyond their designed intent, and how non-Western cultures use and produce with Western and non-Western technologies differently than Westerners do" (281). The inclusion of such texts could have led some respondents to remark on the non-normative nature of his class, as with the comment

Table 2.2. *Rate My Professor* Comments
(http://www.ratemyprofessors.com/ShowRatings.jsp?tid=1235167)

Post-Semester-Long Grading Contract

Poster 1

He conducts class on a "choose your own grade" basis, in which the syllabus lays out a list of assignments for each letter grade, and the students decide what they want to literally "earn." He is quirky and eccentric, all the better for making lessons memorable. However, he tends to be too friendly, even with the disrespectful asshats in class.

Poster 2

Literally is the chillest teacher at [University X]. You choose the grade you want and deserve and he ends up giving it to you if you prove it to him.

Post-Research Assignment Grading Contract

Poster 1

He's a nice guy, but his class can be frustrating. All my friends [. . .] that are in [this course] have less homework and their essays don't have to be as long as his. While pick your own grade sounds good, the requirement for an A are quite extensive. Certainly different than the average English class.

Poster 2

Fun guy really nice however the papers require 8 sources 6 of which are [sic] acedemic. Too much work for [this course]. This is like a upper division research class.

Poster 3

For pick your own grade I picked A. Here's what I had to do. 1) 7–8 pages 2) 8 academic sources 3) read out loud and record onto youtube 4) Annotated BIB 5) citations 6) Presentation 7) 8 minimal blogs this is for ONE paper! Find another teacher its [a] ridiculous amount of work for that.

Poster 4

Nice and funny guy, but he goes over the top with the essays he assigns. 7–8 pages, 6 academic and 2 popular source annotated bib. Digital story for Public Assignment. This is too much work for [this class]. DO NOT TAKE HIS CLASS.

Poster 5

The best English teacher at [University X] and the coolest professor you'll ever meet. He uses pop culture references when explaining rhetorical analysis. The workload is very reasonable and he basically let us choose our grade for one of the essay's. He is an easy grader and is always there if you need help. I actually enjoyed going to class!!

that expressed that his course is "certainly different from the average." However, the inclusion of scholars who address issues such as race in technical communication help make visible the assumption that writers and audiences are accurately represented by a white, middle-class voice (Medina 2014).

Because the association between an instructor of color, grading as an exercise of power, and technological mastery cannot be separated, student resistance is most clearly found in resistance to the contract itself—at least initially viewed as "a huge joke"—and to the "tedious and unnecessary" workload or the "way too much b.s. work" comments in relation to the course's technological orientation. However, the privilege of students can also be detected in the below average ratings for "Usefulness of outside assignments" and "Usefulness of course materials," which does not simply critique the numerous exercises students could perform; rather, the dismissive nature of the course material hints at how students value texts by writers of color who discussed policy such as affirmative action and bilingual education. Students resist what Haas and Scott advocate in terms of integrating critical issues at the core of the curriculum and critically examining the documents and technologies that assume and ascribe certain levels of privilege to the construction of students as the intended audiences (Haas 2012; Scott 2004). Even still, to obviate concerns over usefulness, consequential validity in the use of grading contracts requires the necessary negotiation or Freirian dialogue that avoids re-instantiating oppressive curricula through critical pedagogy.

Medina's tale is emblematic and latent for technical communication in terms of race and pedagogies using digital technologies. Gesturing to grading contracts in and of themselves is not emancipatory and justified. Instead, those values have to be found in the process orientation toward the goals, outcomes, and the kinds of reflective potentialities we frame for our students. Potentially what the grading contract affords is an explicit conversation about the values associated with the goals of a social justice-inflected pedagogy. What is socially just will be found in the process of producing quality writing in relationship to the kinds of assignments typical of technical communication curricula.

MODEL 2: CONSEQUENTIAL VALIDITY AND GRADING CONTRACTS IN TECHNICAL COMMUNICATION SERVICE LEARNING REDESIGN PROJECTS

Designing grading contracts through the frameworks of consequential validity provides one way to integrate critical evaluations into service learning pedagogy in order to preempt co-optation by hyperpragmatic forces (Matthews and Zimmerman 1999; Sapp and Crabtree 2002; Scott 2004; Turnley 2007; Youngblood and Mackiewicz 2013). Walker's experience began in an introductory technical communication course that integrated social justice and a service learning project, particularly at the site grading contracts. For Walker, the fundamental question for using consequential validity in the design of the grading contract was what kinds of agencies, competencies, and performances do I value, and how can these align with the values of social justice to advocate for the socially, politically, and/or culturally under-resourced? With this framing, Walker's grading contract was designed to align social justice values with course assessments of complex and collaborative projects.

The key elements of Walker's redesign project are that it challenges students to establish a relationship with a community partner who identifies with social justice values (developing this relationship either on their own or through the instructor's contacts) and to negotiate a technical redesign project that is needed by the community partner and is manageable for a collaborative student project in under two months.[3] Walker designed the grading contract for this assignment as a substantial portion of the entire course grade (30% in this case). Once student teams secure a community partner and are given the grading contract, they are tasked with discussing their commitment to the project and choosing their own grade. As you can see from table 2.3, this approach allows the instructor to foreground the values upon which grades will be negotiated—equitable collaboration, team initiative, just relationships, attention to process, embodied experience, and usability. Thus, a few of the standard commonplaces of privilege, such as individual effort and cultural capital, are recontextualized within a framework that values relations, processes, and collaboration in addition to labor. The discussion of course grade creates the space to negotiate workload as a function of the consequences that they choose or contract. In this assignment design, social justice values are embedded into the community-based projects and into the evaluation process based on consequential validity—students must work equitably, fairly, and collaboratively among themselves and with community partners in order to achieve the highest grades.[4]

The projects that students in Walker's class produced are emblematic of the kinds of community-based social justice projects that have the potential to encourage and/or complicate student understandings of social justice issues beyond the classroom and in community relationships. For example, one team worked with the Arizona Superior Court to develop, design, and user-test a screencast to assist those seeking to file for a divorce without an attorney. Because racial and ethnic minority women make up the largest percentage of this group, the student team had to maintain a relationship with the court to carefully consider how their information design and delivery could best serve this under-resourced group. Not only did the team work equitably together to produce a quality product, but the reflective evaluation component of the grading contract allowed the student team to consider how the process of the project led all of us to consider more carefully the kinds of structural changes necessary to serve this population. While the goals and outcomes of the project were successful and encouraging, the reflective potentials of the project complicated any simple notions of effective social justice pedagogy. Instead, we all reflected on how even successful socially-justice community projects revealed further systemic inequalities that these kinds of projects are unable to address.

Other projects in the course showed that the majority of student teams were able to use the grading contract as a reference point for working collaboratively and equitably on projects with either explicit or implicit possibilities for social justice work. For example, another project developed a proto website for the Yavapai Health Clinic, thus allowing this team to consider, along with the community partner, how information design and delivery might both reflect indigenous knowledge and promote access to health services. Other projects had more implicit ties to social justice possibilities. One group redesigned the brochure for a local farm seeking to advertise to Co-op shoppers. They reflected that access to low-cost and high-quality local food is a pressing issue for social justice that their technical documentation helped facilitate. Another group worked with a local nonprofit to redesign a homeowner's guide on how to install a DIY storm water storage system at home. This team reflected that storm water storage has the potential to connect a scarce resource to resource-scarce populations. Beyond the range of these more tangible outcomes, Walker and his students found the grading contract usefully made the values embedded in the process of social justice pedagogy more explicit and therefore it was more clear which groups were more or less successful in both the process and products of a socially just-infused redesign project.

Used in these ways, grading contracts make the values of social justice explicit within the processes of service-learning projects, and this has potential to destabilize some of the privileges students have when entering the course. Using the grading contract to collaboratively reflect on the process and to provide the community partner an opportunity to reflect and assess the teams' redesign work results in much more than a grade. At best, the consequences lead to deep reflection on the role of technical communication in community-based social justice projects and the processes used to successfully complete them. Still, an important consideration when working with traditionally marginalized populations is to avoid promising too much, so that failed student collaborations become little more than another stage in the continuum of

Table 2.3. Grading Contract (Walker): Collaborative Service-Learning Redesign Project

In order to earn an A, you must satisfactorily complete the following:	• Using the instructor's list, or on your own, make contact through an email of inquiry with one community partner who identifies with social justice values, and set up a meeting to negotiate an appropriate project scope. • Electronic introduction of instructor to the community partner, which includes discussion of the project's scope. Use of partner's feedback required for contracting group's grade. • Collaboratively develop a redesign project proposal that includes a partner description, a needs analysis, a discussion of possible document designs, a justification for the selected design, and an appendix with a storyboard/design template and style guide. • Collaboratively design, user test, and integrate technical documentation into a community partner's workplace. • Prepare a team presentation that introduces your community partner, reports on your redesign project and the results of your usability testing, identifies the social justice element of your work with the community partner, and draws conclusions for students who might conduct similar projects in the future. • Work equally and collaboratively as a team to fairly distribute the workload, and appropriately use each individual's skills to design the best documentation. • Collaboratively compose an email, with your instructor cc'ed, thanking your community partner, offering a reflective evaluation on your partnership and the product you designed and, using these negotiating points, make a case for why your group deserves the grade you decide on.
In order to earn a B, you must satisfactorily complete the following:	• Using the instructors list, or on your own, make contact through an email of inquiry with one community partner who identifies with social justice values and set up a meeting to negotiate an appropriate project scope. • Send an email introducing your instructor to the community partner so that we may negotiate the scope of your redesign project and use your partner's feedback when negotiating your group's grade. • Collaboratively develop a redesign project proposal that includes a partner description, a needs analysis, a discussion of possible document designs, a justification for the selected design and an appendix with a storyboard/design template (NO STYLE GUIDE). • Collaboratively design, user test, and integrate technical documentation into a community partner's workplace (FEWER REQUIREMENTS FOR USER TEST). • Work collaboratively as a team to distribute the workload, and appropriately use each individual's skills to design the best documentation and establish a professional relationship with a community organization (NOT NECESSARILY EQUALLY OR FAIRLY). • Prepare a team presentation with visuals that introduces your community partner, reports on your redesign project and the results of your usability testing, identifies the social justice element of your work with the community partner, and draws conclusions for students who might conduct similar projects with similar partners in the future. • Collaboratively compose an email, with your instructor cc'ed, thanking your community partner, offering a reflective evaluation on your partnership and the product you designed and, using these negotiating points, make a case for why your group deserves the grade you decide on.

continued on next page

Table 2.3—*continued*

In order to earn a C, you must satisfactorily complete the following:	• Using the instructors list or through simulation, make contact through an email of inquiry with one community partner who identifies with social justice values and set up a meeting to negotiate an appropriate scope for the redesign project (POTENITAL FOR SIMULATION). • Send an email introducing your instructor to the community partner so that we may negotiate the scope of your redesign project and use your partner's feedback when negotiating your group's grade. • Collaboratively develop a redesign project proposal that includes a partner description, a needs analysis, a discussion of possible document designs, and a justification for the selected design (NO STYLE GUIDE; NO DRAFTS OR TEMPLATES). • Collaboratively design, user test, and simulate the integration of technical documentation into a community partner's workplace (FEWER REQUIREMENTS FOR USER TEST). • Work collaboratively as a team to distribute the workload, and appropriately use each individual's skills to design the best documentation (NOT NECESSARILY EQUALLY OR FAIRLY; NO RELATIONSHIP). • Prepare a group presentation with visuals that introduces your community partner, reports on your redesign project and identifies the social justice element of your work, and draws conclusions for students who might conduct similar projects in the future (LIGHTER REQUIREMENTS). • Collaboratively compose an email, with your instructor cc'ed, thanking your community partner, offering a reflective evaluation on the product you designed and, using these negotiating points, make a case for why your group deserves the grade you decide on.
In order to earn a D, you must satisfactorily complete the following:	• Through simulation make contact through an email of inquiry with one community partner who identifies with social justice values (ONLY SIMULATION). • Send an email introducing your instructor to the community partner so that we may negotiate the scope of your redesign project (NO PARTNER RELATIONSHIP). • Collaboratively develop a redesign project proposal that includes a partner description, a needs analysis, a discussion of possible document designs, a justification for the selected design (NO STYLE GUIDE; NO DRAFTS OR TEMPLATES). • Collaboratively design and implement technical documentation into a community partner's workplace (NOT NECESSARILY SUCCESSFUL; NO USER TESTING). • Work collaboratively as a team to distribute the workload, and appropriately use each individual's skills to design the best documentation (NOT NECESSARILY EQUALLY OR FAIRLY; NO RELATIONSHIP; NO PRESENTATION). • Collaboratively compose an email, with your instructor cc'ed, thanking your community partner, offering a reflective evaluation on the product you designed and, using these negotiating points, make a case for why your group deserves the grade you decide on. F grades will be earned by those students who fail to satisfactorily complete more than one of the assignments in the D requirement.

hegemonic institutions shirking responsibilities with these groups. It is in the ability of the instructor and the students to be flexible with the contingencies of each project and to be in constant contact with the team and the community partner that holds the most promise for realizing just consequences in these projects.

CONCLUSION: THE POTENTIALS AND LIMITS OF CONSEQUENCES IN SOCIAL JUSTICE PEDAGOGY

In this chapter we have sought to further the influence of cultural, racial, and social justice theories in technical communication by providing models that contribute to critical practices in course-based assessments. Grading contracts have rarely been discussed with regard to countering institutional and social inequality. By acknowledging the fundamentally rhetorical nature of validity, and by integrating social justice values into our course-based assessments, these models begin the work of acknowledging and correcting for the uneven distribution of knowledge, power, and access to resources found among student populations. The intended and unintended consequences of our pedagogies for systemically marginalized students matter. Using explicit social justice values in our course evaluation systems is one way technical communication instructors can better attend to these consequences for these students. At the programmatic-level, future research might study how consequential validity can help negotiate the inherent tension between programmatic coherence and student potentiality (Gallagher 2012). At the classroom-level, consequential validity leaves us with questions that appear during the process of teaching and aspiring to a critical consciousness: how do educators respond to the racialized social dynamics in our technical communication classrooms that allow entitled white students to continue to perform their privilege, partly by challenging the decisions of an instructor of color, as they might with female, queer, working-class, or disabled instructors? How might these instructors critically rearticulate privilege to support a more just and more equitable distribution of knowledge, power, and access?

It is problematic to assume that a single class can provoke critical consciousness for all students about the many issues impacted by institutional inequality. In chapter 7 of this collection, Marcos Del Hierro conversely problematizes the guiding principle that education serves to civilize poor populations who have been described as "barbarians" in Open Admissions institutions (Horner 1996). Del Hierro warns that the enduring assumption that "the educational process converts young people from wild and misbehaved children into educated and refined citizens has dangerous consequences for non-white students" (174). Our attention to grading contracts underscores a pedagogical change that actively undermines the privilege that normalizes assumptions about underrepresented and under-resourced students and communities. We can neither be afraid nor actively avoid pushback from students when assignments and practices highlight the equitable distribution of labor in groups and for individual students.

Scholarly attention to grading contracts no doubt persists because instructors remain skeptical of the validity of a traditional grading system. The field of technical communication should recognize this exigency because of the growing body of scholarship highlighting why social justice matters with regard to students, educators, businesses, nonprofit organizations, and the environment. Grading contracts offer a system that possesses the potential to make grades more transparent to students; however, grading contracts potentially carry with them misgivings about workload and resentment because of what some students expect because of accrued cultural capital. To still advocate for grading contracts as tools that effect change, we should note that *RMP* posts should be seen as a call to forefront issues of workload, which can be an outcome of valuing labor above privilege. Likewise, the conflicted perspectives on the difficulty of a class as a result of social justice curricula reflect the very same conflict that instructors experience when deciding what and how to challenge students to become critical of the intended outcomes of their writing. While it is certainly true that the race of students can impact the efficacy of grading contracts (Inoue 2012), Medina's experiences as an instructor of color in predominantly white classrooms suggests the reverse: the race of the instructor can also impact the efficacy of grading contracts and lead to messy consequences for the instructor.

As our opening example of Tucson High School's Mexican American Studies Program suggests, if critical educators mean to overturn the contents of privilege's invisible knapsack (McIntosh 1989), then we must be prepared for the mess of sorting it out. Consequential validity provides a framework for making course values explicit, but the decisions and approaches to integrate social justice into classroom curriculum remain rhetorical in that institutional context. Student population, instructor positionality, and departmental support should all be factored into the decision-making that can affect how an instructor is viewed by their students, future students, colleagues, and future hiring committees. Our hope is that by attending to notions of consequential validity, critical pedagogues can disrupt the exercise of privilege and advocate for the marginalized through the course-based assessment tool of grading contracts. But grading contracts alone do not do this. Consequential validity alone does not do this. It is only in the process of attending to consequential validity in grading contracts that we *might* discover workable social justice pedagogies and evaluations in technical communication.

Notes

1. We view Freire's critical pedagogy as possessing a social justice ethos; however, we do not necessarily correlate a direct one-to-one relationship between critical pedagogy and methodology for the theories and practices of social justice because of the farther-reaching possibilities of social justice work that cannot be reduced to critical pedagogy.
2. Students cannot receive a passing grade in first-year composition unless they have submitted drafts and final versions for all major assignments and the final exam. Incompletes are awarded in cases of extreme emergency if and only if 70 percent of the course work has been completed at the semester's end.
3. The literature on grading contracts suggests that instructors use grading contracts for major assignments and/or for the class as a whole.
4. A key part of explicitly valuing community relationships is that the instructor must have dependable relationships with select organizations to ensure that students have the best possible opportunity to succeed. Instructors should always have contingency plans for community partners who do not meet the expectations of their role in the project.

References

Agboka, G. Y. 2013. "Participatory Localization: A Social Justice Approach to Navigating Unenfranchised/Disenfranchised Cultural Sites." *Technical Communication Quarterly* 22 (1): 28–49. https://doi.org/10.1080/10572252.2013.730966.

Allen, Jo. 1993. "The Role(s) of Assessment in Technical Communication: A Review of the Literature." *Technical Communication Quarterly* 2 (4): 365–388. https://doi.org/10.1080/10572259309364548.

Ball, Cheryl E. 2012. "Assessing Scholarly Multimedia: A Rhetorical Genre Studies Approach." *Technical Communication Quarterly* 21 (1): 61–77. https://doi.org/10.1080/10572252.2012.626390.

Banks, Adam J. 2006. *Race, Rhetoric, and Technology: Searching for Higher Ground.* Mahwah, NJ: Lawrence Erlbaum.

Barton, Matthew D., and James R. Heiman. 2012. "Process, Product, and Potential: The Archaeological Assessment of Collaborative, Wiki-Based Student Projects in the Technical Communication Classroom." *Technical Communication Quarterly* 21 (1): 46–60. https://doi.org/10.1080/10572252.2012.626391.

Beard, J. D., J. Rymer, and D. L. Williams. 1989. "An Assessment System for Collaborative-Writing Groups: Theory and Empirical Evaluation." *Journal of Business and Technical Communication* 3 (2): 29–51. https://doi.org/10.1177/105065198900300203.

Cabrera, Nolan L., Jeffrey Milem, and Ronald W. Marx. 2012. "An Empirical Analysis of the Effects of Mexican American Studies Participation on Student Achievement within Tucson Unified School District." UA College of Education, June 20, 2012. Accessed August 27, 2014. https://works.bepress.com/nolan_l_cabrera/17/.

Coppola, Nancy W. 1999. "Setting the Discourse Community: Tasks and Assessment for the New Technical Communication Service Course." *Technical Communication Quarterly* 8 (3): 249–267. https://doi.org/10.1080/10572259909364666.

Corbett, Edward P. J., Nancy Myers, and Gary Tate. 1999. *The Writing Teacher's Sourcebook.* New York: Oxford University Press.

Danielewicz, Jane, and Peter Elbow. 2009. "A Unilateral Grading Contract to Improve Learning and Teaching." *College Composition and Communication* 61 (2): 244–68.

Farber, Jerry. 1990. "Learning How to Teach: A Progress Report." *College English* 52 (2): 135–141. https://doi.org/10.2307/377440.

Freire, Paulo. 1970. *Pedagogy of the Oppressed.* New York: Continuum.

Gallagher, Chris W. 2012. "The Trouble with Outcomes: Pragmatic Inquiry and Educational Aims." *College English* 75 (1): 42–60.

Haas, Angela M. 2012. "Race, Rhetoric, and Technology: A Case Study of Decolonial Technical Communication Theory, Methodology, and Pedagogy." *Journal of Business and Technical Communication* 26 (3): 277–310. https://doi.org/10.1177/1050651912439539.

Herndl, Carl G. 2004. "Introduction to the Special Issue: The Legacy of Critique and the Promise of Practice." *Journal of Business and Technical Communication* 18 (1): 3–8. https://doi.org/10.1177/1050651903258143.

Horner, Bruce. 1996. "Discoursing Basic Writing." *College Composition and Communication* 47 (2): 199–222. https://doi.org/10.2307/358793.

Inoue, Asao. B. 2009. "The Technology of Writing Assessment and Racial Validity." In *Handbook of Research on Assessment Technologies, Methods, and Applications in Higher Education*, ed. C. Schreiner, 97–120. Hershey, PA: Information Science Reference. https://doi.org/10.4018/978-1-60566-667-9.ch006.

Inoue, Asao. B. 2012. "Grading Contracts: Assessing Their Effectiveness on Different Racial Formations." In *Race and Writing Assessment*, ed. A. B. Inoue and M. Poe, 78–93. New York: P. Lang.

Johnson, Jennifer Ramirez, Octavio Pimentel, and Charise Pimentel. 2008. "Writing New Mexico White." *Journal of Business and Technical Communication* 22 (2): 211–236. https://doi.org/10.1177/1050651907311928.

Longo, B. 1998. "An Approach for Applying Cultural Study Theory to Technical Writing Research." *Technical Communication Quarterly* 7 (1): 53–73. https://doi.org/10.1080/10572259809364617.

Manion, Christopher E., and Richard "Dickie" Selfe. 2012. "Sharing an Assessment Ecology: Digital Media, Wikis, and the Social Work of Knowledge." *Technical Communication Quarterly* 21 (1): 25–45. https://doi.org/10.1080/10572252.2012.626756.

Matthews, Catherine E., and Beverly B. Zimmerman. 1999. "Integrating Service Learning and Technical Communication: Benefits and Challenges." *Technical Communication Quarterly* 8 (4): 383–404. https://doi.org/10.1080/10572259909364676.

McIntosh, Peggy. 1989. "White Privilege: Unpacking the Invisible Knapsack." *Peace and Freedom* 49 (4): 10–12.

Medina, Cruz. 2014. "Tweeting Collaborative Identity: Race, ICTs and Performing Latinidad." In *Communicating Race, Ethnicity, and Identity in Technical Communication*, ed. Miriam Williamson and Octavio Pimentel, 63–86. Amityville: Baywood Publishing.

Messick, Samuel. 1989. "Meaning and Values in Test Validation: The Science and Ethics of Assessment." *Educational Researcher* 18 (2): 5–11. https://doi.org/10.3102/0013189X018002005.

Monroe, Barbara Jean. 2004. *Crossing the Digital Divide: Race, Writing, and Technology in the Classroom.* New York: Teachers College Press.

Morain, Matt, and Jason Swarts. 2012. "YouTutorial: A Framework for Assessing Instructional Online Video." *Technical Communication Quarterly* 21 (1): 6–24. https://doi.org/10.1080/10572252.2012.626690.

Moreno-Lopez, Isabel. 2005. "Sharing Power with Students: The Critical Language Classroom." *Radical Pedagogy* 7 (2). http://www.radicalpedagogy.org/radicalpedagogy/Sharing_Power_with_Students__The_Critical_Language_Classroom.html.

Poppen, William A., and Charles L. Thompson. 1971. "The Effect of Grade Contracts on Student Performance." *Journal of Educational Research* 64 (9): 420–423. https://doi.org/10.1080/00220671.1971.10884209.

Ritter, Kelly. 2008. "E-Valuating Learning: Rate My Professors and Public Rhetorics of Pedagogy." *Rhetoric Review* 27 (3): 259–280. https://doi.org/10.1080/07350190802126177.

Sapp, David Alan, and Robbin D. Crabtree. 2002. "A Laboratory in Citizenship: Service Learning in the Technical Communication Classroom." *Technical Communication Quarterly* 11 (4): 411–432. https://doi.org/10.1207/s15427625tcq1104_3.

Scott, J. Blake. 2004. "Rearticulating Civic Engagement through Cultural Studies and Service-Learning." *Technical Communication Quarterly* 13 (3): 289–306. https://doi.org/10.1207/s15427625tcq1303_4.

Scott, J. Blake, and Bernadette Longo. 2006. "Guest Editors' Introduction: Making the Cultural Turn." *Technical Communication Quarterly* 15 (1): 3–7. https://doi.org/10.1207/s15427625tcq1501_2.

Scott, J. Blake, Bernadette Longo, and Katherine V. Wills. 2006. *Critical Power Tools: Technical Communication and Cultural Studies*. Albany: State University of New York Press.

Shor, Ira. 2009. "Critical Pedagogy Is Too Big to Fail." (CUNY) *Journal of Basic Writing* 28 (2): 6–27.

Spidell, Cathy, and William H. Thelin. 2006. "Not Ready to Let Go: A Study of Resistance to Grading Contracts." *Composition Studies* 34 (1): 35–69.

Taylor, Hugh. 1971. "Student Reaction to the Grade Contract." *Journal of Educational Research* 64 (7): 311–314. https://doi.org/10.1080/00220671.1971.10884172.

Turnley, Melinda. 2007. "Integrating Critical Approaches to Technology and Service-Learning Projects." *Technical Communication Quarterly* 16 (1): 103–123. https://doi.org/10.1080/10572250709336579.

Williams, Miriam F., and Octavio Pimentel. 2012. "Introduction: Race, Ethnicity, and Technical Communication." *Journal of Business and Technical Communication* 26 (3): 271–276. https://doi.org/10.1177/1050651912439535.

Wolvin, A. D., and D. R. Wolvin 1975. "Contract Grading in Technical Speech Communication." *Speech Teacher* 24: 139–142.

Yarber, William L. 1974. "Retention of Knowledge: Grade Contract Method Compared to the Traditional Grading Method." *Journal of Experimental Education* 43 (1): 92–96. https://doi.org/10.1080/00220973.1974.10806310.

Yosso, Tara J. 2006. *Critical Race Counterstories along the Chicana/Chicano Educational Pipeline*. New York: Routledge.

Youngblood, Susan A., and Jo Mackiewicz. 2013. "Lessons in Service Learning: Developing the Service Learning Opportunities in Technical Communication (SLOT-C) Database." *Technical Communication Quarterly* 22 (3): 260–283. https://doi.org/10.1080/10572252.2013.775542.

Yu, Han. 2008. "Contextualize Technical Writing Assessment to Better Prepare Students for Workplace Writing: Student-Centered Assessment Instruments." *Journal of Technical Writing and Communication* 38 (3): 265–284. https://doi.org/10.2190/TW.38.3.e.

Yu, Han. 2012. "Intercultural Competence in Technical Communication: A Working Definition and Review of Assessment Methods." *Technical Communication Quarterly* 21 (2): 168–186. https://doi.org/10.1080/10572252.2012.643443.

3

THE UNIVERSITY REQUIRED ACCOMMODATIONS STATEMENT
What "Accommodation" Teaches Technical Communication Students and Educators

Barbi Smyser-Fauble

> *Discussions of access should include more than the obvious exclusion of wheel chair users and individualized accommodation processes for those with learning differences. It should be reasonable to assume that discussions of disability and access will include a consideration of all sorts of embodiments, health issues, or ethnic, religious, or sexuality commitments, and these interlock with issues of race, class and gender.*
> *(Titchkosky 2011, xii)*

When preparing a technical communication course syllabus, how often do we, as instructors, consider the impact of the information contained within this technical document? Do we ask ourselves about how accessible the information within the syllabus is to the students within our courses (the users)? And, as instructors, do we really consider how the construction of a legal statement about accommodation, in reference to the Americans with Disabilities Act (ADA) legislation, is perceived by all students in our courses—including those with documented categorized disabilities? This should be a priority for technical communication instructors, as the courses we teach emphasize the importance of constructing documents that are considered usable and accessible to a wide variety of audiences; our classrooms should be no different.

This chapter offers four major discussion areas. I commence with a brief overview of current discussions of legal literacies and regulatory rhetorics in technical communication studies and put them into conversation with ADA legislation. Second, I discuss contemporary usability and accessibility scholarship and critiques from scholars working at the intersections of disability studies and technical communication. I then

DOI: 10.7330/9781607327585.c003

build a feminist disability framework that incorporates feminist theory and disability studies to extend usability and accessibility in ways that further social justice. I then apply the feminist disability framework to a rhetorical analysis of the Illinois State University (ISU) required Accommodations statement, as well as the information conveyed on the website of the ISU Disability Concerns Office. The analysis will examine: (1) the impacts of ambiguous language within technical documents on the construction of student bodies and identities, (2) how this language also impacts the definition of what is perceived as being "accessible" or "reasonable accommodation," and (3) how this language can dictate who is perceived as being worthy of consideration when addressing issues of accessibility and inclusion in the construction of both technical documents and technical communication course design. Ultimately this analysis brings to light the pedagogical and curricular implications of applying a feminist disability studies framework to the field of technical communication by calling for the construction of accommodation statements and course materials that reflect a more proactive and inclusive approach for accessibility.

INTERPRETING ADA LEGISLATION AS A TECHNICAL COMMUNICATOR

The ADA is an "equal opportunity" law for people with disabilities. Its legislation was "[m]odeled after the Civil Rights Act of 1964, which prohibits discrimination on the basis of race, color, religion, sex, or national origin" (US Department of Justice Civil Rights Division 2009, n.p.). The ADA legislation was meant to extend Section 504 of the Rehabilitation Act of 1973, which stated:

> no qualified individual with a disability in the United States shall be excluded from, denied the benefits of, or be subjected to discrimination under any program or activity that either receives Federal financial assistance or is conducted by any Executive agency or the United States Postal Service. (US Department of Justice Civil Rights Division 2009, n.p.)

Essentially, ADA legislation "prohibits discrimination and guarantees that people with disabilities have the same opportunities as everyone else to participate in the mainstream of American life—to enjoy employment opportunities, to purchase goods and services, and to participate in State and local government programs and services" (US Department of Justice Civil Rights Division 2009, n.p.).

Postsecondary institutions are bound to adhere to the ADA federal mandate passed July 26, 1990 when considering the needs of students and employees with medical diagnoses of categorized disabilities. Universities are legally required to provide "reasonable accommodation." However, ADA legislation does not specify *how* an institution defines or distributes accommodations. Therefore, the legislation appears to be open for interpretation and, thus, is ambiguous for the very entities responsible for adhering to it, including universities. To demonstrate my concerns with this ambiguous language, I rhetorically analyze a case study of ISU's localized response to ADA regulation and compliance.

Due to the ambiguity of the ADA legislation, technical communication instructors need to have a more thorough understanding of how legislation is created and implemented. In this particular analysis, technical communication educators should first focus upon how the ADA legislation impacts their roles and responsibilities. As we know, traditionally, technical communicators (both within academic and non-academic settings) have perceived laws and legislation as "an after-the-fact-phenomenon, something over which they have no influence"; to follow, but have no influence or input as to their construction or interpretation (Hannah 2010, 6). As such, Hannah identifies how, traditionally, texts for technical communicators concerning legal literacies often provide details as to how technical communicators can adhere to laws and regulations (providing guidelines) when constructing technical documents. However, what is not made explicit to technical communicators is how to address issues of ethics and social responsibilities when constructing these documents. As Williams (2006) notes, "historically produced legal and regulatory writings, such as the constitution that states 'we the people,' were constructed in such a way as to only include those within the perceived positions of power (white, educated, economically advantaged, males) as the 'we', while excluding certain communities based on race, ethnicity, gender, ability, sexuality, etc . . ." (141–142). Thus, technical communicators,[1] both within academic and non-academic settings, should not only understand what laws and regulations are required to reference within their constructed documents but to also understand how the laws/regulations were initially conceived and implemented.

Developing legal literacies that take into account social injustices help technical communicators answer Bowdon's (2004) call that we act as "public intellectuals" (325). Bowdon (2004) explicates the dual roles of being both a technical communicator and "public intellectual" by stating that "these roles require that we recognize and embrace the deeply rhetorical nature of our work as shapers of documents, particularly when those documents affect public policy" (326). What is also interesting to note about Bowdon's call to act as a "public intellectual," is that it stems from her work as a technical communicator within both the academic setting (as an instructor and scholar), as well as in a non-academic setting as a writing consultant for the AIDS Prevention Collaboration (APC) of Tucson, Arizona. Specifically, Bowdon's technical communication work with this organization was to assist with the development of specific HIV prevention awareness literature for the community of "gay young males" (GYM), while not overlooking the ethical responsibilities related to the representation of the referenced community. Even though her role was initially acquired to provide technical editing to documents that were already constructed with established conclusions, Bowdon quickly realized that she also needed to be a "responsible rhetorical activist regardless of what stage of development the document was in" (326). In addition, as a technical communication instructor, Bowdon also brought her realizations and work experiences into the classroom to both educate and collaborate with students. Specifically, Bowdon sought the feedback from students concerning the potential "pathological tone of survey questions" (337). The feedback from the students helped to ensure that the survey was more positively framed, with the removal of accusing or judgmental tones. As a result, Bowdon provided feedback to the APC that was beyond surface editorial comments and also redressed rhetoric that perpetuated stigmatizing representations of the GYM community—a result that could undermine the efforts of providing important healthcare information. Ultimately, both she and her students recognized the potential risks these documents posed for the GYM community and took action to amend the stereotypical and negative connotations.

Bowdon's (2004) work provides a model for redressing technical documentation that works to construct community identities in problematic ways. In this chapter then, I seek to examine how university accommodations statements in technical communication course syllabi can impact all students—not just those with categorized disabilities. In fact, I will demonstrate that the mandated accommodations statement can actually work against technical communicators' responsibility to produce documents that are usable and accessible—and that accommodate as many users (students) as possible. This chapter, thus, asks technical communication instructors to understand that a syllabus is a technical document that includes technical, legal, and cultural specifications for a course and that it calls for instructors to adopt a critical legal literacy approach to composing syllabi, especially the portions informed by legislative rhetorics.

USABILITY AND ACCESSIBILITY IN TECHNICAL COMMUNICATION

Current usability and accessibility work in technical communication places a stronger focus on deconstructing the more physical barriers of exclusion. This is evidenced in how processes are developed to ensure technical texts are accessible to a variety of audiences to consume and engage with, while using language that is considered to be clear and easily understood. This position is furthered by technical communication scholar Lisa Meloncon (2013a) within her definition of accessibility within technical communication. According to Meloncon, "accessibility is the material practice of making social and technical environments and texts as readily available, easy to use, and understandable to as many people as possible, including those with disabilities" (5). Meloncon also identifies how "current scholarship within technical communication (related to accessibility) has yet to engage fully the complex intersections of technology and disability" (5).

Technical communicators also need to be more aware of how a disability studies approach can make them more aware of the rhetorical effects of language. The work of technical communication scholars Elmore 2013; Gutsell and Hulgin 2013; Meloncon 2013a, 2013b; Palmeri 2006; and Walters 2010; also demonstrates how the field can benefit from the incorporation of a disability studies perspective. Palmeri critiques current usability and accessibility teaching practices that "segregate assistive technologies" as being a method of providing accommodation or ensuring access (readability) of a text that further stigmatizes not only the technology, but those who use the technology (58). I would add that Palmeri's work also illustrates the impact of using language like "assistive technology." This language appears to only be applied to a technology when it is used by an individual with a medically categorized disability; someone who utilizes a cell phone to "assist" them in communicating with individuals who live a great distance away are perceived as using a technology and not an "assistive technology," even though the technology is meant to literally assist them in their communication efforts. Thus, begging the question of why not all technology is perceived of as being "assistive," and why the label of being an "assistive technology" carries with it the baggage of stigmatizing negativity in the form of being "othered."

The incorporation of a disability studies lens can also help technical communicators identify how current usability practices may actually perpetuate forms of exclusion. For instance, technical communication scholars Elmore (2013) and Meloncon (2013b) focus on how to expand current usability practices that appear to privilege able-bodied (normal) users of technical documents and technologies. Elmore states that "user experience research would benefit from the incorporation of ideas from disability studies, such as the view of disabilities as human differences in abilities that should be accommodated in society to the benefit of all" (32). Meloncon adds that "technical communicators can benefit from a focus on the actual users rather than the idealized 'normal' bodied user that are often the focus of the user-centered experience (user research)" (75). However, while both of these lines identify an affordance of a disability studies framework in the field of technical communication in relation to expanding current usability and accessibility practices, they also appear to perpetuate forms of unconscious, negative othering that furthers the us/them dichotomy. For example, why is there a need to state that "there is a benefit to all" or that "normal" users are perceived of as being idealized? While I am by no means saying that the moves that both of these scholars are making are not warranted or valuable, I am merely pointing out that even within these progressive movements there are elements of silent "othering" that can work to perpetuate the us/them dichotomy, including how "them" (the "othered") can benefit "us" (the norm).

Therefore, technical communicators need to be more aware of the rhetorical effects of language used within technical informational texts. For instance, as identified by technical communication scholars Gutsell and Hulgin (2013), technical communicators need to identify how unconscious forms of oppression within language choices can continue to "distance and avoid" disability (87). For example, Gutsell and Hulgin identify the need for technical communicators to use "'people-first'[2] language that refers to people in terms of their characteristics rather than in terms of disability labels (e.g., Person who uses a wheelchair vs. wheelchair bound), and to refrain from language that appears to undermine an achievement by using the phrase 'despite having a disability'" (86–87). Thus, there is need for the application and consideration of disability studies theory when composing technical documents to better ensure that they are "more inclusive of everyday experiences through language by involving more critical examinations of identities and approaches" (88). Walters (2010) applies disability studies to critique issues of accessibility related to both document construction and the physical space of the technical communication classroom. Her critique identifies how classroom space accommodations tend to be driven by a more retrofit approach. This approach usually focuses accommodating a single type of different ability as a fix for a specific student, rather than an exploration of proactive approaches to make the classroom inclusive to a variety of student abilities by means of "universal design."

Universal design is currently being explored as a potential process for promoting usability, accessibility, and inclusivity in both technical communication and disability studies. The term universal design originated in the field of architecture to promote practices for "creating buildings (blue-print development and actual construction), development of products, and the construction of environments that are accessible to all" (88). Disability rhetorics scholar Jay Dolmage identifies universal design as "talking about the ways of changing space to accommodate the broadest range of users" (Dolmage 2009, 172). However, Dolmage also critiques universal design approaches for "tending to overlook opportunities for feedback from the users of the space as to how effective it is in accommodating the needs of all users" (172). Thus, while universal design has the potential to bring issues of physical accessibility to the forefront, it must provide a mechanism for securing feedback from users to ensure that the space meets and continues to meet the needs of the users. Moreover, while universal design may construct spaces or texts that are accessible by the "broad range of users," it doesn't appear to consider other types of barriers.

Thus, a limitation of a universal design approach is that it is not explicitly concerned with dismantling social barriers of exclusion. For example, a person with a physical disability may have physical access to a classroom or building by means of having to enter through the back door, while other "able-bodied" individuals have the privilege of entering through the front or back door. In response to this type of situation, disability studies scholar Titchkosky (2011) asks, "If this is access, what have we been given access to? And when we get in, what are we in for?" (149). In classroom spaces, the physically accessible desk may be positioned at the front or the very back of the classroom, which can stigmatize the individual needing to use the desk by marking them as "other." Given this, technical communicators need to move beyond understanding accessibility and usability in terms of just physical access to a space or the ability to consume a text in ways that work toward a redressing of socially constructed barriers of exclusion.

HOW DOES A FEMINIST DISABILITY METHODOLOGY WORK?

In order to incorporate a framework that addresses issues of socially constructed barriers of exclusion, I recommend combining theoretical lenses from feminist and disability studies as they often work to dismantle oppression (including social exclusion), value experiential knowledges, and construct more socially responsible and more accurate identities and representations. While there appears to have been more feminist work done within technical communication studies to investigate issues of audience inclusion and representation than there has of disability studies work, little scholarship has tapped the promises that an approach that works at the intersections of these areas of inquiry affords our discipline.

A feminist disability framework combines aspects of feminist and disability theories to make more apparent the value of embodied experiential knowledges, including those who have been traditionally silenced as both audience(s) and resources, by focusing on the location of connections rather than the identification of and focus on differences. Disability studies scholars (Brueggemann et al. 2008; Brueggemann, Garland-Thomson, and Kleege 2005; Davis 2008 and 2013; Dolmage, 2008, 2009, and 2014; Dunn and De Mers 2002; Hamraie 2013; Lott 2001; Meyer 2013; Price 2008; Vidali 2008; Wilson 2000; Wilson and Lewiecki-Wilson 2001) establish more explicit connections between feminist theory and disability studies that extends both areas of study. For example, Rosemarie Garland-Thomson (2002) dubs "feminist disability theory" as one that brings together feminist and disability theories—and is "distinguished from other critical paradigms in that it scrutinizes a wide range of material practices involving the lived body" (10). She justifies the connection between the two areas of inquiry by asserting that, "as with gender, race, sexuality, and class: to understand how disability operates is to understand what it is to be fully human" (28). In fact, "disability—like gender (feminist theory)—is a concept that pervades all aspects of culture: its structuring institutions, social identities, cultural practices, political positions, historical communities and the shared human experiences of embodiment" (4).

It is this "shared human experiences of embodiment" that I wish to focus upon when constructing a combination of feminist and disability theory. Further, I bring together feminist theory's focus on interrogating power structures that privilege the voice of the dominant or expert with that of disability studies' focus on interrogating social constructions of disability. In this way, feminist disability studies illustrates how certain individuals are marked by difference in a way that appears to rationalize their exclusion. This process can demonstrate how the "aggregate of cultural systems operates together, yet distinctly, to support an imaginary norm and structure the relations that grant power, privilege, and status to that norm" (Garland-Thomson 2002, 4). Thus, it is the emphasis of feminist and disability studies' deconstruction of the social barriers of exclusion that I am recommending.

When applying a feminist disability studies lens to interrogate how identities and representations of individuals can be constructed negatively by language, students can become more aware of and, possibly, appreciate different embodied experiential knowledges. From the application of this lens, students may start "understanding how disability operates as an identity category and cultural concept" and how this may "enhance how we understand what it is to be human, our relationships with one another and the experience of embodiment" (Garland-Thomson 2002, 5). In other words, this application may help students connect to people previously marked as "other," giving them a reason to locate, value, and learn from these experiential knowledges—knowledges that, in the past, have been silenced or ignored for being considered not important or valuable to the "norm." In addition, this type of learning can transfer to other socially and culturally constructed identities beyond gender and ability, such as race, ethnicity, age, class, and sexuality to further pursuits of social justice and equality. For, "feminist disability studies all of us, not only women with disabilities: disability is the most human of experiences, touching every family and—if we live long enough—touching us all" (Garland-Thomson 2002, 5). Ultimately, the application of a feminist disability studies framework in technical communication studies and pedagogy can initiate critically reflexive conversations that identify the importance of establishing connections with all people, for we are all constructing what it means to be humans in relation to our experiences with science, technology, and the rhetorics about us and them—albeit some of us more than others.

For this particular case study, I apply a feminist disability framework to conduct a usability analysis of the language used in ISU's required accommodation statement for course syllabi. In particular, I identify how language of this accommodation statement shapes the identities of students with and without categorized disabilities and, in the process, works against the interests of technical communicators to accommodate and advocate for users (in this case, the students) by perpetuating social barriers of exclusion for some users of our syllabi. Although this case study focuses on rhetorically analyzing the accommodations statement at one university, I posit that this case study has implications for technical communication studies writ large, as well as other accommodations statements sponsored by other institutions.

ANALYSIS OF THE UNIVERSITY REQUIRED ACCOMMODATIONS STATEMENT: WHAT IS REASONABLE ACCOMMODATION?

According to ISU's Office of Equal Opportunity, Ethics, and Access website (retrieved on May 19, 2014):

> The University is committed to providing reasonable accommodations to ensure equal employment opportunities and access to University academic programs, services, and facilities in accordance with the requirements of the Americans with Disabilities Act of 1990 (ADA), Section 504 of the Rehabilitation Act of 1973, and other applicable federal and state regulations. (Illinois State University n.d.b.)

The mandated accommodations statement at Illinois State University (ISU), is as follows:

> Any student needing to arrange a reasonable accommodation for a documented disability should contact Student Access and Accommodation Services at 350 Fell Hall, 309-438-5853, or visit studentaccess.Illinoisstate.edu. (Illinois State University 2018)

This accommodation statement is required on all ISU course syllabi, and, according to the Disability Concerns website and letter from the Provost, instructors "should not alter or add to this statement." On the surface, this statement, and the fact that it is now being made a required addition to all syllabi, could be perceived as a move in the right direction. However, the language used within the construction of this statement makes it apparent that this is mandated to cover a legal obligation rather than enact what the ADA legislation intended: the inclusion of all individuals regardless of perceived differences in abilities.

In 1983, a report issued by the US Commission on Civil Rights attempted to set out a "working definition of reasonable accommodation." This definition stated that "reasonable accommodation means providing or modifying devices, services or facilities, or changing practices or procedures in order to match a particular person with a particular program/activity" (Gadacz 1994, 235). This definition was established with two underlying assumptions or principles, that (1) "The disabled individual would otherwise be qualified if it were not for the physical/mental limitation," and (2) that each "reasonable accommodation was to be individually assessed" (235–36).

In addition, it is also important to recognize that, as stated by Jung (2007), "according to the ADA, accommodations are 'reasonable' when they do not cause institutions 'undue hardship,' that is, when they do not 'require significant difficulty or expense'" (162). Jung agrees with scholars like Karen E. Jung who finds that the "undue hardship is always defined from the institutional point of view and it constitutes the technical means . . . by which the university protects itself from legal obligation for failure to accommodate" (162). Hardship can then be seen as nothing more than "unfortunate circumstances," and this type of "misfortune as not having the means to supply accommodation is simply a misfortune and not an injustice" (Michalko and Titchkosky 2001, 208).

This lack of direction and guidelines leads to "placing the actual responsibility of pursuing and obtaining accommodation onto the shoulders of those needing the accommodations" (Jung 2007, 162). Students with a disability are therefore expected to play the roles of educator, self-advocate, and "normal" student. They are required to perform the work of accommodation as follows:

> educating their instructors about their various form(s) of disability and need for accommodation, learning to work in alternate media, seeking better types of accommodation, 'coming up with a plan' for accommodation, and maneuvering through the bureaucracy of obtaining the accommodation. (Jung 2007, 162)

By placing the sole responsibility of obtaining accommodation on the shoulders of those impacted, the accommodation statement "shifts the obligation for change from the university, as an institution, to individual students and individual faculty members', thereby eliminating opportunities for genuine institutional change" (Jung 2007, 162). Individuals with a medical diagnosis of a disability are asked to "adapt themselves to this environment so that they are able to 'fit in' and make a place for themselves in this environment that surrounds them, regardless of how unnatural this environment is for them" (Michalko and Titchkosky 2001, 209). Therefore, students are being asked to "make a place for themselves, by the means available to them; their adaptation" (209). As a result, these students are forced to perform the "work" of obtaining accommodation in an institution that appears to promote practices of silencing. These students are given an unspoken promise that their instructors will not *ask* them to reveal their disability. While this promise also provides an assurance that students will not be stereotyped by their disability, it does not afford them the assurance of actually receiving any accommodation(s). And even when students ask, there is no guarantee they will receive it; someone else, usually those in a position of power who themselves may not need consideration(s) for accommodation, decides who is deserving of accommodation and how freedoms, in the form of rights to accessibility and accommodation, are allocated and/or received.

This type of educational discourse appears to be reiterated and supported within the walls of the technical communication classroom.

Here, instructors can be perceived as role models for appropriate behavioral responses to individuals within traditionally underrepresented communities, including those who have a medical diagnosis of a disability; the instructors enact and model the socially constructed appropriate responses to and interpretations of the ADA legislation. As such, by revising the language within a technical communication course syllabus to better support the purpose of ADA legislation and establish a tone of being more inclusive of a wider variety of users (students), instructors are better positioned to promote more complex student understandings and appreciation of inclusion. Ultimately, by modeling more socially inclusive and responsible behavior within the construction of technical documents and the classroom space, instructors set an example for their students to follow in their own composing practices and behaviors.

For example, an ISU instructor's syllabus, itself a form of a technical document about the instructor's course, is required to include the mandated accommodations statement (indicated at the onset of this analysis). This statement articulates the expectation that students who desire or need accommodation for a medical disability should contact the office of Disability Concerns themselves. Once at this office, these students must provide legal documentation that proves that they have a medical disability that meets the pre-defined categories for eligibility of accommodation. In addition, the required accommodations statement within the syllabus, as informed by the FAQ section of ISU's website on "Disability Concerns," also implies that instructors actually have only two options when it comes to any student's request for accommodation (Illinois State University n.d.a.). The first option centers on the need for students to provide evidence, in the form of an ID card from Disability Concerns Office detailing what accommodation(s) are necessary and that have been pre-approved for the student to receive. Within the FAQ section of the ISU Disability Concerns website it states that "only when a student provides faculty with this ID card are faculty to provide accommodation." The FAQ section goes on to suggest that once a student self-identifies as needing accommodation by displaying their ID card, instructors should "make an appointment to discuss the accommodations required and how best to facilitate those," thus assuming that instructors may not be capable of administering the accommodation in the approved manner. So, instead of Disability Concerns, aware of a student's accommodations needs, reaching out to the instructor in advance (prior to the start of the semester) to assist instructors in constructing a more accessible and inclusive course, this office waits to see if the student will *ask* their instructor for accommodation (i.e., "out" themselves) and initiate the process of actually receiving the approved accommodation. While I recognize that the Disability Concerns Office wants to respect the rights of their users (students with an approved form of a documented disability that has been proven to require accommodation), by not identifying the student to the instructor prior to the start of the semester, I find the process of forcing students to "out" themselves as needing accommodation in order to receive what has already been proven necessary to be problematic. Thus, I argue that this process and its rhetoric is another barrier to inclusion. By not developing processes that proactively work to construct inclusive environments (physical classroom space or course materials), the institution contributes to the exclusion of certain users from the course because of their difference. Ultimately, by placing the burden solely on the students to seek out accommodations (even those that have already been approved), institutions can appear to perpetuate the idea that these differently abled students are not anticipated users.

The second guideline for ISU instructors focuses on students who request accommodation but do not have a Disability Concerns Office ID Card. The Faculty FAQ page states that "if a student does not have a Disability Concerns ID card and claims to have a disability: do NOT provide any accommodation." Instead, faculty are instructed to "refer the student to Disability Concerns," where the office can collect the "appropriate documentation to verify entitlement to accommodation." Thus, instructors are expected to wait to see if a student's request for accommodation is validated by the Disability Concerns Office (ID card) before offering any accommodation. In the meantime, waiting is not action and thus does nothing to support inclusion and/or accessibility for the student or the course.

Another rhetorical effect of this "waiting" is that instructors may react to student requests of accommodation with some speculation as to the motives of the student's request or fear of getting oneself or the institution into legal trouble. For example, the FAQ section of the ISU Disability Concerns website states that "to provide an accommodation without the Disability Concerns ID card as verification would be to allow that student to a right to which s/he may not be entitled and an advantage over other students in the class." Further, it explains that the verification of a student's request for accommodation for a "qualified disability" is necessary, as it "provides legal protection for you, the faculty member, and the university." Given these guidelines, some instructors may think that the process for "proving" accommodation needs may be due to past abuses of this policy and that the university funding needs protection from expenditures on false claims.

Another concern that arises from the current policy that students are to present Disability Concerns ID cards as proof for accommodation is that it does not provide all instructors with the context necessary to best support students for successfully navigating a specific course. The information on the Disability Concerns ID card provides the following types of information: the student's name and a list of items that have been approved for necessary forms of accommodation. The list of accommodations can be as simple as, "more time to take tests" or "additional time to complete assignments." What is interesting about this Disability Concerns Office ID card system is that instructors are expected to enact what is indicated on the card without ever understanding why this accommodation(s) is necessary and how it is beneficial to the student. By not providing this type of information to instructors, Disability Concerns precludes the instructor from implementing curricular changes that might better support a larger population of users. Ultimately, according to this statement, someone else in a position of power (such as personnel at the Disability Concerns Office) who is usually not an instructor for any course at the institution decides what type of accommodation is warranted and proven necessary to navigate any course successfully, and then grants an ID card that acts as both a student's proof and types of accommodation, as well as the instructors' curriculum guide for the required accommodation(s). Thus, the users of this ID card are very diverse in disability and disciplinary identity, and thus have diverse information needs when it comes to understanding accommodations in relation to disciplinary curricular content delivery and assessment practices. Moreover, the list of accommodations on the card could potentially be perceived as the *only* measures to be taken when perhaps others could be negotiated for the rhetorical and curricular situation at hand.

Essentially, the universal accommodations statement and ID card at ISU help to demonstrate that universal design has its limitations, and technical communicators have a responsibility to interrogate and intervene in them. Although all instructors are required to include and abide by the mandated accommodations statement within their syllabus and to thusly make the courses we teach "accessible to all," there are no guidelines provided to faculty as to *how* to make specific courses "accessible to all." The letter from the ISU Vice President and Provost concerning the legal obligations of the ADA legislation and the Disability Concerns Office "Faculty FAQ" website do provide faculty with common teaching techniques, but they do not offer defined expectations beyond the legal policy and required syllabus statement, such as how instructors might evaluate current course materials in relation to accessibility, as well as how to actually construct and implement materials for a course that are more accessible for all students.

Technical communication instructors should take note that providing a list of common teaching techniques is not enough for audiences to fully understand how to implement new ideas or strategies for (re)designing for usability, usefulness, and accessibility. Not only are the newer teaching techniques not fully explicated, but no examples are provided as to how to implement the new approaches or where to find support in developing these techniques for any courses, not just ones in which a student has asked for an accommodation. For example, one of the techniques listed is to "approach teaching and learning from a multi-sensory perspective," which could be extremely useful and beneficial in developing more accessible and inclusive courses. However, there is no explanation of what "multi-sensory" means and how some faculty are incorporating this technique. Moreover, this tip does not also come with a warning that multi-sensory approaches to teaching may cause physical and/or psychological harm to students with some disabilities (e.g., acute vertigo). In addition, the technique that suggests "using accessible technology such as computers or captioning videos" doesn't explain how computers may not be considered accessible to all users or our courses, much less students with disabilities, nor who is responsible for adding the captions to videos or which assistive technologies the instructor has access to (financially and physically) and IT support for. Thus, these suggestions designed to improve accessibility for some may also exclude it for others. Furthermore, if instructors don't feel confident or able to implement these techniques, not only are the list of techniques not usable or useful for instructors but also not for the students they are intended to support.

The ambiguous rhetoric on the ISU Disability Concerns website—intended to articulate how the University complies with the ADA legislation requirements—could be perceived as alleviating instructors of their responsibility and accountability. To explain, the website states that instructors *should consider* incorporating techniques to improve accessibility, instead of *expecting* instructors to, which can suggest that instructors have no real obligation or definitive expectation to proactively implement techniques that would make their course more accessible to more students. In fact, the only thing that the website appears to expect, in terms of action, is to (1) NOT provide any accommodation to a student who does not supply proof (the ID card), and (2) how to respond to a student who specifically requests and/or is deemed eligible for these accommodations by only accommodating what is listed on the ID card. Both of these expectations appear to further the concept of a retrofit approach (only accommodate after a request) versus a proactive approach (try to construct a more accessible course for all students, while also being open to continuous user feedback about the accessibility of the course).

The Disability Concerns rhetoric also sponsors the "othering" of the disability community. The university expects students to "perform" by "speaking in a voice and through codes of those considered to be in positions of power and wisdom" (Bartholomae 2009, 622). Student users of the Disability Concerns interface are forced to assimilate to hegemonic institutional processes and rhetoric in order to be granted any support—a process that perpetuates perceptions of people with disabilities as "other," different, deficient, and abnormal. Instructor users of the Disability Concerns interface are essentially encouraged to promote the stigmatization of students with disabilities vis-à-vis the perpetuation of problematic disability labels and only providing two options for students with disabilities: (1) *Assimilate* by performing what disability scholars refer to as *passing*, or (2) ask for permission from those in hegemonic positions of power to be received or accommodated as *non-normative* by using *normative* rhetoric in a *normative* atmosphere.

Thus, universities using an accommodation system like ISU's can appear to promote an educational discourse that perpetuates a version of a "don't ask, don't tell" stance on the issues of disability and accommodation—a stance that attempts to support educational equality for all students that inevitably falls short. Institutions problematize the *asking* for accommodation by placing strict guidelines and criteria that identify who is expected to ask for or about accommodation, how they should ask for it, to whom they should inquire, what they should or shouldn't expect concerning accommodation, and even questions *who* is considered "worthy" or "eligible" for accommodation. The *tell (or outing)* requirement expected of students with disabilities begs the question, why? Why should these students have to self-identify themselves as being "other" or a "special needs case" if they have to prove themselves disabled enough to receive accommodation? Why should they have to out themselves and risk possible stereotyping only to *possibly* receive the services that were the driving force of this self-identification in the first place? Not to mention that the amount of paperwork and time necessary to be even considered eligible or "entitled" to accommodation may be a deterrent in itself, as some users have disabilities that hinder their access to the process itself while others must wait several months for accommodation approval, much less the actual accommodation(s). How are we meeting ADA requirements and user needs if the process can take so much time to complete that by the time the student is approved to receive accommodation, the semester is almost over?

Given all this, technical communication instructors should be mindful of how usability incidents that interrupt the promises of universal design and evidence the need for accommodating a wider range of audiences, content, and delivery options—among other things—may be perceived by some as disruptive of, and thus perhaps dismissive of, the norm. For example, in relation to this case study, students receiving accommodations could potentially be subjected to social exclusion by their peers and instructor. This is more likely to happen when instructors are not provided with the resources necessary to rhetorically address any potential concerns that students receiving accommodations have an "unfair advantage" and/or redress any potential repercussions that students receiving accommodations may experience from potentially disgruntled peers who may feel that they are being placed at a disadvantage.

These potential effects of social exclusion for those who ask for accommodation services could also deter students from ever seeking accommodations in the first place. In fact, Marshak et al. (2010) conducted a study at a medium-sized state university in the mid-Atlantic region to identify the reasons why students with medically categorized disabilities did not seek out accommodation services. They found that only "40% of students who received special education services and supports in secondary school identify their disability to their postsecondary institution and 88% of those students actually receive support" (152). Thus, the postsecondary institution that appears to have a desire to construct and "pave a level playing field" for all students enrolled may unconsciously promote the need to "assimilate" to the normative expectation by not instituting a more proactive stance in making all courses (not just physical spaces) more accessible to a wide variety of students.

FURTHER IMPLICATIONS AND A CALL FOR ACTION

Usability and accessibility are critical issues for both technical communication and disability studies. In particular, technical communication's rich history of engaging these issues in tandem appears to focus on the deconstruction of more physical barriers of exclusion such as focusing on ensuring that the construction of a physical space (environment), such as a classroom, and/or the physical text or product is able to be accessed or consumed by the "broadest range of users" (Dolmage 2009,

172). As Meloncon (2013a) explains, "accessibility is the material practice of making social and technical environments and texts as readily available, easy to use, and understandable to as many people as possible, including those with disabilities" (5). However, is a space or a document that is perceived as being physically accessible for being "readily available, easy to use, and understandable" more socially inclusive? Do these concerns work to deconstruct stereotypic or negative representations of particular communities within technical informational texts? I would argue that this focus on physical accessibility, though an important piece of inclusion, is not enough.

Thus, I propose that technical communication instructor-scholars more fully redress social barriers of exclusion as they pertain to accessibility and usability in our scholarship and pedagogy. To do so, we could use our rhetorical expertise and a feminist disability studies framework to identify, deconstruct, and revise the rhetorics and practices which—perhaps unwittingly—work to exclude users of our discipline and courses. For example, mapping Meloncon's (2013b) definition of accessibility onto the case study at hand, the following is a heuristic that demonstrates that the accommodation statement appears to meet each element of her definition:

- "readily available": the accommodation statement is required to appear on every ISU instructor's course syllabus.
- "easy to use": the accommodation statement is a legally approved statement that is provided to every instructor and is expected to be included word for word on every course syllabus.
- "understandable": the accommodation statement clearly instructs students that they must contact the Disability Concerns Office if they have a disability and want to request accommodation.

On the surface, one could say that, technically, this statement is both usable and accessible; however, this statement is not socially accessible and inclusive. Rhetorically placing the burden of proof on the student with a disability to *prove* their need and entitlement to an accommodation can construct an identity of deviancy (e.g., one who seeks an "unfair advantage" and may not be "entitled" to accommodation), thereby further marking an individual as "other." This is especially problematic in a technical communication course, where we teach the value of developing texts that are usable and accessible to diverse users, as students may see the accommodation statement as a justifiable way to exclude certain users.

A feminist disability studies framework can help to not only uncover how rhetoric can be used to further stigmatize or marginalize, but it can also establish important connections that draw attention to embodied human experiences and what it means to be human and work to revise rhetorical—and thus embodied—"othering." I am by no means saying that a feminist disability studies framework is the ultimate answer or resolution to the construction of a more inclusive and accessible course. However, a feminist disability studies framework in this case has helped to reveal that technical communication could benefit from expanding our current investment in physical accommodation to social accommodation as well—as well as our current legal and social justice literacies as they pertain to our students and their access to our course content and the transfer of skills to their future professional lives.

In *Deconstructing the Digital Divide*, Virginia Eubanks (2011) posits that inclusion should extend beyond "bringing people in or together, but in challenging the structures that make their inclusion a paradox: how to get people to include those they disregard" (n.p). For technical communication instructors, this means helping students establish these connections and position them as agents of social change in hegemonic structures—agents with the skill and know-how to design texts with more socially responsible representations of underrepresented and disenfranchised users. As Dolmage (2009) reminds us, "we as teachers are the establishment that often distributes access, regulates relationships with technologies, and can either impose or remove barriers" (185). The documents (and policies therein) that instructors and technical communicators create and distribute—such as institutional accommodation statements in course syllabi—can impose or remove barriers.

Given this pedagogical imperative, as well as our disciplinary responsibility to perform rhetorical action as "public intellectuals" (Bowdon 2004), there are steps we can take to remove institutional barriers to inclusion and accessibility—even in hegemonic structures where agency appears limited. In the case of the required accommodations statement, a feminist disability studies approach to constructing this portion of our syllabi could include providing an additional statement for users of the course that identifies the intentions of the instructor. For example, accommodations are only one element of my broader "Classroom Inclusion Policy," which states:

> I want to ensure that all students are given the tools and resources necessary to be successful within this course. Therefore, I am dedicated to the idea and practice of doing whatever I can to help accommodate all students, regardless of their abilities or needs. I attempt to accomplish this by providing more inclusive strategies for the classroom that focus on ensuring accessibility of information, content, and resources for students that work to equip each student with the tools and resources needed to

successfully navigate and complete the course. Additionally, I am more than willing to work in conjunction with ISU Disability Concerns to provide whatever a student requires or needs. However, I want to encourage all students to feel free to approach me at any time with suggestions and/or requests in regards to accessibility of information, accommodation, etc . . . as I am always interested in suggestions that will not only meet a specific student's needs, but could be something that I proactively employ to make the overall class more accessible and inclusive of more individuals. (Smyser-Fauble 2014)

This policy, while appearing with the required ISU Accommodations statement, works to promote more complex understandings of inclusion and accessibility and more diverse ways of being accommodated. In addition, this policy engages with Erin Frost's "apparent feminist pedagogical approach" by providing an explicit model of a more culturally inclusive classroom philosophy that works to model socially-just behavior.

Another suggestion—inspired by Bowdon (2004)—is for technical communication instructors and students to collaborate on constructing more ethical and responsible cultural representations of communities within technical documents. Bowdon's technical communication students helped to assess and revise the language used in a questionnaire distributed to GYM community members, as well as other report tools and findings. This collaborative approach could be applied to co-developing a university accommodations statement with students—one that responsibly supports the intention of the ADA legislation to construct more inclusive spaces (both physically and socially) in work and educational institution settings. Such an activity would engage students' critical engagement, cultural rhetorical analysis, and cultural usability skills, while fostering new legal literacies and further developing their technical communication skills. Ultimately, like Medina and Walker's use of grading contracts within technical communication courses in chapter 2 of this collection, this activity can provide technical communication instructors an opportunity to enact more inclusive and accessible pedagogical practices. In this way, both students and instructors actively participate in "recognizing the powerful effect for positive and negative change that their (technical communicators') work may have in their communities and in the messy world that we all share" (Bowdon 2004, 339). In addition, technical communication instructors can extend Bowdon's work of collaborating with students to construct more socially responsible and inclusive texts by asking students to provide feedback on the technical communication products designed for the course. Soliciting feedback from the users of our course syllabus, for example, in relation to its accessibility models for our students shows that we don't just talk about audience needs, rather we actually value them.

Feminist disability frameworks are useful and useable for disciplinary research and pedagogy. The work that we do with feminist disability frameworks has the potential to establish us interdisciplinarily as leaders in knowing how to accommodate the "broadest range of users," including students in higher educational contexts. Further, feminist disability studies framework can help us to better demonstrate and incorporate the value of diverse embodied experiential knowledges in the construction of our information products, such as more accurate representations of under- and mis-represented identities and socially responsible relationships with the disenfranchised communities about and with whom we communicate. Ultimately, feminist disability theory provides a framework for using technical communication scholarly and pedagogical practices to work toward social justice.

Notes

1. Throughout this chapter, the term *technical communicator* is used to reference those individuals working in academic (instructors and scholars) and non-academic (professionals in the public sphere) settings. While this chapter focuses on identifying pedagogical approaches and curricular implications of a feminist disability studies framework for technical communication instructors, it also discusses the effects of implementing these approaches with our students (future technical communication professionals). Thus, the term *technical communicator* also indicates a bridging of skill sets and traditions impacting both the field (professional non-academic settings) and discipline (academic settings of technical communication).
2. While the use of "person first language" has been critiqued by disability scholars like Titchkosky (2011) and Linton (2008) for creating what is perceived as perpetuating a "negative" perception of any type of disability (negative perception of the terms affiliated with disability), there is a greater need within technical communication for the use of "person-first language" to ensure that individuals are not equated with their disabilities: disability is not their only identity.

References

Bartholomae, David. 2009. "Inventing the University." In *The Norton Book of Composition Studies*, ed. Susan Miller, 605–30. New York: W.W. Norton.

Bowdon, Melody. 2004. "Technical Communication and the Role of the Public Intellectual: A Community HIV-Prevention Case Study." *Technical Communication Quarterly* 13 (3): 325–340. https://doi.org/10.1207/s15427625tcq1303_6.

Brueggemann, Brenda J., Rosemarie Garland-Thomson, and Georgina Kleege. 2005. "What Her Body Taught (or, Teaching about and with a Disability): A Conversation." *Feminist Studies* 31 (1): 13–33. https://doi.org/10.2307/20459005.

Brueggemann, Brenda J., Linda F. White, Patricia A. Dunn, Barbara Heifferon, and Johnson Chieu. 2008. "From Becoming Visible: Lessons in Disability." In *Disability and*

the Teaching of Writing: A Critical Sourcebook, ed. Cynthia Lewiecki-Wilson and Brenda Jo Brueggemann, 141–46. Boston, MA: Bedford/St. Martins.
Davis, Lennard. 2008. "From the Rule of Normalcy." In *Disability and the Teaching of Writing: A Critical Sourcebook*, ed. Cynthia Lewiecki-Wilson and Brenda Jo Brueggemann, 206–10. Boston, MA: Bedford/St. Martins.
Davis, Lennard. 2013. *The End of Normal: Identity in a Biocultural Era*. Ann Arbor: University of Michigan Press.
Dolmage, Jay. 2008. "Mapping Composition: Inviting Disability in the Front Door." In *Disability and the Teaching of Writing: A Critical Sourcebook*, ed. Cynthia Lewiecki-Wilson and Brenda Jo Brueggemann, 14–27. Boston, MA: Bedford/St. Martins.
Dolmage, Jay. 2009. "Disability, Usability, Universal Design." In *Rhetorically Rethinking Usability: Theories, Practices, and Methodologies*, ed. Susan Miller-Cochran and Rochelle L. Rodrigo, 167–90. Cresskill, NJ: Hampton Press.
Dolmage, Jay. 2014. *Disability Rhetoric*. Syracuse, NY: Syracuse University Press.
Dunn, Patricia A., and Kathleen Dunn De Mers. 2002. "Reversing Notions of Disability and Accommodation: Embracing Universal Design in Writing Pedagogy." *Kairos* 7 (1). http://kairos.technorhetoric.net/7.1/coverweb/dunn_demers/.
Elmore, Kimberly. 2013. "Embracing Interdependence: Technology Developers, Autistic Users, and Technical Communicators." In *Rhetorical Accessibility: At the Intersection of Technical Communication and Disability Studies*, ed. Lisa Meloncon, 15–38. New York: Baywood Publishing. https://doi.org/10.2190/RAAC1.
Eubanks, Virginia. 2011. "Deconstructing the Digital Divide." *Chicago Humanities Festival.* YouTube video, 23:15. December 12, 2011. https://www.youtube.com/watch?v=pJwZcUJQFkk.
Gadacz, Rene. 1994. *Re-thinking Dis-ability*. Edmonton, AB: The University of Alberta Press.
Garland-Thomson, Rosemarie. 2002. "Integrating Disability, Transforming Feminist Theory." *NWSA Journal* 14 (3): 1–32. https://doi.org/10.2979/NWS.2002.14.3.1.
Gutsell, Margarat, and Kathleen Hulgin. 2013. "Supercrips Don't Fly: Technical Communication to Support Ordinary Lives of People with Disabilities." In *Rhetorical Accessibility: At the Intersection of Technical Communication and Disability Studies*, ed. Lisa Meloncon, 83–94. New York: Baywood Publishing. https://doi.org/10.2190/RAAC4.
Hamraie, Aimi. 2013. "Designing Collective Access: A Feminist Disability Theory of Universal Design." *Disability Studies Quarterly* 33 (4). https://doi.org/10.18061/dsq.v33i4.3871 http://dsq-sds.org/article/view/3871/3411.
Hannah, Mark A. 2010. "Legal Literacy: Coproducing the Law in Technical Communication." *Technical Communication Quarterly* 20 (1): 5–24. https://doi.org/10.1080/10572252.2011.528343.
Illinois State University. 2018. "Student Access and Accommodation Services." Accessed June 10, 2018. https://studentaccess.illinoisstate.edu/policies/statement/.
Illinois State University. n.d.a. "Disability Concerns: Faculty FAQs." Accessed October 11, 2011. https://disabilityconcerns.illinoisstate.edu/students/faculty-faq/.
Illinois State University. n.d.b. "Office of Equal Opportunity, Ethics, and Access." Accessed October 11, 2011. https://equalopportunity.illinoisstate.edu/ada/.
Jung, Julie. 2007. "Textual Mainstreaming and Rhetorics of Accommodation." *Rhetoric Review* 26 (2): 160–178. https://doi.org/10.1080/07350190709336707.
Linton, Simi. 2008. "From Reassigning Meaning." In *Disability and the Teaching of Writing: A Critical Sourcebook*, ed. Cynthia Lewiecki-Wilson and Brenda Jo Brueggemann, 174–82. Boston, MA: Bedford/St. Martins.
Lott, Deshae. E. 2001. "Going to Class with (Going to Clash with?) the Disabled Person: Educators, Students, and their Spoken and Unspoken Negotiations." In *Embodied Rhetorics: Disability in Language and Culture*, ed. James C. Wilson and Cynthia Lewiecki-Wilson, 135–53. Carbondale: Southern Illinois University Press.

Marshak, Laura, Todd Van Weren, Diane Ferrell, Lindsay Swiss, and Catherine Dugan. 2010. "Exploring Barriers to College Student Use of Disability Services and Accommodations." *Journal of Postsecondary Education and Disability* 22 (3): 151–65.

Meloncon, Lisa, ed. 2013a. *Rhetorical Accessability: At the Intersection of Technical Communication and Disability Studies*. New York: Baywood Publishing.

Meloncon, Lisa. 2013b. "Toward a Theory of Technological Embodiment." In *Rhetorical Accessibility: At the Intersection of Technical Communication and Disability Studies*, ed. Lisa Meloncon, 67–82. New York: Baywood Publishing.

Meyer, Craig A. 2013. "Disability and Accessibility: Is There an App for That?" *Computers and Composition Online*. http://www2.bgsu.edu/departments/english/cconline/spring 2013_special_issue/Meyer/index.html.

Michalko, Rod, and Tanya Titchkosky. 2001. "Putting Disability in Its Place: It's Not a Joking Matter." In *Embodied Rhetorics: Disability in Language and Culture*, ed. James C. Wilson and Cynthia Lewiecki-Wilson, 200–228. Carbondale: Southern Illinois University Press.

Palmeri, Jason. 2006. "Disability Studies, Cultural Analysis, and the Critical Practice of Technical Communication Pedagogy." *Technical Communication Quarterly* 15 (1): 49–65. https://doi.org/10.1207/s15427625tcq1501_5.

Price, Margaret. 2008. "Writing from Normal: Critical Thinking and Disability in the Composition Classroom." In *Disability and the Teaching of Writing: A Critical Sourcebook*, ed. Cynthia Lewiecki-Wilson and Brenda Jo Brueggemann, 56–73. Boston, MA: Bedford/St. Martins.

Smyser-Fauble, Barbi. 2014. *Syllabus*. Normal: Illinois State University.

Titchkosky, Tanya. 2011. *The Question of Access: Disability, Space, Meaning*. Toronto: University of Toronto Press.

US Department of Justice Civil Rights Division. 2009. "A Guide to Disability Rights Laws." Ada.gov, April 30, 2015. https://www.ada.gov/cguide.htm?.

Vidali, Amy. 2008. "Discourses of Disability and Basic Writing." In *Disability and the Teaching of Writing: A Critical Sourcebook*, ed. Cynthia Lewiecki-Wilson and Brenda Jo Brueggemann, 40–55. Boston, MA: Bedford/St. Martins.

Walters, Shannon. 2010. "Toward an Accessible Pedagogy: Dis/ability, Multimodality, and Universal Design in the Technical Communication Classroom." *Technical Communication Quarterly* 19 (4): 427–454. https://doi.org/10.1080/10572252.2010.5 02090.

Williams, Miriam F. 2006. "Tracing W. E. B. Dubois' 'Color Line' in Government Regulations." *Journal of Technical Writing and Communication* 36 (2): 141–165. https://doi.org/10.2190/67RN-UAWG-4NFF-5HL5.

Wilson, James C. 2000. "Making Disability Visible: How Disability Studies Might Transform the Medical and Science Writing Classroom." *Technical Communication Quarterly* 9 (2): 149–161. https://doi.org/10.1080/10572250009364691.

Wilson, James C., and Cynthia Lewiecki-Wilson, eds. 2001. *Embodied Rhetorics: Disability in Language and Culture*. Carbondale: Southern Illinois University Press.

PART II

Space, (Em)Place, and Dis(Place)Ment

4

SPATIAL ORIENTATIONS
Cultivating Critical Spatial Perspectives in Technical Communication Pedagogy

Elise Verzosa Hurley

One point of studying technical writing as a cultural practice is to make visible what seems invisible in technical writing, to view what seems inevitable as a product of culture (Longo 1998, 65).

We must be insistently aware that space can be made to hide consequences from us, how relations of power and discipline are inscribed into the apparently innocent spatiality of social life (Soja 1989, 6).

In the summer of 2013, I embarked on a cross-country move from the rugged Sonoran desert where I completed my graduate work at the University of Arizona (UA), to the flatlands of the Midwest in order to begin a new position as an assistant professor at Illinois State University (ISU). Having only visited the Midwest once before, during my whirlwind campus visit, my entire spatial orientation changed. Whereas in Tucson I could locate myself cardinally by simply looking to the mountain ranges surrounding the city to the north, south, east, and west, in Bloomington-Normal, I had no geographical landmarks, no sense of direction. The GPS application on my smart phone became my lifeline; I relied (and continue to rely) on it to (mis)direct me and tell me where to go. As a native Texan, my concept of time and distance similarly changed. Whereas I could drive for hours—even days—through parts of Texas and the Southwest, I was amazed that I could cross state lines to Missouri, or to Indiana, in as little as two hours. Geographical changes aside, there were other ways in which the significance of space became apparent to me in the days and months after the move. Marked by overlapping "real and imagined" spatial differences (Soja 1996, 11), the transition from graduate student to assistant professor, from living in the Southwest to the Midwest was, for me, simultaneously material *and* metaphorical.

For example, I now live in a house with a basement, a new-to-me space that became my home office in an effort to keep the piles of papers and books out of sight, hidden away from the main living space. This same house is located in a quintessential suburban neighborhood where, for the first time in my life, I quickly realized that I was the only person of color. Having earned the terminal degree also allowed me access to new spaces and places at my new university. I now have an office all to myself with a door that I can keep closed or opened. I have access to the department's supply room, copy room, and became privy to the meeting rooms where department meetings happen. Although I have been thinking about space in relation to my research and teaching for quite some time (Verzosa and Crump 2012), the transition to a new region, new town, new university, new position, made clear to me that I was/am in the spatial turn: space, place, location, embodiment—all of these things mattered.

I begin this chapter on integrating spatial perspectives in technical communication pedagogy with this narrative in order to underscore the ways in which the workings of our everyday lives—whether personal or professional—are comprised of numerous and overlapping spatial stories, or what de Certeau (1984) calls "spatial practices" (115). While the terms *space* and *place*[1] certainly encompass physical and material geographic sites, they also emerge from discursive and metaphoric constructions; thus, spatialization need not be restricted to geographic space as a fixed, static location. Because space is produced by and productive of social relations, spatial practices are always cultural, rhetorical, and necessarily political. Feminist cultural geographer Doreen Massey (2005) thus argues for an alternative view of human relationships to space and forwards three propositions crucial to its reimagining: that space is "the product of interrelations," that we understand space as "the sphere of possibility of the existence of multiplicity," and that space is "always in the process of being made. It is never finished; never closed. Perhaps we could imagine space as a simultaneity of stories-so-far" (9). Massey grounds these three propositions within a larger critique of the "frameworks of progress, of development and of modernization" (10) that helps to construct and reinforce a socio-spatial imaginary in which concepts like globalization are conceived of and articulated as a predetermined historical queue, a view that reproduces a singular, linear narrative about progress and development: that the developed world is at one advanced point and the rest of the developing world simply need to catch up. According to Massey, only when we reimagine space as an always open collection of multidimensional trajectories and relations—rather than as a flat surface to be filled, or traversed and conquered—can there be a future that is unscripted and open "for a politics which can make a difference" (11).

As the two epigraphs to this chapter suggest, however, notions of transparency, neutrality, and objectivity often elide both technical communication practices as well as spatial practices. When viewed as an already given inevitability, as innocent and uncomplicated, space—whether natural, built, real, imagined, or some combination thereof—can render discursive practices as similarly innocent, static, fixed, and unchangeable. When translated to classroom settings where students all too often perceive technical communication as a neutral and objective practice, a singular, closed view of space limits the possibility for students to understand how the work they do in our classrooms are tied to knowledge work in a multiplicity of other spaces beyond the university. Moreover, focusing too narrowly on the vocational spaces of technical communication, as many others have noted, forestalls the potential for students to consider other spaces in which technical discourse can work in service of social justice.

My purpose in this chapter, then, is to illustrate how attention to rhetorics of space, place, and location may help to cultivate critical spatial perspectives that can enrich our teaching of technical communication. Because all social practices, including communication practices such as those in technical documents, traffic in numerous spaces and places which, in turn, are productive of cultural, ideological, and rhetorical meanings, I advocate for a spatially-oriented pedagogy that asks students to think critically about the ways in which spaces and places interact relationally with subjectivities and communication practices, as well as their potentials and constraints for rhetorical action. Critical spatial perspectives, I argue, can complement existing cultural studies frameworks in professional and technical communication pedagogy that seek to intervene in what Scott, Longo, and Wills (2006) have dubbed the "legacy of hyperpragmatism" (9). In this chapter I offer a discussion of how scholars and teachers have taken up the "spatial turn" in professional and technical communication as well as in rhetoric and writing studies more broadly. Then, I argue that theoretical contributions from cultural geography can further inform, complicate, and extend our discipline's understanding of space and place. Finally, I describe how spatially oriented pedagogical approaches informed by cultural geography can be integrated into a graduate-level course design and illustrate the potentials for critical spatial perspectives to further social justice work in our discipline.

WRITING SPACE: LOCATING CONVERSATIONS IN TECHNICAL COMMUNICATION AND RHETORIC AND WRITING STUDIES

The spatial turn in the disciplines of technical communication and rhetoric and writing studies has gained significant traction in recent years, all of which affirm that *where* communicative practices happen is just as important as *how* and *why* they happen. Scholarship in technical communication has recently been concerned with issues of space, particularly in areas concerning hybrid technological spaces (Swarts 2007), human-technology interactions (Sun 2009; Welch 2005), and distributed work (Pigg 2014; Spinuzzi 2007). Moreover, concerns about space, place, and location are especially evident through scholarship that studies maps (Barton and Barton 1993; Propen 2012) and mapping practices (Johnson-Eilola 1996), most often in the form of "mapping essays" that serve as visual and conceptual tools in charting disciplinary concerns. For example, Patricia Sullivan and James Porter highlighted the "curricular geography" of professional and technical communication and pointed to its spatialization within the terrain of English departments (Porter and Sullivan 2007; Sullivan and Porter 1993). Johndan Johnson-Eilola and Stuart Selber developed a spatial matrix of the locations of usability for pedagogical uses beyond the professional and technical communication classroom (Johnson-Eilola and Selber 2007), and Carolyn Rude (2009) mapped the research questions in technical communication to explore the field's disciplinary status. Indeed, conversations about space and place have provided multiple areas of technical communication with a means of engaging and interrogating the ways in which space is produced and contested through a variety of communication practices in a wide array of settings.

Scholarship in the broad area of rhetoric and writing studies has been similarly infused by spatial concerns, an increasing area of interest evinced by Christopher Keller and Christian Weisser's claim that "nearly all of the conversations in composition studies involve place, space, and location, in one way or another" (Keller and Weisser 2007, 1). Indeed, we need only look to the explanatory spatial metaphors used to describe the teaching of rhetoric and writing—such as trenches, frontiers, landscapes, cities, contact zones, borderlands, architectures, margins, and sites—to provide evidence for Keller and Weisser's claim. After all, it makes sense for the field to engage most readily with representations of space, or space as it is conceived through discourse. To illustrate, Nedra Reynolds (1998) asserted that "writing itself is spatial" and explicitly situated the politics of space within writing and rhetoric's "imagined geographies" that both shape and are shaped by the spatial metaphors used to describe the field (14). Roxanne Mountford (2005) forwarded the notion of rhetorical space as "the geography of a communicative event" in her study of pulpits as a gendered location (42). Roberta Binkley and Marissa Smith critiqued the spatialized origins of Western rhetorical history, arguing that the privileging of Greek rhetorical principles erases "those whose spatial history and context are different from 'mainstream' Western Eurocentric heritage" (Binkley and Smith 2006, n.p.).

Indeed, the spatial turn and its focus on the locations of writing and rhetoric has gained significant traction in recent years, resulting in a growing number of spatially-oriented scholarship in ecocomposition and environmental rhetorics (Dobrin and Weisser 2002; Killingsworth 2005), information and networked spaces (Kimme Hea 2009; Rice 2012), institutional critique (Grego and Thompson 2007; Porter et al. 2000), public discourses and public rhetorics (Ackerman 2003; Fleming 2009), community outreach (Deans 2000), and spatial praxes (Haley-Brown, Holmes, and Kimme Hea 2012) just to name a few lines of inquiry in addition to several edited collections and special journal issues devoted to space and place.

Despite these rich and varied inquiries, however, only a handful of scholars have explicitly engaged perspectives from cultural geography in technical communication research and teaching. Drawing from landscape theory in cultural geography, Lisa Meloncon (2007) developed a heuristic for determining the preparedness of professional and technical communication instructors to teach in online environments. In a later book chapter on teaching visual communication in the professional and technical communication classroom, Meloncon (2013), building on the work of geographer Pierce Lewis, again argued for the relevance of landscape theory in teaching students to read, interpret, and produce visual documents. Employing a critical cartographic perspective, Amy Propen (2012) articulated a visual-material framework in studying the spatial and embodied implications of navigation technologies and built spaces. A primary purpose of this chapter, then, is a call for further intellectual alliances among cultural geography, rhetoric, and technical communication to enrich our classroom pedagogies. Most compelling, in my view, is the transformative potential critical spatial perspectives may provide technical communication students.

SOCIAL SPACE: LOCATING CONVERSATIONS IN CULTURAL GEOGRAPHY

Cultural geographers and spatial theorists have made apparent that understanding the spatiality of everyday life is integral to understanding how the spaces and places we inhabit always already produce—and are produced by—social, cultural, ideological, and rhetorical meanings. As a branch of human geography that studies how people impact the physical world and vice versa, cultural geography focuses on the relationships among identity, meaning, and place. And because bodies produce space materially (by creating built environments), perceptually (by making meaning about our experiences), and conceptually (by imposing and negotiating socially constructed codes), cultivating a critical spatial consciousness can be potentially transformative for our students when viewed as a spatial practice that can build coalitions across disciplinary domains in order to intervene in various forms of oppression.

The theoretical work of Henri LeFebvre and his assertion that we view space as both a concept and a practice has been pivotal in engaging a more robust understanding of space. Concerned about dominant ideologies that take space for granted and troubled by the Cartesian dualism that views space as either abstract or material, Lefebvre asserts that space is inherently social and relational. Thus, he argues for a triadic understanding of *spatial practice* (perceived space, or the "real" sensory experience of how we see and navigate the world), *representations of space* (conceived space, or how we understand or "imagine" space as it is mediated by the norms of institutional knowledge through codes, symbols, and signs), and *representational space* (lived space, or the lived experiences that emerge from the relations among perceived space and conceived space) (Lefebvre 1991).

Bernadette Longo (1998) cautioned technical communication scholars and practitioners to attend to the ways in which "struggles for knowledge legitimation . . . are influenced by institutional, political, economic, and/or social relationships, pressures, and tensions" (61), and numerous others have argued that we have a responsibility to teach our students to be cognizant of the broader cultural conditions that inform technical discourse (Henry 2001; Herndl 1993; Scott 2004; Scott, Longo, and Wills 2006). I posit that a triadic spatial understanding can help us to think more expansively about the spatialized nature of knowledges generally deemed "technical," "professional," "scientific," or otherwise neutral. For example, the highly expert, important, and "objective" knowledge-work of many scientific laboratories (conceived space), is often supported and legitimized by physical measures, such as security systems and specialized equipment (perceived space), all of which may occlude interlocking power geometries at play on the bodies for whom laboratories are workplaces (lived space). Cultivating critical spatial perspectives in technical communication curricula can provide students not only with a means of exploring and interrogating space—in its perceived, conceived, and lived relational iterations—but also may serve as a starting point in discussing the ways in which spatial configurations can be reimagined through policy work and other initiatives that depend on technical discourse.

Building on the work of Lefebvre, Edward Soja (1996) forwards the notion of *thirdspace* as means of calling attention to the geographies that difference makes, a perspective that provides "a context from which to build communities of resistance and renewal that cross the boundaries and double-cross the binaries of race, gender, class, and all oppressively othering categories" (84). Soja thus asserts that investigations concerning spatiality "must be additionally guided by some form of potentially emancipatory praxis, the translation of knowledge into action in a conscious—and consciously spatial—effort to improve the world in some significant way" (22).[2] Many others in our discipline (Brasseur 1993; Haas 2012; Williams and Pimentel 2014) have sought to explicitly situate the knowledge work of technical communication as deeply concerned with social justice by integrating pedagogical frameworks informed by cultural studies, critical race theories, postcolonial and decolonial theories, and feminist theories. I suggest that pedagogical approaches informed by cultural geography can similarly engage our students in conversations about social *and* spatial justice. I am particularly drawn to Soja's (2010) use of the term spatial justice "as the geography of social justice . . . [for] everything that is social (justice included) is simultaneously and inherently spatial, just as everything spatial . . . is simultaneously and inherently socialized" (4–6). Moreover, the term spatial justice emphasizes the concrete sites and locations from which social justice work can begin.

Cultural geography's call for cultivating a critical spatial perspective and consciousness is, above all, a call to attend to the ways in which spatialization affects human lives. Thus, integrating conversations about the spatialization of everyday life in technical communication pedagogy is especially apt because, as Steven Katz (1992) pointed out, "perhaps even more than other kinds of rhetorical discourse, [technical communication] always leads to action, and thus impacts human life" (259). Further, as Natasha Jones, Gerald Savage, and Han Yu argued in a guest editorial concerning the status of diversity and social justice in technical communication programs, our field shares common interests with those in "economics, agriculture, geography, forestry, and the broadly interdisciplinary area of development studies . . . yet little of our scholarship has paid any attention to such studies" (Jones, Savage, and Yu 2014, 146).

Jones, Savage, and Yu assert that making connections across disciplinary domains is imperative if we are to further social justice work in our field. In the remainder of this chapter, I describe how spatially oriented approaches informed by cultural geography can be integrated into a graduate seminar course design. In so doing, I posit that technical communication scholars and teachers have a responsibility to consciously influence and take action in the spaces we inhabit and that one site where this work can begin is the technical communication classroom.

TEACHING AND LEARNING ABOUT SPACE: A SPATIALLY ORIENTED COURSE DESIGN

I taught my first graduate seminar at ISU during the spring semester of 2014, a topics course in technical writing that I named visual and spatial rhetorics.[3] This course is open to graduate students enrolled in the doctoral program in English Studies and the master's program in professional writing and rhetoric, although it is not a required course for degree completion.

My goals in designing a course that focuses on both visuality and spatiality were threefold. First, the ways in which spaces and places are represented—through maps, images, and other technical discourse—construct specific kinds of knowledge that are always culturally, ideologically, and rhetorically saturated. As Reynolds (2004) noted, "geography is very much a *seeing* discipline, whose premises and proofs, methodologies and conclusions, stem from visual evidence" (252). Further, what we see or don't see is shaped, in part, by the spaces and places in which viewing happens. Finally, I also wanted to call attention to the seeming transparency of space—that even as we experience and see spaces and places, we tend to view them as static, inevitable, and unchangeable. In other words, I wanted to make apparent that how we view, design, and experience spaces and places, particularly as they are mediated by technologies and texts, have the potential to be revised and transformed.

Because I only had one semester of teaching experience at ISU prior to teaching the course and had limited encounters with the graduate students who would enroll in it—in addition to the fact that a major premise of the course assumes that space is open, dynamic, and produced by social relations—I wanted the course to be an intellectual space where multiple research trajectories could be engaged. Thus, despite the official designation of the course as a topics course in technical writing, I designed the course with a broader interdisciplinary focus that not only drew from scholarship in technical communication but those from other fields as well. In so doing, the course design supported the English Studies model valued by ISU's English Department, while also making room for the potential interests and needs of the course's primary users. In sum, I wanted students to be able to: build and gain facility with both visual and spatial theories, methodologies, and praxes; interrogate the relationships among rhetoric, culture, and power, in order to have a critical understanding of how both visuality and spatiality are articulated in our contemporary culture; explore the ways in which we can construct, deconstruct, articulate, and re-envision new possibilities for visuality and spatiality in ways that account for varying lived-experiences and contexts; and consider the ways in which visual and spatial rhetorics can enrich our pedagogies and professional practices.

In order to support the goals of the course, I designed the course so that we would have the opportunity to engage inquiries related to both visuality and spatiality, beginning with a focus on spatial theories and readings during the first half of the semester, and concluding with a focus on visual theories and readings during the second half of the sixteen-week semester, knowing that the boundaries between these two areas are fluid and overlapping.[4]

I have already mentioned some of the primary texts I used in the course to provide the theoretical grounding and framing for exploring space and spatiality from perspectives informed by cultural geography (Lefebvre 1991; Massey 2005; Soja 1996), cultural theory (de Certeau 1984; Foucault 1985, 1986), technical communication (Killingsworth 2005; Meloncon 2007; Meloncon 2013; Propen 2012; Swarts 2007), and writing and rhetoric (Mountford 2005; Reynolds 2004). Because I wanted students to make connections about the ways in which spatiality and visuality are imbricated in the construction of difference and inequality, I supplemented the texts with readings concerning socio-spatial exclusion that are often enforced through various kinds of technical discourse such as policy work, institutional critique, and cartographic practices (Barton and Barton 1993; Dolmage 2008; Ewalt 2011; Porter et al. 2000; Sibley 1995; Wallace 2009). To further engage the transformative potential of a critical spatial perspective, I also included readings informed by feminist theories about marginality and borderlands as a location, practice, and worldview that opens up possibilities for transgressing real-and-imagined boundaries (Anzaldúa 1987; Blunt and Rose 1994; hooks 2000). Finally, I included technical communication scholarship that were not explicitly concerned with space or spatiality in their original purpose and intent, though after engaging with the spatial theories and frameworks mentioned above, could be read and understood through a critical spatial lens (Jeyaraj 2004; Kitalong, Moody, Middlebrook, and Ancheta 2009; Koerber 2006; Popham 2005; Salvo 2004).[5]

A total of nine students enrolled and remained in the course throughout the duration of the semester: two first-year PhD students interested in the overlapping areas of rhetoric, composition, and technical communication; one second-year PhD student in rhetoric and composition; five master's students in the professional writing and rhetoric degree program; and one first-year graduate student at-large who was later formally admitted to the PhD program.

After consulting with my colleagues who were much more familiar with the student population and with the course content that had been taught in previous iterations of this particular topics course, I learned that while some of the students had taken varying amounts of coursework in technical communication, rhetorical theory, and/or visual and digital rhetorics, a course with a focus on space, place, and location had not been previously offered in the department. Thus, prior to the beginning of the semester, I assigned the students three articles to read in preparation for our first class meeting: Sullivan and Porter's original and retrospective pieces on the curricular geography of professional and technical communication (Porter and Sullivan 2007; Sullivan and Porter 1993) and Dobrin's book chapter on the occupation of composition within the university (Dobrin 2007). In so doing, I aimed to provide students with a shared (albeit, brief) introduction to disciplinary and curricular historical perspectives (Malone and Wright 2012) while also cognitively orienting them to the ways in which various issues—including curricular, disciplinary, and historical concerns—can be engaged through spatial approaches.

To further make the intersections among spatialities, visualities, technologies, subjectivities, and communication practices more apparent, on the first day of class I asked students to introduce themselves to one another through a mapping activity. Using Google Maps, I asked students to plot the spaces, places, and pathways they inhabit by prompting them to consider:

- personally significant physical and conceptual spaces and places
- spaces and places you inhabit regularly
- spaces and places you rarely frequent and/or avoid
- commonly traveled routes
- landmarks or monuments (literal and otherwise) that are important to you or that help you to "locate" yourself

Students had free-reign over the visual design of their maps. Some added multiple layers to the map and made use of features such as satellite or topographical views; some used different icons to represent different types of spaces and places; others plotted routes and provided directions while noting differences in distance and time. Once their maps were complete, students then drafted a set of instructions that served to guide our class members in interpreting their individual maps. After looking at each other's maps in small groups, we engaged in a whole-class discussion about the spaces and places they mapped, the reasons why they mapped them, and the differences in the ways in which class members interpreted their maps versus their original intents.

Beyond the initial purpose of introducing themselves to one another, this first-day mapping activity served other purposes throughout the duration of the semester. First, it established from the outset of the course that even seemingly straightforward, neutral, and objective technical documents, such as maps, are always partial, contingent, culturally situated, and open to multiple interpretations and understandings. Most important, it allowed the students to consider the connections among the spaces and places they inhabit—as well as to those that they don't— and prompted them to reflect on the ways in which their relationships to spaces and places intersect with their own embodied experiences and with social practices that are influenced by cultural and ideological beliefs, which are then mediated through an array of technologies and texts. Finally, their maps served as concrete references to spaces and places with which they had some working familiarity that were useful in helping them to grapple with the highly theoretical spatial concepts we engaged in our readings and discussions.

In addition to weekly responses posted to our course learning management system, students were asked to facilitate one class period in which the primary goals are to engage class members with a productive discussion of the assigned readings for the week and to provide visual and/or spatial examples that illustrate related issues or concepts raised by the readings. One week prior to their course facilitation, I asked students to email me a detailed agenda for the class period and invited them to meet with me outside of class to talk through their ideas. My rationale for including the course facilitation component into the course design were motivated just as much by spatial goals as they were by pedagogical goals. If I wanted students to take seriously the course content—that space is socially produced and is always in the process of production—the intellectual and material space of our classroom needed to reflect this in praxis. In sum, I wanted students to have the opportunity to co-construct our classroom space and to be able to engage their own interests and trajectories concerning the production of space.

During one such meeting, a master's student who chose to do her course facilitation on a set of readings concerning the spatialization of cities, states, and institutions used her map from the mapping activity as a starting point for her course facilitation agenda. Noting that numerous spaces and places on her initial map were predominantly plotted around Chicago's South Side where her family is from, the student astutely connected theoretical concepts about institutionalized power and its spatialized consequences to the material distribution of resources. Specifically, she used the week's readings to engage the class in a discussion about the rhetoricity of maps by providing the class with other maps and documents about the area and, during class discussion, raised concerns about the lack of access to basic resources such as grocery stores, adequate housing, and trauma centers in the most impoverished areas of Chicago's South Side, areas deeply impacted by racial tension, violence, and economic struggle. As Miriam Williams (2014) recently noted, our discipline needs to further "uncover those communicative practices used to negatively impact historically marginalized groups and identify new practices that can be used to encourage cultural competence within institutions and communities" (2). The initial mapping exercise thus allowed the student to consider a triadic spatial perspective by problematizing the *conceived* space of a specific locale (the spatial representation of Chicago through maps) in relation to the *perceived* experiences the maps and documents convey (that Chicago's South Side should be avoided because of its negative associations) and the *lived* experiences of those who inhabit it. (In her own words, the student asserted that the unequal spatialization of resources in the area "effectively quarantines the South Side and those who live there from the rest of the city"). I posit that integrating critical spatial perspectives in our classroom pedagogies, when coupled with discursive analyses such as the study of policies and legislation—both of which fall under the purview of technical communication—can make visible the lived material conditions of socio-spatial exclusion across intersecting axes of difference. In so doing, students may come to understand that while maps and the borders represented in them (both material and metaphorical) may be imposing, borders are not static or inevitable; rather, they can be resisted and potentially reshaped through social interaction.

In accordance with my investments in facilitating a classroom space open to multiple possibilities and trajectories, I designed the culminating course project so that it would allow for different kinds of inquiry. Above all, I wanted the final course project to be useful and usable for the students and allowed for an array of potential deliverables such as a thesis/dissertation chapter, a pedagogical project such as a course design and rationale, a researched report/seminar paper, an instantiation of a public project or other creative project. Regardless of the project deliverable, students were required to submit a project proposal along with a tentative bibliography, and give a formal oral presentation with accompanying presentation slide and/or handouts.

Building on our course readings, discussions, and course facilitations, the final course project allowed students to engage their own scholarly, professional, pedagogical, and personal interests concerning spatiality and visuality as they intersect with texts, technologies, and lived experiences. Given the course's dual foci, the students' final course projects engaged issues of spatiality and visuality to varying extents, though nearly all of the students used critical spatial perspectives informed by cultural geography as a primary framework. Included below is a summary of the course projects:

- Drawing from Reynolds's discussion of *flanerie* and Propen's study concerning embodiment and navigation technologies such as Global Positioning Systems (GPS), a part-time MA student conducted fieldwork to explore and theorize issues such as agency and access in relation to her own experiences navigating a space with which she was unfamiliar: Estero Island, a small island in the Gulf of Mexico which she visited during the semester. Her multimodal webtext intersperses video clips, maps, driving and walking routes, photographs, and interviews with island locals, in order to document her two-phase spatial practice in navigating the island (the first phase includes her experience of the space as a flaneur, in contrast to her reliance solely on GPS technology for directions and recommendations during the second phase).
- A PhD student with previous work experience as a land surveyor created maps of signage at a regional state park by using a Geographic Information System (GIS) in order to theorize the material, spatial, and embodied rhetorics of wayfinding documents as boundary genres that help to construct and support a specific narrative about the park as a place.
- As part of the portfolio requirement for the professional writing and rhetoric degree program, an MA student used critical spatial theories in addition to research about the social construction of criminality to inform a web-based white paper analyzing the social, cultural, and spatial implications of digital crime mapping technologies.
- Merging the spatial theories of de Certeau, Lefebvre, and Massey with Foucault's discussions of power, a PhD student interested in digital rhetorics employed a triadic spatial analysis focused on identity politics and user/developer agency in a few case studies concerning augmented reality devices.

- An MA student interested in public rhetorics analyzed maps and GPS images of the Middle East to illustrate the ways spatial images have been used in contemporary news media to justify a primarily Western, nationalist understanding of ongoing socio-spatial conflicts.
- Using case studies of online communities such as "mommy" blogs and forums, a part-time MA student interested in rhetorics of mothering and motherhood explored the ways in which communal spaces are socially produced in digital networked environments.
- An MA student interested in public relations and corporate communication conducted a visual analysis of the annual report narratives of a regionally based, global manufacturer of agricultural machinery and theorized the socio-spatial-cultural implications of its distribution.
- A PhD student with a background in environmental studies conducted a visual analysis of the US Department of Agriculture's (USDA) report on climate change in order to theorize the best practices for communicating climate change discourses to a specific geographic audience of rural Midwestern farmers.
- A first-year, part-time PhD student who worked as a full-time high school English teacher wanted to intervene in the traditional text-based literacies of her school's curriculum. Thus, she used the final course project to design a curriculum complete with lesson plans, readings lists, and classroom activities that focused on integrating spatial, visual, and multimodal literacies for the sophomore English curriculum at her school, in addition to writing a research-based proposal and rationale for her school's administrators about the need for and viability of her proposed curriculum.

As these final projects illustrate, integrating critical spatial perspectives into a graduate-level professional and technical communication course design have the potential to engage students in expanding the purview of the field by looking to the ways in which space, place, and location impact the users of texts, technologies, communities, organizations, and institutions. While the course design I describe in this chapter provides just one example of a localized approach, I maintain that more engagement with critical spatial perspectives informed by cultural geography can further enrich our research and teaching.

OPEN SPACE: A MULTIPLICITY OF TRAJECTORIES

Critical spatial perspectives require a shift in our thinking. Rather than thinking of spaces and places in terms of stasis with pre-determined boundaries, critical spatial perspectives ask us to move beyond the space-as-container model and to focus instead on the relations that produce and are produced by space. The intellectual space of the classroom, as I have suggested here, is just one place where we can begin to do this work. While the course design I described in this chapter may not be feasible across all institutions and programs, insights gleaned from cultural geography are applicable to a range of inquiries within the purview of our discipline. For example, a special topics course focused on social justice in technical communication would do well to include research from cultural geography that examines the material impacts of the spatial distribution of socially valued resources which have historically provided access and advantages to some, while barring and marginalizing others. Such a course might include the historical study of treaties, legislation, and policies in relation to practices of domination, subjugation, resource depletion, and human rights— all of which have been brought to bear on cultures, peoples, and bodies through space and spatialization. (E.g., in the next chapter of this collection, Godwin Agboka calls for our field's sustained engagement with human rights pedagogy, which can be further enriched by critical spatial perspectives.) Conversations about space, place, and location can similarly complement courses focused on intercultural technical communication, where document globalization or localization should take into account the spaces and places within which users will interact with them. More importantly, critical spatial perspectives in intercultural technical communication can allow students to acknowledge, negotiate, and reflect on the challenges of producing technical documents for those whose geographical and sociopolitical realities might be different from their own.

Critical spatial perspectives need not be confined to graduate study, however, and can also be integrated in undergraduate technical communication courses. Mapping activities such as the one I described could be integrated in lessons concerning instructions and wayfinding. Teaching students to conduct triadic spatial analyses that not only looks at how spaces are perceived, but also represented, and experienced could easily be integrated into course designs focused on service-learning, community partnerships, action-research, and relational work. In so doing, students may better understand both the visible and invisible spatial structures that can both shape and constrain rhetorical action.

Most important, critical spatial perspectives can further allow us to continue conversations about social justice issues in our classrooms and in our research because "thinking spatially about justice not only enriches our theoretical understanding, it can uncover significant new insights that extend our practical knowledge into more effective actions to achieve greater justice and democracy" (Soja 2010, 1). Likewise, when trained to consider critical spatial perspectives, our students—future technical communicators—can, in turn, develop a disposition for what Sackey calls cognitive justice (see chapter 6 in this collection), and may provide new insights through spatially informed praxis in organizations, corporations, and nonprofits in local, national, and transnational contexts. Moreover, critical spatial perspectives can be especially useful for technical communicators who may encounter the ways in which "many global businesses and industries seem oblivious to the negative impact of their activities of unenfranchised and disenfranchised populations" (Jones, Savage, and Yu 2014, 146), and may provide a starting point for questioning the potential complicity of technical discourse in injustice. Further alliances among technical communication, cultural geography, development studies, and human rights thus may have even greater potential to intervene in such situations.

As I have demonstrated in this chapter, technical communication and critical cultural geography engage similar issues that are intellectually and politically overlapping and can mutually inform and thereby enrich one another. The chapters in this edited collection represent a growing body of knowledge focused on theoretical frameworks and cultural theories that can enrich technical communication pedagogies and practices, offering us a generative way to reimagine our discipline in ways that promote diversity, equity, and justice. Teaching our students to be mindful and reflective of the spaces and places they inhabit—and the potentials for spaces and places to be redefined—is just one way that we can open space to a multiplicity of trajectories that can effect social change.

Notes

1. Although there is considerable overlap and slippage between the terms "space" and "place," most spatial theorists use the term "space" to refer to the abstract conception of a site and "place" to refer to a site's concrete location, hence de Certeau's assertion that "space is a practiced place" (de Certeau 1984, 117). Massey asserts, however, that "if space is a simultaneity of stories-so-far, then places are collections of those stories, articulations within the wider power-geometries of space" (Massey 2005, 130). The distinctions between space and place thus cannot be viewed in binary terms such as either/or; space and place are mutually constitutive. Additionally, see LeFebvre (1991) and Tuan (1979).
2. A concern for redressing social injustices and oppressions has informed much of the scholarship in cultural geography in recent years. For example, feminist cultural geographers have pointed out the spatialization of patriarchal power and its implications on cities and urban development (Grosz 1992), economic geography (Massey 1994), sexuality and queer studies (Colomina 1992; Valentine 2002), colonialism, postcolonialism, and third world feminisms (Blunt and Rose 1994), critical race studies (Price 2010), and the discipline of geography itself (Rose 1993). In

addition to perspectives informed by feminisms, other critical interventions from cultural geographers have focused on uneven development in specific urban locales (Brenner, Marcuse, and Mayer 2011; Scott and Soja 1998) as well as the effects of globalization in national and transnational contexts (DeBres 2005; Smith 2008).

3. I am deeply indebted to Amy C. Kimme Hea whose course on spatial and visual rhetoric at the University of Arizona has been influential in my research and teaching in countless ways. I can only hope that the course I describe in this chapter will be as intellectually productive for the students enrolled as Amy's class was for me.

4. I found that beginning the semester with a focus on space and spatiality helped the students to better consider the spatial factors and implications involved in visuality.

5. Among the texts I used to inform and support the focus on visual theories and rhetorics during the second half of the semester include readings on visual culture theories: Margarita Dikovitskaya, *Visual Culture: The Study of the Visual after the Cultural Turn* (Cambridge, MA: Massachusetts Institute of Technology Press, 2005); Nicholas Mirzoeff, "The Right to Look," *Critical Inquiry* 37, no. 3 (2011): 473–96; W. J. T. Mitchell, "There are No Visual Media," *Journal of Visual Culture* 4, no. 2 (2005): 257–66; readings on visual rhetoric theories: Sonja Foss, "Theory of Visual Rhetoric," in *Handbook of Visual Communication: Theory, Methods, and Media*, eds. Ken Smith, Sandra Moriarty, Gretchen Barbatsis, and Keith Kenney (Mahwah: Lawrence Erlbaum, 2005), 141–52; Carolyn Handa, ed., *Visual Rhetoric in a Digital World: A Critical Sourcebook* (New York: Bedford/St.Martin, 2004); Charles Hill and Marguerite Helmers, eds., *Defining Visual Rhetorics* (Mahwah: Lawrence Erlbaum, 2004); Carlos Salinas, "Technical Rhetoricians and the Art of Configuring Images," *Technical Communication Quarterly* 11, no. 2 (2002): 165–83; readings on theories of design, design praxes, and multi-literacies: Victor Margolin, ed., *Design Discourse: History, Theory, Criticism* (Chicago: University of Chicago Press, 1986); Gunther Kress and Theo Van Leeuwen, *Reading Images: The Grammar of Visual Design* (New York: Routledge, 2006); readings on communicating visual information: Nancy Allen, "Ethics and Visual Rhetorics: Seeing's Not Believing Anymore," *Technical Communication Quarterly* 5, no. 1 (1996): 87–105; Ben Barton and Marthalee Barton, "Modes of Power in Technical and Professional Visuals," *Journal of Business and Technical Communication* 7, no. 1 (1983): 138–62; Lee Brasseur, *Visualizing Technical Information* (New York: Baywood, 2003); Edward Tufte, *Beautiful Evidence* (Cheshire: Graphics Press, 2006); and readings on visual literacies and pedagogies: Brumberger and Northcut, *Designing Texts.*

References

Ackerman, John. 2003. "The Space for Rhetoric in Everyday Life." In *Towards a Rhetoric of Everyday Life: New Directions in Research on Writing, Text, and Discourse*, ed. Martin Nystrand and John Duffy, 84–117. Madison: University of Wisconsin Press.

Anzaldúa, Gloria. 1987. *Borderlands/La Frontera: The New Mestiza.* San Francisco: Anne Lute.

Barton, Ben, and Marthalee Barton. 1993. "Ideology and the Map: Toward a Postmodern Visual Design Practice." In *Professional Communication: The Social Perspective*, ed. Nancy Roundy Blyler and Charlotte Thralls, 49–79. Newbury Park: Sage Publications.

Binkley, Roberta, and Marissa Smith. 2006. "Re-Composing Space: Composition's Rhetorical Geography." *Composition Forum* 15. http://compositionforum.com/issue/15/.

Blunt, Alison, and Gillian Rose. 1994. *Writing Women and Space: Colonial and Postcolonial Geographies.* New York: Guilford Press.

Brasseur, Lee. 1993. "Contesting the Objectivist Paradigm: Gender Issues in the Technical and Professional Communication Curriculum." *IEEE Transactions on Professional Communication* 36 (3): 114–123. https://doi.org/10.1109/47.238051.

Brenner, Neil, Peter Marcuse, and Margit Mayer. 2011. *Cities for People, Not for Profit: Critical Urban Theory and the Right to the City.* New York: Routledge.
Colomina, Beatriz. 1992. *Sexuality and Space.* New York: Prince Architectural Press.
de Certeau, Michel. 1984. *The Practice of Everyday Life.* Los Angeles: University of California Press.
Deans, Thomas. 2000. *Writing Partnerships: Service Learning in Composition.* Urbana: National Council of Teachers of English.
DeBres, Karen. 2005. "Burgers for Britain: A Cultural Geography of McDonald's UK." *Journal of Cultural Geography* 22 (2): 115–139. https://doi.org/10.1080/088736 30509478241.
Dobrin, Sidney I. 2007. "The Occupation of Composition." In *The Locations of Composition*, ed. Christopher J. Keller and Christian R. Weisser, 15–35. Albany: SUNY Press.
Dobrin, Sidney, and Christian Weisser. 2002. *Natural Discourse: Toward Ecocomposition.* Albany: SUNY Press.
Dolmage, Jay. 2008. "Mapping Composition: Inviting Disability in the Front Door." In *Disability and the Teaching of Writing*, ed. Brenda Jo Bruggemann and Jay Dolmage, 14–27. Boston: Bedford/St.Martin.
Ewalt, Joshua. 2011. "Mapping Injustice: The World is Witness, Place Framing, and the Politics of Viewing on Google Earth." *Communication, Culture & Critique* 4 (4): 333–354. https://doi.org/10.1111/j.1753-9137.2011.01109.x.
Fleming, David. 2009. *City of Rhetoric: Revitalizing the Public Sphere in Metropolitan America.* Albany: SUNY Press.
Foucault, Michel. 1985. "Panopticism." In *Discipline and Punish: The Birth of the Prison*, trans. Alan Sheridan, 195–228. New York: Vintage.
Foucault, Michel. 1986. "Of Other Spaces: Utopias and Heterotopias." *Diacritics* 16 (1): 22–27.
Grego, Rhonda, and Nancy Thompson. 2007. *Teaching/Writing in Thirdspaces: The Studio Approach.* Carbondale: Southern Illinois University Press.
Grosz, Elizabeth. 1992. "Bodies-Cities." In *Sexuality and Space*, ed. Beatriz Colomina, 241–53. New York: Princeton Architectural Press.
Haas, Angela. 2012. "Race, Rhetoric and Technology, A Case Study of Decolonial Technical Communication Theory, Methodology, and Pedagogy." *Journal of Business and Technical Communication* 26 (3): 277–310. https://doi.org/10.1177/1050651 912439539.
Haley-Brown, Jennifer, Ashley Holmes, and Amy Kimme Hea. 2012. "Guest Editor's Introduction: Spatial Praxes: Theories of Space, Place, and Pedagogy." *Kairos: A Journal of Rhetoric, Technology, and Pedagogy* 16 (3). http://kairos.technorhetoric. net/16.3/loggingon/index.html.
Henry, Jim. 2001. "Writing Workplace Cultures." *College Composition and Communication* 53 (1): 1–12.
Herndl, Carl. 1993. "Teaching Discourse and Reproducing Culture: A Critique of Research and Pedagogy in Professional and Non-Academic Writing." *College Composition and Communication* 44 (3): 349–363. https://doi.org/10.2307/358988.
hooks, bell. 2000. "Choosing the Margin as a Space of Radical Openness." In *Gender, Space, Architecture*, ed. J. Rendell, 203–9. London: Routledge.
Jeyaraj, Joseph. 2004. "Liminality and Othering: The Issue of Rhetorical Authority in Technical Discourse." *Journal of Business and Technical Communication* 18 (1): 9–38. https://doi.org/10.1177/1050651903257958.
Johnson-Eilola, Johndan. 1996. "Stories and Maps: Postmodernism and Professional Communication." *Kairos: A Journal of Rhetoric, Technology, and Pedagogy* 1 (1). http://kairos.technorhetoric.net/1.1/binder.html?features/johndan/stories_and _maps_029.html.

Johnson-Eilola, Johndan, and Stuart Selber. 2007. "The Locations of Usability." In *The Locations of Composition*, ed. Christopher Keller and Christian Weisser, 171–94. Albany: SUNY Press.

Jones, Natasha, Gerald Savage, and Han Yu. 2014. "Tracking Our Progress: Diversity in Technical and Professional Communication Programs." *Programmatic Perspectives* 6 (1): 132–152.

Katz, Steven. 1992. "The Ethic of Expediency: Classical Rhetoric, Technology, and the Holocaust." *College English* 54 (3): 255–275. https://doi.org/10.2307/378062.

Keller, Christopher J., and Christian R. Weisser. 2007. *The Locations of Composition*. Albany: SUNY Press.

Killingsworth, M. Jimme. 2005. "From Environmental Rhetoric to Ecocomposition and Ecopoetics: Finding a Place for Professional Communication." *Technical Communication Quarterly* 14 (4): 359–373. https://doi.org/10.1207/s15427625tcq1404_1.

Kimme Hea, Amy C. 2009. "Perpetual Contact: Re-articulating the Anywhere, Anytime Pedagogical Model of Mobile and Wireless Composing." In *Going Wireless: A Critical Exploration of Wireless and Mobile Technologies for Composition Teachers and Researchers*, ed. Amy C. Kimme Hea, 199–221. Cresskill: Hampton Press.

Kitalong, Karla, Jane Moody, Rebecca Helminen Middlebrook, and Gary Saldana Ancheta. 2009. "Beyond the Screen: Narrative Mapping as a Tool for Evaluating a Mixed-Reality Science Museum Exhibit." *Technical Communication Quarterly* 18 (2): 142–165. https://doi.org/10.1080/10572250802706349.

Koerber, Amy. 2006. "Rhetorical Agency, Resistance, and the Disciplinary Rhetorics of Breastfeeding." *Technical Communication Quarterly* 15 (1): 87–101. https://doi.org/10.1207/s15427625tcq1501_7.

Lefebvre, Henri. 1991. *Production of Space*. Cambridge: Blackwell Press.

Longo, Bernadette. 1998. "An Approach for Applying Cultural Study Theory to Technical Writing Research." *Technical Communication Quarterly* 7 (1): 53–73. https://doi.org/10.1080/10572259809364617.

Malone, Edward, and David Wright. 2012. "The Role of Historical Study in Technical Communication Curricula." *Programmatic Perspectives* 4 (1): 42–82.

Massey, Doreen. 1994. *Space, Place, and Gender*. Minneapolis: University of Minnesota Press.

Massey, Doreen. 2005. *For Space*. London: Sage Publications.

Meloncon, Lisa. 2007. "Exploring Electronic Landscapes: Technical Communication, Online Learning, and Instructor Preparedness." *Technical Communication Quarterly* 16 (1): 31–53. https://doi.org/10.1080/10572250709336576.

Meloncon, Lisa. 2013. "How to Read Landscapes: A Method for Integrating Visual Communication in the Technical Communication Classroom." In *Designing Texts: Teaching Visual Communication*, ed. Eva Brumberger and Kathryn Northcut, 13–32. Amityville: Baywood. https://doi.org/10.2190/DTTC1.

Mountford, Roxanne. 2005. "On Gender and Rhetorical Space." *Rhetoric Society Quarterly* 31 (1): 42.

Pigg, Stacey. 2014. "Coordinating Constant Invention: Social Media's Role in Distributed Work." *Technical Communication Quarterly* 23 (2): 69–87. https://doi.org/10.1080/10572252.2013.796545.

Popham, Susan. 2005. "Forms as Boundary Genres in Medicine, Science, and Business." *Journal of Business and Technical Communication* 19 (3): 279–303. https://doi.org/10.1177/1050651905275624.

Porter, James, and Patricia Sullivan. 2007. "Remapping Curricular Geography: A Retrospection." *Journal of Business and Technical Communication* 21 (1): 15–20. https://doi.org/10.1177/1050651906293507.

Porter, James, Patricia Sullivan, Stuart Blythe, Jeffrey Grabill, and Libby Miles. 2000. "Institutional Critique: A Rhetorical Methodology for Change." *College Composition and Communication* 51 (4): 610–641. https://doi.org/10.2307/358914.

Price, Patricia. 2010. "At the Crossroads: Critical Race Theory and Critical Geographies of Race." *Progress in Human Geography* 34 (2): 147–174. https://doi.org/10.11 77/0309132509339005.

Propen, Amy. 2012. *Locating Visual-Material Rhetorics: The Map, the Mill, and the GPS.* Anderson: Parlor Press.

Reynolds, Nedra. 1998. "Composition's Imagined Geographies: The Politics of Space in the Frontier, City, and Cyber Space." *College Composition and Communication* 50 (1): 12–35. https://doi.org/10.2307/358350.

Reynolds, Nedra. 2004. *Geographies of Writing: Inhabiting Places and Encountering Difference.* Carbondale: Southern Illinois University Press.

Rice, Jeff. 2012. *Digital Detroit: Rhetoric and Space in the Age of the Network.* Carbondale: Southern Illinois University Press.

Rose, Gillian. 1993. *Feminism and Geography: The Limits of Geographical Knowledge.* Minneapolis: University of Minnesota Press.

Rude, Carolyn. 2009. "Mapping the Research Questions in Technical Communication." *Journal of Business and Technical Communication* 23 (2): 174–215. https://doi.org/10 .1177/1050651908329562.

Salvo, Michael. 2004. "Rhetorical Action in Professional Space: Information Architecture as Critical Practice." *Journal of Business and Technical Communication* 18 (1): 39–66. https://doi.org/10.1177/1050651903258129.

Scott, Allen, and Edward Soja. 1998. *The City: Los Angeles and Urban Theory at the End of the Twentieth Century.* Oakland: University of California Press.

Scott, J. Blake. 2004. "Tracking Rapid HIV Testing through the Cultural Circuit: Implications for Technical Communication." *Journal of Business and Technical Communication* 18 (2): 198–219. https://doi.org/10.1177/1050651903260836.

Scott, J. Blake, Bernadette Longo, and Katherine V. Wills, eds. 2006. *Critical Power Tools: Technical Communication and Cultural Studies.* New York: SUNY Press.

Sibley, David. 1995. *Geographies of Exclusion: Society and Difference in the West.* London: Routledge. https://doi.org/10.4324/9780203430545.

Smith, Neil. 2008. *Uneven Development: Nature, Capital and the Production of Space.* Athens: University of Georgia Press.

Soja, Edward. 1989. *Postmodern Geographies: The Reassertion of Space in Critical Social Theory.* London: Verso.

Soja, Edward. 1996. *Thirdspace: Journeys to Los Angeles and Other Real-and-Imagined Places.* Malden, MA: Blackwell Press.

Soja, Edward. 2010. *Seeking Spatial Justice.* Minneapolis: University of Minnesota Press. https://doi.org/10.5749/minnesota/9780816666676.001.0001.

Spinuzzi, Clay. 2007. "Guest Editor's Introduction: Technical Communication in the Age of Distributed Work." *Technical Communication Quarterly* 16 (3): 265–277. https://doi .org/10.1080/10572250701290998.

Sullivan, Patricia, and James Porter. 1993. "Remapping Curricular Geography: Professional Writing in/and English." *Journal of Business and Technical Communication* 7 (4): 389–422. https://doi.org/10.1177/1050651993007004001.

Sun, Huatong. 2009. "Toward a Rhetoric of Locale: Localizing Mobile Messaging Technology into Everyday Life." *Journal of Technical Writing and Communication* 39 (3): 245–261. https://doi.org/10.2190/TW.39.3.c.

Swarts, Jason. 2007. "Mobility and Composition: The Architecture of Coherence in Non-Places." *Technical Communication Quarterly* 16 (3): 279–309. https://doi.org/10.1080 /10572250701291020.

Tuan, Y. F. 1979. "Space and Place: Humanistic Perspective." *Philosophy in Geography: Theory and Decision Library*, ed. S. Gale and G. Olsson. An International Series in the Philosophy and Methodology of the Social and Behavioral Sciences, vol. 20. Dordrecht: Springer.

Valentine, Gil. 2002. "Queer Bodies and the Production of Space." In *Handbook of Lesbian and Gay Studies*, ed. D. Richardson and S. Seidman, 145–60. London: Sage Publications. https://doi.org/10.4135/9781848608269.n10.

Verzosa, Elise, and Adrienne Crump. 2012. "Visualizing Writing Space: A Reflection." *Kairos: A Journal of Rhetoric, Technology, and Pedagogy* 16 (3). http://kairos.technorhetoric.net/16.3/praxis/hea-et-al/crump-verzosa/index.html.

Wallace, Aurora. 2009. "Mapping City Crime and the New Aesthetic of Danger." *Journal of Visual Culture* 8 (1): 5–24. https://doi.org/10.1177/1470412908100900.

Welch, Kathleen. 2005. "Technical Communication and Physical Location: Topoi and Architecture in Computer Classroom." *Technical Communication Quarterly* 14 (3): 335–344. https://doi.org/10.1207/s15427625tcq1403_12.

Williams, Miriam. 2014. "Introduction." In *Communicating Race, Ethnicity, and Identity in Technical Communication*, ed. Miriam Williams and Octavio Pimentel, 1–4. Amityville: Baywood.

Williams, Miriam, and Octavio Pimentel, eds. 2014. *Communicating Race, Ethnicity, and Identity in Technical Communication*. Amityville: Baywood Press.

5
INDIGENOUS CONTEXTS, NEW QUESTIONS
Integrating Human Rights Perspectives in Technical Communication

Godwin Y. Agboka

> *Business enterprises should respect human rights. This means that they should avoid infringing on the human rights of others and should address adverse human rights impacts with which they are involved. (Office of the High Commissioner for Human Rights 2011, 9)*

Our technical communication program plans to start a certificate program in *energy industry writing*. The decision was motivated by two overarching considerations: ethics and industry needs. Following developments after the Enron scandal (2001) and the Texas City plant explosion (2005), there were growing concerns, mainly in business schools, for a paradigmatic shift in pedagogy to capture the centrality of ethics. Although not necessarily a direct response to these concerns, our program developed a set of ethics modules, with support from an NEH grant to inform pedagogical practices, particularly at the graduate level. The grant allowed us to organize a series of workshops—what we called the "NEH Ethics Workshops"—the consequence of which led to our developing case studies that highlighted the ethical ramifications of corporate decision-making and its impact on quality of life. Also considered, Houston, the site of my university, is "headquarters for the American oil industry" (Gilmer 2014), providing a haven for about 40 of America's 145 publicly traded *oil and gas* exploration and production firms (e.g., ExxonMobil, Shell, Valero Energy Corp, etc.). Basically, energy has been the primary factor in the Houston economy over the years, employing over 100,000 workers in oil and production services and paying over $140,000 in wages, salaries, and employer benefits in 2011 (Gilmer 2014).

DOI: 10.7330/9781607327585.c005

The certificate program, when it begins, will attempt to educate students to be effective, functional writers within the energy sector. To be sure, technical communication features prominently in the energy industry, as those employed perform a range of functions that support the industry's operations. Depending on the contexts and vision of the organizations in which they work, technical communicators serve as support professionals or are part of the management of those organizations.[1] Essentially, the energy industry (like others such as medical, engineering, etc.) can provide potential opportunities for technical communicators.

However, as a social justice scholar, I am concerned about the potential problems that the industry also poses for local communities, especially those who[m] the activities of the industry might affect in destructive ways. I am also concerned about the complicity of technical communicators, whether directly or indirectly, given that technical communication artifacts are central in the operations of the industry. Globally, oil discovery, production, and extraction have been associated with mixed fortunes. For example, of the major oil producing countries, the oil sector (i.e., oil and gas production, processing and refining) alone contributes some 30 percent to 60 percent of the gross domestic products (GDP) of those economies (Al-Moneef 2006), leading, in many cases, to economic growth and independence. Elsewhere, aside the reported social responsibility of oil and gas companies in many local communities (Aghalino 2004), the industry is also implicated in some of the worst excesses of human rights abuses globally—sometimes with the support of governments (Ebeku 2003; "Briefing Note" 2013). As Collinson and MacLeod argue, "the poor of oil-rich countries suffer from displacement, pollution, corruption, poor public services, and a lack of livelihood options" (Collinson and McLeod 2010, 2). Unfortunately, most of these reported human rights abuses have been documented in developing, unenfranchised sites—although there are some documented cases of abuse in the USA and Canada as well.

The few scientific studies available suggest that these appear to be the result of the careless attitude of oil companies to environmental issues (Ebeku 2003; United Nations Environmental Programme 2011). From a human rights perspective, such infractions infringe on people's right to the environment, safety, mental and physical health, land tenure, and overall quality of life. The International Bill of Human Rights (IBHR) [made up of the Universal Declaration of Human Rights (UDHR) (United Nations 1948), the International Covenant on Civil and Political Rights (ICCPR) (United Nations 1966a), and the International Covenant on Economic, Social and Cultural Rights

(ICESCR) (United Nations 1966b)], along with several protocols and conventions, prescribes a set of precepts and guidelines that link human well-being, quality of life, and dignity to a clean, safe environment. For example, recalling Principle 1 of the Stockholm Declaration on the Human Environment, the 1998 Aarhus Convention on access to justice in environmental matters recognized "that adequate protection of the environment is essential to human well-being and the enjoyment of basic human rights, including the right to life itself" and "that every person has the right to live in an environment adequate to his or her health and well-being" (United Nations Economic Commission for Europe 1998, 2).[2] In essence, we have an obligation as scholars, teachers, and practitioners to address all the human rights issues that intersect with our work—particularly in areas where populations might lose more than they might gain.

Let me state that the abuses described here are not particularly a peculiar problem in technical communication, because in many organizations only few technical communicators are part of the decision-making bodies. More so, technical communication, I believe, is a virtuous field, because we have been involved with issues of ethics for many years (Allen and Voss 1997; Appelbaum and Lawton 1990; Bowdon 2002; Dombrowski 2000; Dragga 1996; Faber 2001; Sullivan and Martin 2001). However, because our discursive practices might be implicated in these abuses, we must aggressively begin to interrogate ways in which we *may* be complicit in human rights abuses if we are employed in, work for, and shape practices in specific organizations and cultural sites. As our students become professionals and transition into the workplace, they—directly or indirectly—share in the successes of the organizations as they do the negative effects of their practices on target audiences. Writing about the effects of globalization, for instance, Savage and Mattson (2011) argued "insofar as technical communication as a practice and as a discipline participates in and seeks to benefit from globalization, it also shares responsibility for globalization's effects, whether good or ill" (5).

As a sequel to chapter 4, in which Hurley discusses the merits of "spatial orientations," this chapter takes up the notion of "cultural geography" (136), by exploring the intersections of technical communication and human rights, particularly how theories of human rights can and/or should inform technical communication research and pedagogy. It crystallizes the connections between cultural geography and identity. Thus I use, as a case study, the Ogoni oil crisis in the Niger Delta of Nigeria in which many lives were lost and displaced as a result of political unrest, gas flares, oil spills, and pollution. In discussing this case, I connect with and draw from Stephen Katz's (1992) earlier work on the ethic expediency by conducting a brief rhetorical analysis of a memo, mainly couched in militaristic language, that was used by the Nigerian government and Shell Petroleum Development Company (SPDC) to induce the Army and to plan attacks on residents of Ogoniland. I begin the chapter by examining human rights research in the field of technical communication and then interrogate why specific research on human rights is largely missing in our field. Next, I trace the trajectory of human rights discourse, examine the foundational concepts of human rights theories, and then examine notions of right to environment in the context of the Ogoni crisis. Then, I discuss how a memo was used as an instrument to perpetrate human rights abuses on residents of Ogoniland in the Niger Delta area of Nigeria. Finally, I discuss the implications of this case on technical communication research and pedagogy.

HUMAN RIGHTS RESEARCH IN TECHNICAL COMMUNICATION

Although scholarship that specifically addresses how technical communication can promote or inhibit human rights is only beginning, the thematic strands of human rights discourse have been consistent and progressive scholarly interests. In our field, many aspects of human rights and social justice, including issues of gender, race, age, disability, nationality, sexuality, and class, etc. are increasingly being explored in the technical communication literature and in many facets of curriculum design (Agboka 2013; Haas 2012; Huckin 2002; McKee and Porter 2010; Palmeri 2006; Sapp 2004; Savage 2001; Savage and Mattson 2011; St.Amant and Rife 2010; St. Germaine-McDaniel 2010; Walton and DeRenzi 2009 Walton, Price, and Zraly 2013). For example, in response to the upsurge in scholarship over the limitations of the "modernistic assumptions that permeate[d] technical communication discourse" (Wilson 2001, 73), scholarship has shifted to a more democratic and political posture that has sought to acknowledge the political and ideological function of discourse (Herndl 1993, 350). Champions of these progressive, sometimes radical, approaches (Agboka 2013; Blyler 1995; Dragga and Voss 2001; Faigley 1986; Herndl 1996; Huckin 2002; Miller 1979; Ornatowski and Bekins 2004; Savage 2001; Sullivan 1990; Thralls and Blyler 1993) have often tended to not only make a case for the field to disengage from its beginnings, but "open up the field to civic advocacy and action and make it responsive to progressive political agendas focused on individual and collective emancipation and empowerment" (Ornatowski and Bekins 2004, 252). A common feature of all this scholarship has been the centrality of humanity. Thus, working through the lens of visual rhetoric, Dragga and Voss (2001) were worried that "[a]gain and again in . . . visual images people are deprived of their humanity and objectified for purposes of statistical manipulation" (262). Savage (2001) also called on technical communicators to recognize the relevance of their fields in global practices in terms of their social, economic, and environmental consequences (94). Similarly, Huckin (2002, 1) wondered why technical communicators were not interrogating the effects of corporate practices on the environment, while Sapp (2004) was worried that our failure to engage these issues might lead to further colonizing practices in specific sites.

Furthermore, in their introduction to the most recent collection on human rights and professional writing (the only notable collection that investigates the impacts of technical communication on human rights), Sapp, Savage, and Mattson (2013) note that our field cannot pretend that "human rights has no significance for scholarship, teaching, and practice of professional communication, especially where it concerns developing nations and marginalized populations" (1). They encourage us, then, to acknowledge responsibility for our work in all contexts and also "develop curricula and courses that offer much more to students than jobs in corporations in which their career paths may depend heavily on subservience to that singular culture that places competition and profit ahead of more basic core responsibilities . . ." (6). Besides carefully laying out how corporate structures may often be used in ways that abuse human rights, they also call on us to embrace practices that "respect human rights, socio-cultural conventions, and the sovereignty of people and governments where they do business" (3). Fortunately, though, other scholars do show that, indeed, technical communication can be used in ways to liberate populations from abuses (Ding and Pitts 2013; Dura, Singhal, and Elias 2013).

Evidently, a common strand in the literature reviewed shows that technical communication may be used for good or bad purposes. But, another worrying thread is the paucity of research in this area—particularly in areas that are disadvantaged. But, it's fair to say that the limited research in this area of inquiry has legitimate sources. First, the scholarship in the field, so far, shows that "many teachers have a commitment to human rights principles such as justice and equity, and are skilled in the art of socializing students," (Osler and Starkey 1994, 350) but they do not have the expertise to apply specific concepts of human rights to their pedagogy and research because of the specialized nature of human rights. Also, from the field's scholarly trajectory, it appears that the impact of technical communication on users in many "low" and "medium" development-index sites has been largely unexplored in the field of technical communication—of course, that is not to say that human rights is only a concern in these sites. Still, the controversial nature of human rights and the difficult dynamics of addressing those topics remain, for a lot of scholars, a major obstacle to incorporating and implementing human rights objectives. For example, the problematic dynamics of the tensions involved in Universalist and Relativist notions and applications of human rights tend to complicate discourses on human rights. This leads Lyon and Olson (2011) to ask: "Are certain human rights so fundamental as to be 'universal,' as the United Nations proclaimed in 1948? Does this ostensibly 'universal' character of human rights mean that certain human rights principles can never be compromised and, if so, how have communities dealt constructively with collisions among conflicting human rights?" (203).

However, for a field that interfaces with diverse populations, by the unique virtue of our communication practices, these challenges of applying these human rights objectives present a unique opportunity for us to position ourselves as ambassadors and activists for the audiences we engage. Ultimately, the significance of these rights to contemporary social, cultural, and political struggles of many sites where technical communication intersects means that technical communicators must develop strategies to address such issues. In the next section I discuss the discourses of human rights and also outline the key principles that motivate these discourses.

HUMAN RIGHTS: DISCOURSES AND THEORETICAL FRAMEWORK

Although there is no unified definition for these rights, as the scope varies from one context to another, human rights scholars generally agree that they refer to:

> Those rights, which are inherent in our nature and without which we cannot live as human beings. Human rights and fundamental freedoms allow us to fully develop and use our human qualities, our intelligence, our talents and our conscience and to satisfy our spiritual and other needs. Human rights are based on mankind's increasing demands for a life in which the inherent dignity and worth of each human being will receive respect and protection. (United Nations 1987, 4)

The human rights discourse, over the years, has become an authoritative platform for articulating improved human worlds and incorporating new visions of society (Coombe 2010, 235). For example, over the years the United Nations (UN), mainly through its development-centered frameworks (e.g., the Human Development Index (HDI), the Millennium Development Goals (MDGs), the Global Compact, etc.), and many donor agencies (e.g., DANIDA, etc.) have made consistent efforts to promote rights-based approaches to development that are centered on recognizing the centrality of human dignity and human well-being in development objectives. Thus, these human rights have become *the* de facto moral and legal framework for protecting individuals and groups around the world from various cultural, political, economic, and environmental rights violations. The human rights framework enjoins states and other social actors (e.g., corporations, non-governmental organizations, the media, civil society, rights-groups, etc.) to make the protection of human rights a central feature of their agenda—although the framework doesn't necessarily have the force of law.

It is inspired by the International Bill of Human Rights (IBHR), which is made up of three legal frameworks [i.e., Universal Declaration of Human Rights (UDHR) (United Nations 1948), the International Covenant on Economic, Social and Cultural Rights (ICESCR) (United Nations 1966b), and the International Covenant on Civil and Political Rights (ICCPR) (United Nations 1966a)]. The International Covenant on Civil and Political Rights (United Nations 1966a) contains a list of ambitious guidelines that capture civil and political rights, also known as first generation rights (articles 1–21); economic and socio-cultural rights, also known as second generation rights (articles 22–27); and solidarity rights and obligations, also known as third generation rights (articles 28–30). The first generation of rights addresses general issues of freedom (i.e., protection from state power) relating to discrimination, freedom of thought and conscience, and freedom of speech, association, and religion. Over the years the second generation of rights has sought to address issues of linguistic and cultural participation, social security, education, and employment. Then, solidarity rights and obligations, known as the third generation of rights, capture environmental rights and rights to natural resources, among others.

The UDHR (United Nations 1948) came into effect after mounting "international pressure to develop meaningful guarantees of human rights which intensified following World War II, particularly in light of the massive human rights violations committed by the Axis Power, as well as Nazi Germany's policies of genocide in Europe (the Holocaust)" (Gaudelli and Fernekes 2004, 17). Many years after its adoption, the UDHR was a non-legally binding document, but in 1966 the United

Nations transformed the provisions of the UDHR into legally binding obligations by adopting two separate, but interdependent, treaties: the International Covenant on Civil and Political Rights (ICCPR) (United Nations 1966a) and the International Covenant on Economic, Social, and Cultural Rights (ICESCR) (United Nations 1966b), which have been ratified by about 155 parties. These two treaties, together with the UDHR and the 1946 UN Charter, form the International Bill of Rights (Chamberlain 2001, 211).

Essentially, from a theoretical standpoint, "A human rights framework facilitates a practice . . . that enables transgressions in an institutionalized environment fraught with . . . challenges" (Falcon and Jacob 2011, 30). These challenges manifest themselves in discriminatory policies and practices; gender relations; class dispositions; political contexts (i.e., government-citizen relations); issues of sexuality; race; disability; health; communication policies; among all other practices that seek to impede the realization of quality of life. Such challenges can be a concern within institutional contexts such as academic institutions (i.e., within or outside the classroom); health contexts (e.g., abuse of patients, etc.); states (e.g., physical, emotional, sexual, and mental abuse through the use of state resources, etc.), all of which can impede the development of human qualities and talent. Thus, human rights discourses seek to understand—and in a lot of ways, rupture—the historical, structural, cultural, and socio-economic " . . . forces underpinning the obstacles [that impede human rights] as well as put in place strategies for the removal of such obstacles in the processes of social change and transformation" (International Consultation on the Pedagogical Foundations of Human Rights Education 1996). In many ways, these challenges are institutional and systemic, meant to maintain and sustain a socio-political order of injustice; more so, they see cultural territories as sites of struggle and contestation and therefore enjoin social actors to work to liberate victims from injustices through equitable distribution of resources in these sites. They seek to not only create opportunities for cultural critique, but also provide academic opportunities for action both in and outside academic institutions.

Overarchingly, the process of attaining human rights is a political endeavor—that simply suggests that we are all affected by and responsible for injustices in our society, so we should work to address all such issues. Thus, much like the decolonial project in many postcolonial contexts, human rights discourses highlight attempts at and the need for self-determination and liberation as well as the struggles against injustices and colonialism—in whatever forms they take.

TECHNICAL COMMUNICATION, CONTEXTS, AND THE ENVIRONMENT

Technical communicators produce works and artifacts for action (i.e., Environmental Impact Statements, etc.), and such works may influence these rights discussed above. For example, "technical communicators play important roles in developing work that contributes to democratic interaction locally, and may produce work that influences human rights globally" (Herrington 2013, 17). They also work in governmental, intergovernmental, nongovernmental, and civil society organizations," (Sapp, Savage, and Mattson 2013, 5) which may specifically impact rights of populations. Because the case study I describe in this chapter relates to how technical communication may impact the environmental and overall quality of life, I describe, briefly, the right to environment under human rights law.

Under human rights law, the issue of whether there is a specific "right to environment" is contested (Ebeku 2003), but under the second and third generations of human rights, there are various protocols, charters, and treaties that describe a range of environmental rights to be enjoyed by people. For example, in its preamble, the UDHR acknowledges that the "inherent dignity and of the equal and inalienable rights of all members of the human family" should be "the foundation of freedom, justice and peace in the world" (United Nations 1948). It emphasizes in its Article 25 the "right to a standard of living adequate for the health and well-being . . . " at the center of which is the environment (United Nations 1948). The environment is considered to be the most relevant to the realization of all types of rights. Consequently, during the 1998 Aarhus Convention in Denmark, the members present recognized the need to "protect, preserve and improve the state of the environment . . . " which is "essential to human well-being and the enjoyment of basic human rights, including the right to life itself" (United Nations Economic Commission for Europe 1998, 2). Although there were several before it, the Aarhus Convention became the first authoritative document that imposed on its parties a duty toward its citizens. Similarly, in the Declaration of the United Nations Conference on the Human Environment (United Nations 1972) held in Stockholm, the international community declared in Principle 1 of the document that "Man has the fundamental right to freedom, equality and adequate conditions of life, in an environment of a quality that permits a life of dignity and well-being, and [he] bears a solemn responsibility to protect and improve the environment for present and future generations." Ultimately, all these documents place at the center of life, human dignity. In an earlier document that was adopted by an international group of experts on human rights and environmental protection, which was convened at the United Nations in Geneva, a direct link is made between quality of life and "an ecologically sound environment." The document underscores the right "to protection and preservation of the air, soil, water, sea-ice, flora and fauna . . . "; "to the highest attainable standard of health free from environmental and industrial hygiene"; to safe and healthy food and water adequate to their well-being"; "to a safe and healthy working environment"; and "to adequate housing, land tenure and living conditions in a secure, healthy and ecologically sound environment" (Draft Principles 1994). However, the most important statement yet on the environment is put forward by the African Commission in its judgment on a case brought to it by two non-governmental organizations (NGOs) on behalf of the Ogoni people against the Shell Petroleum Development Company (SPDC). In that judgement the Commission concluded that "an environment degraded by pollution and defaced by the destruction of all beauty and variety is as contrary to satisfactory living conditions and development as the breakdown of the fundamental ecologic equilibria is harmful to physical and moral health" (United Nations 1966b). To be sure, the African Charter establishes that every "individual shall have the right to enjoy the best attainable state of physical and mental health" (African Commission on Human and Peoples' Rights 1987, Article 16(1)); and that "All peoples shall have the right to a general satisfactory environment favourable to their development" (African Commission on Human and Peoples' Rights 1987, Article 24).

Unfortunately, these universal conventions, treaties, and declarations do not have the force of law in the same way as national constitutions do. As a result, weaker populations who are impacted negatively by non-compliance cannot legally make a case that these actions be enforced. In many cases, the United Nations recognizes the relevance of other social actors in safeguarding these rights, although the biggest stakeholder in the protection and fulfillment of these rights is national governments—obviously because they are the single biggest party with the resources to do so. But, making the state the chief protector of rights can be problematic because individuals are too vulnerable in many developing, "Third World" countries. As Linden (1999) has noted, for example, implementing human rights in many "Third World" countries is cumbersome, given that human rights are violated widely.

In the following section I map out how these rights may be violated by highlighting a case study involving the Ogoni people of Nigeria, inhabitants of an oil-rich area, who suffered human rights violations from Shell Development Corporation and the Nigerian government. I will end this section with an analysis of a memo that was used as an instrument of abuse.

CASE STUDY: OGONI OIL CRISIS, SHELL, AND HUMAN RIGHTS ABUSES

The Ogoni communities are an indigenous group who inhabit an oil-rich area in the northeastern part of the Niger Delta of Nigeria. Since oil exploration began in Nigeria, the Ogonis alone have contributed about $30 billion to the Nigerian economy ("Factsheet on the Ogoni Struggle" n.d.). As a largely agricultural and fishing society, their survival has depended on their lands (Boele, Fabig, and Wheeler 2001, 76). Nigeria is the sixth largest oil-producing nation, with oil contributing about 90 percent of the country's foreign exchange, 20 percent of gross domestic product, and 88 percent of government's revenue (Coble 2014). SPDC, a subsidiary of Royal Dutch Shell, began oil exploration in the Ogoni area since 1958, when then British Colonial government granted it "exclusive exploration and prospecting rights" (1). The SPDC's operations in Nigeria account for 14 percent of their worldwide crude oil (the greatest outside the United States), producing half of Nigeria's total daily production of 2 million barrels (Howarth 1997, 384). However, between 1990 and 1995, the people of Ogoni embarked on a series of non-violent protests against Shell after they accused Shell of destroying the ecosystem of the area. For example, between 1976 and 1991, almost three thousand separate oil spills, averaging seven hundred barrels each, occurred in the Niger Delta ("Factsheet on the Ogoni Struggle" n.d.).

The Ogoni people united under the Movement for the Survival of the Ogoni People (MOSOP), led by author, activist, and Nobel Prize nominee, Ken Saro-Wiwa. In the process, MOSOP developed a bill, the Ogoni Bill of Rights, which sought to demand environmental protection, self-determination, cultural rights, equal share of the oil revenue, and social justice (Boele, Fabig, and Wheeler 2001). Although the activities of MOSOP were non-violent, about thirty villages were destroyed in the process, 600 people detained, 750 people killed, and 30,000 people left homeless (Corby 2011). Saro-Wiwa, the leader of the group, was hanged and nine other leaders of the group killed. A three-week investigation by *Human Rights Watch/Africa* implicated the government and Shell for using the army to oppress the local people. The report highlighted "flagrant human rights abuses" (Corby 2011). As a consequence, *Human Rights Watch/ Africa* argued that:

Because the abuses set in motion by Shell's reliance on military protection in Ogoniland continue, Shell cannot absolve itself of responsibility for the acts of the military. [. . .] The Nigerian military's defence [*sic*] of Shell's installations had become so intertwined with its repression of minorities in the oil-producing areas that Shell cannot reasonably sever the two. (Human Rights Watch/Africa 1995)

Although both the government and Shell denied involvement in the attacks, often citing ethnic clashes, memos exchanged between them indicate a deliberate attempt to induce the Army and plan attacks on the people (see figure 5.1 below). More so, although Shell attributed the oil spills to sabotage, a report by Amnesty International found it guilty of malpractice and argued that Shell was ". . . overstating the case in an effort to deflect attention from many oil spills . . ." (Amnesty International 2013). Also important, in 1996 two NGOs brought a complaint before the African Commission on behalf of the people of the Niger Delta. The commission, in its decision, found the Nigerian government in violation of:

> Articles 16 (i.e. *(1) Every individual shall have the right to enjoy the best attainable state of physical and mental health (2) States Parties to the present Charter shall take the necessary measures to protect the health of their people and to ensure that they receive medical attention when they are sick)* and Articles 24 (*All peoples shall have the right to a general satisfactory environment favourable to their development.*) of the African Charter.

It recommended that the government not only stop attacks on the people but ensure that appropriate environmental assessment impacts were conducted for future oil development. More recently, a 2011 United Nations Environmental Programme (UNEP) report, the first scientific investigations of the state of environmental destruction in the Ogoni area, cited destruction to aquatic life, vegetation, public health, soil and groundwater, adding that even though the oil industry is no longer active in the Ogoni area, oil spills continue to occur with alarming regularity (United Nations Environmental Programme, 2011).

Although it has denied involvement for many years, Shell, in 2009, agreed to pay $15.5 million in settlement after a court case was brought against it for collaborating in the execution of Ken Saro-Wiwa (Pilkington 2009). Shell, however, indicated that it made the payment to focus on the future for the Ogoni people (Pilkington 2009). Ironically, Shell, in 2009, posted a YouTube video on its webpage in which the Vice-President of the Ethical Affairs Committee of Royal Dutch Shell acknowledged responsibility and apologized for the abuses in the Niger Delta area.

RHETORICAL ANALYSIS OF MEMO BETWEEN SHELL AND NIGERIAN ARMY

The memo presented below (see figure 5.1) is a document used to communicate administrative procedures between the Nigerian Army and Shell, leading to attacks on the Ogoni people. The analysis shows how innocent technical communication artifacts and rhetoric may be central in human rights practices. Let me begin with a brief definition of rhetoric.

"Rhetoric is the use of language to inform and persuade" (Evans 2011). It's used to facilitate action through language, as the goal of the rhetor is to have something done by way of action. The rhetor attempts to "construct a persuasive argument . . . through the five canons of rhetoric: invention, organization, style, delivery, and memory" (Foss 2009). In a rhetorical analysis, then, a writer engages in "a criticism (or close reading) that employs the principles of rhetoric to examine the interactions between a text, an author, and an audience" (Nordquist n.d.). In many respects, technical communication artifacts (i.e., resumes, memos, letters, etc.) are rhetorical documents because their primary objective is to persuade a specific audience toward a specific action. Thus, memos are not only technical communication documents; they are also rhetorical artifacts that use language to achieve rhetorical ends. They are texts that rely on language to state their intent. For example, in "The Ethic of Expediency: Classical Rhetoric, Technology, and the Holocaust," Katz (1992) smartly describes how a regular, innocuous rhetorical document (i.e., a memo) was used to make a case for exterminating people during the Holocaust. Similarly, this memo under analysis, mainly using militaristic language, makes a persuasive argument toward a course of action.

Although there is no date in the "to" "from" "subject" area at the top left of the memo, and while the document's all capitals posture may be distracting, it passes for a technical communication document. The memo has a specific purpose, in the "subject" line, and a clear message, in the bulleted format. Its intent is to galvanize the Army to use violence to stop the Ogoni people from further protests against Shell's operations. It also has a clear, specific audience: Shell Petroleum Development Company (SPDC). The memo itself is written from an objective, detached point of view, giving an indication of the distance between the writer and the action being recommended. It's no wonder that the author's name is not indicated in the "from" area of the document. Only the author's title appears, and unless readers are able to locate the year of publication of the memo, it will be difficult for them to trace it to Major Paul Okuntimo, who was the Chairman of the

```
RIVERS STATE GOVERNMENT HOUSE OF NIGERIA FACTS SHEET
RESTRICTED RESTRICTED RESTRICTED
RIVERS STATE INTERNAL SECURITY TASK FORCE, GOVERNMENT HOUSE, PH.

MEMO
TO: HIS EXCELLENCY THE MILITARY ADMINISTRATOR RIVERS STATE
FROM: THE CHAIRMAN RIVERS STATE INTERNAL SECURITY (RSIS)
SUBJECT: RSIS OPERATIONS: LAW AND ORDER IN OGONI, ETC

OBSERVATIONS:
* POLICE IN OGONI REMAIN INEFFECTIVE SINCE 1993.
* SHELL OPERATIONS STILL IMPOSSIBLE UNLESS RUTHLESS MILITARY OPERATIONS ARE UNDERTAKEN FOR SMOOTH
ECONOMIC ACTIVITIES TO COMMENCE.
* WA IBOM AND OPOBO BORDERS INADVISABLE BECAUSE OF INACCESSIBILITY. ADDED TO DISAGREEMENT BETWEEN
OPOBO/ANDONI MAKING COOPERATION BY THE FORMER UNREALISABLE.
* DIVISION BETWEEN THE ELITIST OGONI LEADERSHIP EXISTS.
* EITHER BLOC LEADERSHIP LACKS ADEQUATE INFLUENCE TO DEFY NYCOP DECISIVE RESISTANCE TO OIL PRODUCTION
UNLESS REPARATION OF 400 MILLION DOLLARS PAID WITH ARREARS OF INTEREST TO MOSOP AND KEN SARO-WIWA.
RECOMMENDATIONS/STRATEGIES:
* INTRA-COMMUNAL/KINGDOM FORMULAE ALTERNATIVE AS DISCUSSED TO APPLY.
* WASTING OPERATIONS DURING MOSOP AND OTHER GATERINGS MAKING CONSTANT MILITARY PRESENCE JUSTIFIABLE.
* WASTING TARGETS CUTTING ACROSS COMMUNITIES AND LEADERSHIP CADRES ESPECIALLY VOCAL INDIVIDUALS IN
VARIOUS GROUPS.
* DEPLOYMENT OF 400 MILITARY PERSONNEL (OFFICERS AND MEN).
* NEW CHECKPOINTS SLIGHTLY DIFFERENT FROM OPERATION ORDER NO. 4/94 DATED 21/4/94 BY COMMISSIONER OF
POLICE RIVERS STATE COMMAND.
* DIRECT DAILY REPORT TO MILAD.
* WASTING OPERATIONS COUPLED WITH PSYCHOLOGICAL TACTICS OF DISPLACEMENT/WASTING AS NOTED ABOVE.
* PRESS MONITOR AND LOBBY.
* RESTRICTION OF UNAUTHORISED VISITORS ESPECIALLY THOSE FROM EUROPE TO THE OGONI.
* MONTHLY PRESS PRIEFING BY CHAIRMAN, RIVERS STATE INTERNAL SECURITY (RSIS).

FINANCIAL IMPLICATIONS (ESTIMATES/FUNDING):
* INITIAL DISBURSEMENT OF 50 MILLION NAIRA AS ADVANCED ALLOWANCES TO OFFICERS AND MEN AND FOR
LOGISTICS TO COMMENCE OPERATIONS WITH IMMEDIATE EFFECT AS AGREED.
* ECOMOG ALLOWANCE RATES APPLICABLE AS EARLIER DISCUSSED.
* PRESSURE ON OIL COMPANIES FOR PROMPT REGULAR IMPUTS AS DISCUSSED.
* OMPADEC STANDS BY AS ARRANGED.

REMARKS:
* THE IKWERRE-IJAW-AHOADA (OBAGI) AGENDA FOR SKELETAL OPERATIONS UNTIL FULL ECONOMIC ACTIVITIES COMMENCE
IN OGONI.
* SURVEILLANCE ON OGONI LEADERS CONSIDERED AS SECURITY RISKS/MOSOP PROPELLERS.
* PRESENT SSG OBVIOUSLY SENSITIVE (OGBAKOR/IKWERRE CONNECTION).
* MOSIEND AND MORETO IN IJAWS TERRITORY AS TARGETS FOR CLAMP DOWN.
* MODIFICATIONS OF PROGRAMME CONTINUOUSLY.
* RUTHLESS OPERATIONS AND HIGH LEVEL AUTHORITY FOR THE TASK FORCE EFFECTIVENESS.
* DIRECT SUPERVISION BY MILAD TO AVOID UNRULY INTERFERENCE BY OTHER SUPERIOR OFFICERS.
* RSIS INDEPENDENCE NECESSARY DESPITE SOME MOPOL INPUTS.

12/05/94
```

Figure 5.1. Nigeria government's memo to Shell on military crackdown on Ogoni residents to enable Shell to resume oil operations.

River State Internal Security at the time these abuses took place. The document also does not consider restoration of life or property after the brutality, and so there is no concern for life and for property. One of its primary goals is to source for funds to execute its intentions. In the "Observations" section of the memo, for example, the originator of the memo notes that "SHELL OPERATIONS STILL IMPOSSIBLE UNLESS RUTHLESS MILITARY OPERATIONS ARE UNDERTAKEN FOR SMOOTH ECONOMIC ACTIVITIES TO COMMENCE." By this, he sanctions government brutality against the Ogoni people so that Shell can begin oil exploration. The statement is also necessitated because the originator of the document believes that "POLICE IN OGONI REMAIN INEFFECTIVE SINCE 1993," warranting military action against the unarmed, mostly defenseless, people. Thus, it recommends "WASTING OPERATIONS" against Ogoni campaigners, and "WASTING TARGETS" such as "VOCAL INDIVIDUALS" mainly made up of Ogoni leaders. Just as in Katz's (1992) analysis, only a reader with contextual information will be able to decode the militaristic terms such as "wasting operations" and "wasting targets," which are simply code terms for using ammunitions to maim and kill (257). The document also described, as a strategy, using " . . . PSYCHOLOGICAL TACTICS OF DISPLACEMENT/WASTING . . . " Thus, it is obvious that the author knows his intended audience because there is no attempt, at any time, to define any of the terms used, which might be misunderstood by an "outsider."

This analysis points to two important observations. First, it's obvious that the attacks on the MOSOP leadership were calculated and preplanned. Second, rhetoric was instrumental in planning the attacks. We clearly see the use of the memo through rhetoric to normalize the dehumanization of humans to make violence appear acceptable and regular. More so, there is a deliberate attempt to make those who were to embark on the violence to see the actions as "normal." In essence, by dehumanizing the Ogonis, the writer of the memo "renders the requisite horrors of [violence] tolerable" (Elliott 2004, 99). He also makes sure that "no moral relationship with the [enemy] inhibits the victimizer's violent behavior" (Haslam 2006, 254). Writing in *A Rhetoric of Motives*, Burke (1969) argues, for example, that "On every hand, we find men . . . preparing themselves for the slaughter, even to the extent of manipulating the profoundest grammatical, rhetorical, and symbolic resources of human thought to this end" (264).

Obviously, the RSIS and SPDC used the memo as a tool to achieve a specific economic end by finding ways to continue the exploration of oil. As I argued earlier in this chapter, rhetoric and technical communication are virtuous enterprises, although they may be implicated in excesses as well. It must be said that before the release of this memo, both Shell and the Nigerian government had denied involvement in the deaths of the Ogoni people, although other documents had also pointed to inducements from Shell to the Nigerian Army.

IMPLICATIONS FOR TECHNICAL COMMUNICATION

Technical communicators add value by making information, otherwise specialized and technical, usable and accessible to audiences. In the process they also mediate the activities of the organizations they work for. As ambassadors, they take the credit for the successes and the blame for the failures of those organizations. But how are technical communicators responsible for the actions of their employers, when they have good intentions for composing the artifacts that support practices in organizations? The analysis of the memo suggests that our documents—no matter their organization and design—can raise ethical or human rights issues, even if the technical communicator has good intentions, or even if the technical communicator is merely a writer—not necessarily an author. Pedagogically, this raises a number of questions: can we just teach about formats, style, and design without much consideration for the ethical considerations of these elements? Will our writing pedagogies be more helpful to our students if we teach them that technical communication artifacts can be merely administrative documents, but also instruments of oppression and abuse? The analysis of the memo suggests that we can teach technical communication artifacts in other ways by encouraging our students to acknowledge how technical communication artifacts are not innocent or value-neutral, or even how their rhetorical practices could sustain negative practices or promote human rights. Our students will benefit from our pedagogical practices if we provide opportunities in the classroom for analysis of these artifacts, particularly those that may potentially raise ethical issues. Such analysis can make connections between all the elements of good technical communication and ethics or, more precisely, how these elements manifest themselves unjustly in the composing process. We can no longer take it for granted that our students are aware of these challenges or will act justly when presented with situations in corporate contexts.

Essentially, Shell's posture in the Ogoni crisis crystallizes how corporate structures can impose on the cultural and socio-economic systems of indigenous peoples and therefore lead to human rights abuses. The complex dynamics in the SPDC-Nigerian government-Ogoni relationship point to ways in which human rights may be impacted, although, more important, this relationship also suggests that technical communicators may participate in human rights abuses—whether directly or indirectly—because of our work in and for organizations. In many instances these organizations "wield economic power—and sometimes political power—equal to or greater than the power or governments in the nation states where they conduct business" (Sapp, Savage, and Mattson 2013, 1). Trivett writes that, "If Wal-Mart were a country, its revenues would make it on par with the GDP of the 25th largest economy in the world by, surpassing 157 smaller countries" (Trivett 2011). Thus, as agents of power and action, technical communicators are not only transmitters of information but decision makers through their discursive practices.

Because pedagogical practices create opportunities for action and learning, I suggest further, in the following section, ways in which our field can address human rights issues. We have to recognize that our classroom is a space that is home to students with diverse perspectives, experiences, socializations, who, in turn, will impact global contexts because of the nature of their work. If we consider these issues, we can create learning opportunities that will situate them either physically or spatially.

Article 26(2) of the UDHR mandates that education should be directed " . . . to the strengthening of respect for human rights and fundamental freedoms" (United Nations 1948). To actualize these, the United Nations, in its manifesto on the fundamentals of teaching human rights, suggests a predominantly three-prong approach: incorporation of human rights in national legislations; revision of curricula and textbooks; and development of educational materials (United Nations 2004, 7). A good starting point, for us, is to interrogate ways in which power, ideology, cultural and economic domination, and disfranchisement shape complex relationships in global contexts. More so, in light of the challenges of unenfranchised communities, another useful approach will be to understand the human rights struggles of the communities we engage to inform the content and methods of our pedagogical practices.

So, what should technical communication pedagogical practices look like? First of all, I believe that programmatic avenues offer one of the most practical platforms for enacting change and action. We can begin by making some programmatic adjustments—both big and small. Sometimes small, small changes can lead to big, big results. As Falcon and Jacob (2011) argue, "Human rights pedagogy means that we are all responsible for injustice and we all have a role to combat it daily—in both small and substantial ways" (31). These adjustments could come in forms such as altering programmatic perspectives by designing specific courses on human rights, to small changes such as introducing human rights components in our courses. Of course, the danger in introducing units in courses or designing individual courses in the curriculum to address specific needs is the fear of what Smith and Mikelonis (2011) call "ghettoization" of content (90) which itself makes students think that human rights issues are separate from the technical communication issues they will undertake in the "real" world. However, we can make these components meaningful by effectively connecting our practices and discourses with ways in which technical communication impacts with rights, ethics, and issues of disenfranchisement or oppression. As I highlighted in my discussions of the analysis of the Shell memo, our writing pedagogies can focus on traditional assignments (i.e., resumes, letters, memos, etc.) but also highlight how these traditional documents may be instruments of liberation or abuse in specific contexts. Then, in focusing on the power dynamics of these artifacts, we may expose them to how corporate institutional structures can work to promote or inhibit human rights. Emphasis on documents that are situated in specific contexts can help our students to make meaningful connections with problems that occur on the global scale. As Spilka (1989) pointed out, "We have noticed that when given "real" communication tasks, our students seem to evaluate their work more carefully before, during, and after the writing process, and to place more value on what they are learning in the class" (151).

Also important, our field must begin to engage the various human rights treaties, protocols, and conventions—what I call the genre of human rights pedagogy—and interrogate how they impact our work. For example, when we teach about issues of gender and/or disability studies, we can situate them within the broader contexts of UN conventions (e.g., the Convention on the Elimination of All Forms of Discrimination Against Women (CEDAW) or the Convention on the Rights of Persons with Disabilities (CRPD) and discuss how they fit into the global narrative. Doing this may provide opportunities for our students to appreciate the broader issues surrounding human rights and also make them better citizens who will know about the extent of human rights violations within the academic institution, nationally, and globally.

I should add that courses that include human rights components may involve some community initiatives in which students can reach out and make connections between the human rights concepts they learn in the course and specific "real" world experiences. These service-learning initiatives provide students rare opportunities to look beyond the traditional institution by directly engaging or supporting cultural sites and spaces that are confronted with some inequalities. Furthermore, these opportunities may provide both students and the populations in those sites unique opportunities to benefit from each other's experiences. For example, students may learn about forms of injustice and how to contribute to changing the narrative, while the communities may also learn from the educational experiences of students and how the educational system can be a powerful tool in addressing issues of inequality. Students can engage migrant communities, prisons, refugee camps, civil society groups, and/or other non-governmental organizations. This list is not exhaustive, but these non-traditional experiences provide innovative approaches to learning that help students to engage, first-hand, in the narrative about issues of (in)justice and (in)equality.

From a pedagogical perspective, these approaches will attune students to responsible citizenship, self-awareness and consciousness, and critical thinking. As Grabill put it, "These habits of mind will allow our students the intellectual autonomy to make good personal decisions and become "good" writers for global organizations of all kinds—organizations that localize products thoughtfully, that design diversity into systems that create safe, entertaining, and beautiful systems and documents, and treat people and the environment justly" (Grabill 2005, 274). Proposing assignments that disengage students offer students fewer opportunities and resources to think about the impact of these assignments on the world. Yet another strategy is to assign students to read works from postcolonial (e.g., Gaudelli and Fernekes 2004; Grabill 2005; Perez 1999; Said 1978, 1993; Smith 1999; Spivak 1999; wa Thiong'o 1986, etc.) and decolonial (e.g., Haas 2012; Perez 1999; Smith 1999) literature that foreground issues of colonialism, socio-economic imbalances, and various issues of inequalities in geopolitical discourse, so that they can interrogate their roles in the world as well as foster equality and change. More so, they can read entire sections of various instruments of the International Bill of Human Rights and either critique them or discuss how they can be applied in technical communication. Still, students may also discuss specific case studies of unethical behaviors by corporations in some sites. While human rights education alone cannot eliminate human rights violations, it can certainly be instrumental toward that end (Gaudelli and Fernekes 2004, 25).

The classroom is a site for political, cultural, and economic negotiation and therefore a good site for broaching some of the most critical issues. As Sullivan (1990) noted, "Those of us who teach . . . are really placed in a situation that allows us to be political agents for change" (377). Technical communication is a rights profession. Thus, we are policy makers at the micro level and can use specific approaches to help students become aware of their rights and responsibilities, and help safeguard the rights of others.

CONCLUSION

This chapter has examined the interstices of human rights and technical communication. While the case of Shell in Ogoni may be an extreme one, the practices of many organizations do have the potential to impact human rights in many contexts—and technical communicators have important roles to play. Because technical communicators are connected to the global economy in many ways, and because they have an important role to play, they must make a sustained effort to address such issues. Not all organizational contexts or sites may raise human rights issues because the scope of work and research in every context is different. However, even if these issues do not occur everywhere, the expanding nature of our research requires that human rights issues be addressed. Technical communication is already global, and the fact that many of our graduates already work (and will continue to work) for multinational corporations who design communication and content for international audiences means that we cannot continue to ignore the human rights implications of our work. I encourage more research in other contexts to enrich this conversation.

Notes

1. See Slack et al.'s (1993) discussions on the transmission, translation, and articulation models of communication in "The Technical Communicator as Author: Meaning, Power, Authority"
2. See pages 170–176 for a full discussion of these rights.

References

African Commission on Human and Peoples' Rights. 1987. "African Charter on Human and Peoples' Rights." Accessed December 16, 2014. http://www.achpr.org/files/instruments/achpr/banjul_charter.pdf.

Agboka, Godwin. 2013. "Participatory Localization: A Social Justice Approach to Navigating Unenfranchised/Disenfranchised Cultural Sites." *Technical Communication Quarterly* 42:1–21.

Aghalino, Samuel O. 2004. "Oil Firms and Corporate Social Responsibility in Nigeria: The Case of Shell Petroleum Development Company." *Journal of History and International Studies* 42:1–17.

Al-Moneef, Majid. 2006. "The Contribution of the Oil Sector to the Arab Economic Development." Paper presented at the High-level Roundtable *Partnership for Arab Development: A Window of Opportunity*, OFID, May 5, 2006.

Allen, Lori, and Dan Voss. 1997. *Ethics in Technical Communication: Shades of Gray*. New York: Wiley.

Amnesty International. 2013. *Bad Information: Oil Spill Investigations in the Niger Delta*. London: Amnesty International Publications.

Appelbaum, David, and Sarah V. Lawton, eds. 1990. *Ethics and the Profession*. Englewood Cliffs, NJ: Prentice Hall.

Blyler, Nancy. 1995. "Research as Ideology in Professional Communication." *Technical Communication Quarterly* 4 (3): 285–313. https://doi.org/10.1080/105722595093 v64602.
Boele, Richard, Heike Fabig, and David Wheeler. 2001. "Shell, Nigeria and the Ogoni. A Study in Unsustainable Development: I. The Story of Shell, Nigeria and the Ogoni People—Environment, Economy, Relationships: Conflict and the Prospects for Resolution." *Sustainable Development* 9 (2): 74–86. https://doi.org/10.1002/sd.161.
Bowdon, Melody. 2002. "A Practical Ethics for Professional and Technical Writing Teachers, or a Millers' Tale." *Technical Communication Quarterly* 11 (2): 222–224. https://doi.org/10.1207/s15427625tcq1102_11.
"Briefing Note." 2013. Business and Human Rights Resource Center. https://www.business-humanrights.org/media/documents/africa-oil-week-briefing-note-nov-2013.pdf.
Burke, Kenneth. 1969. *A Rhetoric of Motives*. Berkeley: University of California Press.
Chamberlain, Mark. 2001. "Human Rights Education for Nursing Students." *Nursing Ethics* 8 (3): 211–222. https://doi.org/10.1177/096973300100800306.
Coble, Breane. 2014. "Shell's Corporate Social Responsibility in the Niger Delta." Accessed April 12, 2014. http://www.inasp.info/uploads/filer_public/2013/04/03/3_handout_4.pdf.
Collinson, Helen, and Rod MacLeod. 2010. "Striking Oil: Blessing or Curse? Supporting Civil Society Advocacy to Ensure That the Benefits Are Shared." Accessed May 16, 2014. https://www.intrac.org/data/files/resources/684/Praxis-Note-52-Striking-Oil-Blessing-or-Curse.pdf.
Coombe, Rosemary J. 2010. "Honing a Critical Cultural Study of Human Rights." *Communication and Critical/Cultural Studies* 7 (3): 230–246. https://doi.org/10.1080/147914 20.2010.504594.
Corby, Elorwyn. 2011. "Ogoni People Struggle with Shell, Nigeria, 1990–1995." November 3, 2011.
Ding, Huiling, and Elizabeth Pitts. 2013. "Singapore's Quarantine Rhetoric and Human Rights in Emergency Health Risks." *Rhetoric, Professional Communication, and Globalization* 4:56–77.
Dombrowski, Paul. 2000. *Ethics in Technical Communication*. Needham Heights, MS: Allyn & Bacon.
"Draft Principles on Human Rights and the Environment." 1994. June 16, 2011. https://www1.umn.edu/humanrts/instree/1994-dec.htm.
Dragga, Sam. 1996. "Is this Ethical? A Survey of Opinion on Principles and Practices of Document Design." *Technical Communication (Washington)* 43:255–65.
Dragga, Sam, and Dan Voss. 2001. "Cruel Pies: The Humanity of Technical Illustrations." *Technical Communication (Washington)* 3:265–74.
Dragga, Sam and Dan Voss. 2001. "Cruel Pies: The Inhumanity of Technical Illustrations." *Technical Communication* 48 (5): 265–274.
Dura, Lucia, Arvind Singhal, and Eliana Elias. 2013. "Minga Peru's Strategy for Social Change in the Peruvian Amazon: A Rhetorical Model for Participatory, Intercultural Practice to Advance Human Rights." *Rhetoric, Professional Communication, and Globalization* 4:33–54.
Ebeku, Kaniye S. A. 2003. "The Right to a Satisfactory Environment and the African Commission." *African Human Rights Law Journal* 3:149–66.
Elliott, Kimberly C. 2004. "Subverting the Rhetorical Construction of Enemies through Worldwide Enfoldment." *Women & Language* 27:98–103.
Evans, Robert. 2011. "The Pentad and the EIS: Using Burke's Pentad to Analyze Environmental Impact Statements Issued by the U.S. Military." Paper presented at the Professional Communication Conference (IPCC), IEEE International, Cincinnati, Ohio, October 17–19, 2011. https://doi.org/10.1109/IPCC.2011.6087202.

Faber, Brenton D. 2001. "Gen/Ethics? Organizational Ethics and Student and Instructor Conflicts in Workplace Training." *Technical Communication Quarterly* 10 (3): 291–318. https://doi.org/10.1207/s15427625tcq1003_4.

"Factsheet on the Ogoni Struggle." n.d. Accessed March 4, 2014. https://www.ratical.org/corporations/OgoniFactS.html.

Faigley, Lester. 1986. "Competing Theories of Process: A Critique and a Proposal." *College English* 48 (6): 527–542. https://doi.org/10.2307/376707.

Falcon, Sylvia M., and Michelle M. Jacob. 2011. "Human Rights Pedagogies in the Classroom: Social Justice, US Indigenous Communities, and the CSL Projects." *Societies without Borders* 6:23–30.

Foss, Karen A., ed. 2009. *Encyclopedia of Communication Theory*. Thousand Oaks, CA: SAGE.

Gaudelli, William, and William R. Fernekes. 2004. "Teaching about Global Human Rights for Global Citizenship." *Social Studies* 95 (1): 16–26. https://doi.org/10.3200/TSSS.95.1.16-26.

Gilmer, Robert W. 2014. "Houston: American Oil Headquarters." *Tierra Grande*, January 2014. https://recenter.tamu.edu/pdf/2051.pdf.

Grabill, Jeffrey T. 2005. "Globalization and the Internationalization of Technical Communication Programs: Issues for Program Design." In *2005 IEEE International Professional Communication Conference Proceedings*, ed. Vasundara V. Varadan. Piscataway, NJ: IEEE. https://doi.org/10.1109/IPCC.2005.1494200.

Haas, Angela M. 2012. "Race, Rhetoric, and Technology: A Case Study of Decolonial Technical Communication Theory, Methodology and Pedagogy." *Journal of Business and Technical Communication* 26 (3): 277–310. https://doi.org/10.1177/1050651912439539.

Haslam, Nick. 2006. "Dehumanization: An Integrative Review." *Personality and Social Psychology Review* 10 (3): 252–264. https://doi.org/10.1207/s15327957pspr1003_4.

Herndl, Carl G. 1993. "Teaching Discourse and Reproducing Culture: A Critique of Research and Pedagogy in Professional and Non-Academic Writing." *College Composition and Communication* 44 (3): 349–363. https://doi.org/10.2307/358988.

Herndl, Carl G. 1996. "Tactics and the Quotation: Resistance and Professional Discourse." *Journal of Advanced Composition* 16:455–70.

Herrington, TyAnna. 2013. "Global Intellectual Property Law, Human Rights, and Technical Communication." *Rhetorical, Professional Communication and Globalization* 4: 13–32.

Howarth, Stephen. 1997. *A Century in Oil: The "Shell Transport and Trading Company 1897–1997."* London: Weidenfeld and Nicolson.

Huckin, Thomas. 2002. "Globalization and Critical Consciousness in Technical and Professional Communication." Paper presented at the Conference of Council of Programs in Scientific and Technical Communication, Logan, UT, October 3–5.

Human Rights Watch/Africa. 1995. "The Ogoni Crisis: A Case Study of Military Repression in South East Nigeria." *Africa: Journal of the International Africa Institute* 7 (5). http://www.refworld.org/docid/3ae6a7d8c.html.

International Consultation on the Pedagogical Foundations of Human Rights Education. 1996. *"Towards a Pedagogy of Human Rights Education."* La Catalina, Costa Rica, July 22–26, 1996. http://www.pdhre.org/dialogue/costarica.html.

Katz, Steven B. 1992. "The Ethic of Expediency: Classical Rhetoric, Technology, and the Holocaust." *College English* 54 (3): 255–275. https://doi.org/10.2307/378062.

Linden, A. 1999. "Communicating the Right to Development: Towards Human Rights-Based Communication Policies in Third World Countries." *International Communication Gazette* 61 (5): 411–432.

Lyon, Arabella, and Lester C. Olson. 2011. "Special Issue on Human Rights Rhetoric: Traditions of Testifying and Witnessing." *Rhetoric Society Quarterly* 41 (3): 203–212. https://doi.org/10.1080/02773945.2011.575321.

McKee, Heidi A., and James E. Porter. 2010. "Legal and Regulatory Issues for Technical Communicators Conducting Global Internet Research." *Technical Communication (Washington)* 57:282–99.

Miller, Carolyn R. 1979. "A Humanistic Rationale for Technical Writing." *College English* 40 (6): 610–617. https://doi.org/10.2307/375964.

Nordquist, Richard. n.d. "Grammar & Composition: Rhetorical Analysis." Accessed April 12, 2014 https://www.thoughtco.com/rhetorical-analysis-1691916.

Office of the High Commissioner for Human Rights. 2011. "Guiding Principles on Business and Human Rights." June 16, 2011. http://www.ohchr.org/Documents/Publications/GuidingPrinciplesBusinessHR_EN.pdf.

Ornatowski, Cezar M., and Linn K. Bekins. 2004. "What's Civic about Technical Communication? Technical Communication and the Rhetoric of 'Community.'" *Technical Communication Quarterly* 13 (3): 251–269. https://doi.org/10.1207/s15427625tcq1303_2.

Osler, Audrey, and Hugh Starkey. 1994. "Fundamental Issues in Teacher Education for Human Rights." *Journal of Moral Education* 23 (3): 349–359. https://doi.org/10.1080/0305724940230311.

Palmeri, Jason. 2006. "Disability Studies, Cultural Analysis, and the Critical Practice of Technical Communication Pedagogy." *Technical Communication Quarterly* 15 (1): 49–65. https://doi.org/10.1207/s15427625tcq1501_5.

Perez, Emma. 1999. *The Decolonial Imaginary: Writing Chicanas into History*. Bloomington: Indiana University Press.

Pilkington, Ed. 2009. "Shell Pays out $15.5m over Saro-Wiwa Killing." *The Guardian*, June 8, 2009. https://www.theguardian.com/world/2009/jun/08/nigeria-usa.

Said, Edward. 1978. *Orientalism*. New York: Random House.

Said, Edward. 1993. *Culture and Imperialism*. New York: Alfred Knopf.

Sapp, David A. 2004. "Global Partnerships in Business Communication: An Institutional Collaboration between the United States and Cuba." *Business Communication Quarterly* 67 (3): 267–280. https://doi.org/10.1177/1080569904268051.

Sapp, David, Gerald Savage, and Kyle Mattson. 2013. "After the International Bill of Human Rights (IBHR): Introduction to Special Issue on Human Rights and Professional Communication." *Rhetoric, Professional Communication, and Globalization* 4:1–12.

Savage, Gerald J. 2001. "International Technical Communication Programs and Global Ethics." In *CPTSC Proceedings 2001: Managing Change and Growth in Technical and Scientific Communication*, ed. Bruce Maylath. Pittsburgh, PA: Council for Programs in Technical and Scientific Communication Proceedings.

Savage, Gerald, and Kyle Mattson. 2011. "Perceptions of Racial and Ethnic Diversity in Technical Communication Programs." *Programmatic Perspectives* 3:5–57.

Slack, Jennifer D., David J. Miller, and Jeffrey Doak. 1993. "The Technical Communicator as Author: Meaning, Power, Authority." *Journal of Business and Technical Communication* 7 (1): 12–36. https://doi.org/10.1177/1050651993007001002.

Smith, Shelley L., and Victoria M. Mikelonis. 2011. "Incorporating 'Shock and Aha!' into Curriculum Design: Internationalizing Technical Communication Courses." In *Teaching Intercultural Rhetoric and Technical Communication: Theories, Curriculum, Pedagogies and Practices*, ed. Barry Thatcher and Kirk St.Amant, 89–112. Amityville, NY: Baywood Publishing Company. https://doi.org/10.2190/TIRC5.

Smith, Tuhiwai L. 1999. *Decolonizing Methodologies: Research and Indigenous Peoples*. London: Zed Books.

Spilka, Rachel. 1989. "The Audience Continuum." *Technical Writing Teacher* 2:147–52.

Spivak, Gaytri C. 1999. *A Critique of Postcolonial Reason: Towards a History of the Vanishing Present*. Cambridge, MA: Harvard University Press.

St.Amant, Kirk, and Martine C. Rife. 2010. "Legal Issues in Global Contexts: Reconsidering Content in an Age of Globalization." *Technical Communication (Washington)* 57:249–50.

St. Germaine-McDaniel, Nicole. 2010. "Technical Communication in the Health Fields: Executive Order 13166 and Its Impact on Translation and Localization." *Technical Communication (Washington)* 57:251–65.

Sullivan, Dale. 1990. "Political-Ethical Implications of Defining Technical Communication as Practice." *Journal of Advanced Composition* 10:375–86.

Sullivan, Dale L., and Michael S. Martin. 2001. "Habit Formation and Storytelling: A Theory for Guiding Ethical Action." *Technical Communication Quarterly* 10 (3): 251–272. https://doi.org/10.1207/s15427625tcq1003_2.

Thralls, Charlotte, and Nancy R. Blyler. 1993. "The Social Perspective and Professional Communication: Diversity and Directions in Research." In *Professional Communication: The Social Perspective*, ed. Nancy. R. Blyler and Charlotte Thralls, 3–34. Newbury Park, CA: Sage.

Trivett, Vincent. 2011. "25 US Mega Corporations: Where They Rank If They Were Countries." *Business Insider*, June 27, 2011. Accessed April 4, 2014. http://www.businessinsider.com/25-corporations-bigger-tan-countries-2011-6?op=1.

United Nations. 1948. *Universal Declaration of Human Rights* (UDHR). Accessed March 15, 2014. http://www.un.org/en/documents/udhr/.

United Nations. 1966a. "International Covenant on Civil and Political Rights." Accessed March 15, 2014. http://www2.ohchr.org/english/bodies/hrc/docs/ngos/DCI_Israel99.pdf.

United Nations. 1966b. "International Covenant on Economic, Social and Cultural Rights." Accessed December 16, 2014. http://www.ohchr.org/EN/ProfessionalInterest/Pages/CESCR.aspx.

United Nations. 1972. "Declaration of the United Nations Conference on the Human Environment." Accessed December 16, 2014. http://www.unenvironment.org/Documents.Multilingual/Default.Print.asp?documentid=97&articleid=1503.

United Nations. 1987. *Human Rights: Questions and Answers*. New York: UN.

United Nations. 2004. *Teaching Human Rights: Practical Activities for Primary and Secondary Schools*. New York: United Nations Publications.

United Nations Economic Commission for Europe (UNECE). 1998. "Convention on Access to Information, Public Participation in Decision-making and Access to Justice in Environmental Matters." Accessed March 15, 2014. http://www.unece.org/fileadmin/DAM/env/pp/documents/cep43e.pdf.

United Nations Environmental Programme. "Environmental Assessment of Ogoniland." 2011. Accessed March 15, 2014. http://www.unenvironment.org/disastersandconflicts/CountryOperations/Nigeria/EnvironmentalAssessmentofOgonilandreport/tabid/54419/Default.aspx.

wa Thiong'o, Ngugi. 1986. *Decolonizing the Mind: The Politics of Language in African Literature*. Portsmouth, NH: Heinemann.

Walton, Rebecca, and Brian DeRenzi. 2009. "Value-Sensitive Design and Health Care in Africa." *IEEE Transactions on Professional Communication* 52 (4): 346–358. https://doi.org/10.1109/TPC.2009.2034075.

Walton, Rebecca, Ryan Price, and Maggie Zraly. 2013. "Rhetorically Navigating Rwandan Research Review: A Fantasy Theme Analysis." *Rhetoric, Professional Communication, and Globalization* 4:78–102.

Wilson, Greg. 2001. "Technical Communication and Late Capitalism: Considering a Postmodern Technical Communication." *Journal of Business and Technical Communication* 15 (1): 72–99. https://doi.org/10.1177/105065190101500104.

6
AN ENVIRONMENTAL JUSTICE PARADIGM FOR TECHNICAL COMMUNICATION

Donnie Johnson Sackey

This chapter examines the role of several environmental perspectives as approaches to social justice. What do these theories do? What influence might they have upon the type of work technical communicators do and will be expected to do in the future? More importantly, how might they affect the ways in which we design not only our courses but also build programs that engender the principles of environmental justice? Most technical communicators understand theory as an important part of the work that we do. Recently, several scholars within technical and professional communication increasingly have called for a shift in attention toward social justice through institutional diversity, teaching, research, and service (Haas 2012; Savage and Mattson 2011). Thus, there is a need for theories and approaches to social justice. This is not to suggest that this shift is entirely new. For example, Nancy Blyler offered the *critical perspective* as a political re-orientation of technical professional communication practice that empowers research participants to actively participate in driving the goals and outcomes of research by working from participants' epistemological positions rather than their own. The principle motivation for incorporating this perspective in pedagogy was the belief we should "enable students to learn about [and work among] the network of social relations on the job" (Wells quoted in Blyler 1998, 34). Much in the same way Jeffery Grabill and Michelle Simmons were concerned with how traditional risk communication practices are arhetorical in that they decontextualize risk and fail to take into account the various factors that influence a public's perception of risk. Their understanding of a *critical rhetoric* breaks down the barriers between risk assessment and risk communication by incorporating scientific interests *and* public concerns over uncertainty as a means of changing the flow of information and subsequently the way we make public policy (Grabill and Simmons 1998). Key to their approach is the analysis of power dynamics in order to critically assess the ethical involvement of a public.

This concern has only increased over the years as globalization has pushed technical and professional communication pedagogy to provide more consideration to the international contexts of our practices. In an age that calls increasingly for global rhetorical citizenship, we should be apt to consider repeatedly what constitutes good rhetorical practice. Much of these calls have been geared toward helping students become effective communicators in contemporary corporate settings; however, we should also consider "is it our responsibility [as US-based educators], and our students' responsibility, to question and respond to organizational values and practices that we may find troubling? Or is it our job to prepare students to be effective writers and meet, uncritically, the needs of employing organizations?" (Grabill 2005, 3). These are not necessarily questions that can and should be answered easily; however, they are questions that we should consider. These are questions that point to tensions between local-global divides that both raise social justice concerns and underscore why technical and professional communication should be committed to social justice. Godwin Agboka (2013) is one of a handful of researchers who have argued increasingly for technical communicators to attend to the local. His work highlights the way in which technical communication's sole focus on linguacultural concerns often "overlooks the broader contexts within which these linguacultural factors operate" (29). This is an argument for technical communicators to be attuned to other cultural phenomena that form the social conditions that comprise the everyday in situated locales. Therefore, as a corrective toward engendering social justice, he posits that technical communicators "understand ideology, power, economics, knowledge, politics, law, and ethics all as dimensions of a locale, not separate from it" (29). In this chapter I argue that theories of environment should factor heavily in the ways in which we approach the mitigation of communication issues surrounding environmental concerns. I offer four environmental perspectives and consider their role in the work technical communicators might do. These heuristics draw attention to the fact that while social justice is a question of power, we should also draw our attention to the emplaced, material, and environmental nature of social justice concerns. I do want to raise an important caveat. I do not intend to suggest that there is a line of demarcation between social justice and environmental justice. Instead, I only wish to underscore that while environmental justice approaches are concerned with achieving some level of social justice, the latter is not necessarily dedicated toward exploring environmental factors. My goal here is to stress that environment always matters and should be foregrounded in analysis.

LOOKING FOR JUSTICE IN 48217

In 1989, the Ford Motor Company moved out of the steel-making business and divested its assets in the industry. The company renamed its steelmaking operation the Rouge Steel Company in 1981, which soon began to operate as an independent subsidiary. In December 1989, the Marico Acquisition Corporation purchased 80 percent of the company and then bought the final 20 percent in 1992 from Ford. This transferred 45 percent of the Rouge Complex (a site that sits on 1,200 acres—roughly 1.8750 square miles—between the Rouge and Detroit Rivers in Dearborn, MI, and at one time hosted 93 buildings with 15,767,708 square feet of factory floor space) out of Ford's possession. The company was renamed Rouge Steel Industries in 1996 and grew to become the fifth largest steel producer in the United States. Although it was no longer a part of the Ford Motor Company, Ford accounted for about a third of the company's sales. This is why Ford was able to provide Rouge Steel with a $75 million loan in 2002, a period when US-based steelmakers struggled against imports. Despite the fact that there were cheaper steel prices elsewhere, the relationship between Ford, Rouge, and the community was a sentimental one. Nevertheless, Rouge Steel Industries filed for bankruptcy in October 2003. Its assets were made publicly available in a competitive bidding process. Severstal, a Russian company, purchased Rouge Steel and its assets in 2004 for $285 million.

Severstal specializes in manufacturing flat-rolled carbon steel products used primarily in the automotive industry. In January 2006, the Michigan Department of Environmental Quality (MDEQ) provided Severstal with a permit to make modifications to the plant. This included the installation of baghouses used to capture particles as a means of reducing emissions. (This specifically included both a C Blast Furnace and Desulfurization Baghouse.) When Severstal underwent the required smokestack compliance testing in 2008, the results showed that the plant's release of manganese, mercury, lead, sulfur dioxide, carbon monoxide, and particulate matter exceeded the allowable levels authorized by MDEQ and the US Environmental Protection Agency (EPA). Based on this information, MDEQ issued a "letter of violation" to Severstal Dearborn in February 2009. It was at this time that Severstal officials petitioned MDEQ to alter the permit to allow for pollution emissions in excess to the permit limit. In a March 2009 letter to the department, the company indicated that it wanted to make corrections to its existing permit rather than filing an application for a new permit. Nevertheless, in a meeting that month with MDEQ's Air Quality Division (AQD) regarding the permit limit violations, Severstal agreed to submit an evaluation of available technologies that might enable them to achieve current emission limits allowed by the permit limits. The company responded with a white-paper in May 2009 indicating that it could not meet the emission levels they accepted in 2006, because there was no feasible technology options available to achieve compliance. In December 2010, the company submitted its application to modify. During this five-year span, ADQ Enforcement issued twenty violation notices. This included a notice for exceeding manganese and lead emission limits based on stackhouse testing for the C Blast Furnace and Desulfurization Baghouse in January 2011 and an observed kish graphite fallout in August 2011, for which the company denied responsibility. It is also important to note that these violations were not merely limited to a lack of technology. There were issues with the company's "failure (or refusal) to properly install/maintain/operate equipment." It was not until June 2012 that the EPA stepped in and issued a notice of violation based upon the ADQ violation notices. No substantial action took place to address the violations from 2006 to 2014. Severstal, state, and federal officials knowingly were aware that the facility was not working to comply with emission standards.

So far this story has been about the relationship between corporations and regulatory bodies fighting over how much pollution is too much pollution. On the other side of this story is the world that resides outside of the plant, which comprises many of the plant's workers, residents of Dearborn, Detroit (especially communities residing in the 48217 zip code) and Windsor, Ontario, Canada. Between when the company submitted its application to modify its permit and the present (May 5, 2009 to May 2014), there have been 117 complaints alleging fallout and opacity from various processes at the facility (Lamb and Koster 2012). Many of these complaints have originated in Detroit's 48217 zip code, which many have identified as an environmental justice area. Of particular concern has been Dearborn's South End neighborhood, which is a predominantly Arab-American community. Many of these residents have recently immigrated to the United States. Another concern is the close proximity not only to Severstal but also to the Marathon petroleum refinery and other industrial plants. Residents have to live with continual haze and thick smoke, as well as a variety of petrochemical smells throughout the day. Many residents have claimed that the smells and smoke appear to be stronger at night. It is almost as if the factories purposefully release it in the evening. Within the 48217 zip code, asthma and respiratory-related health problems continue to rise at disconcerting levels. A 2010 study between the University of Michigan and the *Detroit Free Press* ranked postal zones in southeast Michigan based on chemical releases from the EPA's 2006 Toxic Release Inventory. Detroit's 48217 ranked the highest toxicity with a score of 2,576 (Tanner-White and Lam 2010). Moreover, five of the surrounding zip codes were represented in the top ten most toxic areas. It is important to note that the 48217-area's ranking was forty-five times more than the statewide average for Michigan. It was so bad that officials at the Sierra Club dubbed the area a "sacrifice zone for energy production" where regulators appear lax to deliver punishments against industries, and low-income and minority communities have to bear the highest burdens (McArdle 2011). Residents' ability to seek justice and secure the safety of the communities has mostly been stymied by an inability to pinpoint the exact causality between a single polluter (or polluters) and health and/or environmental problems. Companies within the area either blame other entities, point to a lack of existing research that would demonstrate links, or deny culpability altogether. Moreover, residents often have to deal with seemingly impenetrable layers of bureaucracy.

This brief story of industrial pollution in this part of southeast Michigan calls for justice in many ways. There is a need for economic justice, which entails participative, distributive, and social justice (Kelso and Adler 2011). This idea calls for the radical redress of an individual or group's position in a social order and emphasizes their right to participate in decision-making within and receive the benefits of the economic institutions that will directly and indirectly affect the quality of their lives. Regarding the history of the Rouge Complex, to what extent did community members willingly become signatories to an economic contract that would reduce their quality of life over several decades? We also cannot separate economic injustice from racism. Patterns of wealth distribution and economic development have often come at the expense of people of color and poor whites, who marginally have enjoyed the privileges of whiteness. Racial justice touches upon the close proximity of minority communities to hazardous waste sites and locations of urban decay. This of course is compounded by the historic and continual oppression of poor black and brown communities, who reside in southwest Detroit and Dearborn pre- and post- the civil rights era and 9/11, in the form of exploitation, marginalization, powerlessness, cultural dominance, and violence (Young 1990, 1999). Often of lesser concern is justice for nonhuman others who share the environment. The ways in which we bring nonhumans into deliberation of environmental concerns reflects what several theorists have referred to as the "domination of nature" through the lens of reductionist science (Shiva 2010, 29). Nature is reduced to an object of study or use. Accordingly, Vandana Shiva (2010) bases the ontological and epistemological assumptions of reductionist science as seeing "all systems as made up of the same basic constituents, discrete, unrelated and atomistic, and [as assuming] that all basic processes are mechanical" (22). Nature becomes a machine that produces for humanity. This sets up a relationship whereby people can take from nature and exploit its resources. Many contend that the conception of nature as object has been particularly violent and destructive, as there exists a "nexus between reductionist science, patriarchy, violence, and profits" (Shiva 2010, 23). A large segment of scientific research has a close relationship with the military industrial complex and private industry. In terms of the environment, science's reduction of nature (forestry and agriculture) as an object of study and exploitation is rarely about human welfare and more about commercial profit. The truth is that threats against nonhumans are threats against us, as we are well aware that assaults against biodiversity and changes in the environment present many unforeseen consequences on human (especially indigenous peoples) and nonhuman communities.

Of course there is also a concern for cognitive justice. This regards the function of knowledge as it traces between scientific, governmental, and lay communities. Concern centers upon the primacy that scientific communities afford to certain methods and languages as means of making knowledge and the purposes that drive scientific inquiry. The same problem characterized as science's domination of nature also functions as an issue of cognitive justice. Scientific research seems to follow a model of science for the sake of science or research for the sake of research. This means that most scientific research is not necessarily driven from the ground up but instead from the perspective of businesses for the purpose of profit, science for the sake of science, or regulatory bodies who have to juggle environmental regulation with economic interests. The latter of the three represents an extreme conflict of interest. We cannot even begin to weigh economic and environmental costs equally because the cost of the latter often take time to manifest, linger for indeterminate periods of time, and present their own sets of economic hardships.

Evelyn Fox Keller articulates that *doing science* requires the adoption of disciplinary languages and the deployment of specific syntactical structures in writing. If you do not have the specialized language required in science, you cannot participate in scientific communities. In this regard, feminist critiques of science have questioned access in scientific communities by pointing to either the relative absence of women and minority scientists or the lack of interaction between scientific communities and citizens within local communities. If we want to ensure justice so that people on the ground can truly benefit and become the driving force behind research, science must realize that "sharing a language means sharing a conceptual universe" (Keller 2001, 136). Therefore, science has to be willing to translate its research into forms that local communities can more readily understand and use for purposes that suit their collective goals. This follows closely with anti-globalization activists' demands that the scientific epistemology of the project be written within the vernacular of the people (Visvanathan 2007, 84). Yet, it is not enough that science translate its work into local vernaculars. The largest problem with science and knowledge is the lack of collaboration between scientific and local communities. Much of this has resulted from science's inability to value local epistemologies. Therefore, community-based/non-scientific methods of knowledge building (e.g., intuition and everyday experience) remain unvalued within science. In critiquing acts of *doing science*, I am articulating that (1) the scientific epistemology of a research project be written and understood in terms of local epistemologies, and (2) science open itself to more collaborative research with local communities (Verchick 2004, 66–69; Visvanathan 2007, 84). The goal is to reform science so that it is not only more participatory but, more important, so that it makes people the center of research. The objective of changing the way we *do science* in pursuit of justice is a larger multidisciplinary goal. The primary focus of this chapter, however, regards what theoretical lenses technical and professional communicators can deploy in order to help solve environmental problems with justice in mind. Theory gives us a language for understanding and navigating the world. Theory is problem-posing and problem-solving in nature. It is impossible to do work as a technical communicator without theories as guides. If we want to address environmental problems in ethical and responsible ways, it makes sense that we train technical communicators to have (1) a theory of the environment, and (2) a theory of justice.

A FEW ENVIRONMENTAL APPROACHES TO JUSTICE

With the story above in mind, I want to dedicate the rest of this chapter toward exploring the affordances that some justice-oriented approaches hold for the work that technical communicators might do in approaching environmental problems. Specifically, I present feminist materialism, feminist political ecology, ecofeminism, and environmental justice. These theories come with a set of affordances and limitations. Despite environmental justice being the only framework that foregrounds *justice* in its name, there is nothing specific that makes the other perspectives less justice-oriented. They all emerge from varying traditions that have different approaches to addressing injustice. Nevertheless, their usefulness in providing insight into how technical communicators can work with and within communities like 48217 highlights their greater utility. So I do want to present them as potential tools for technical communicators by examining them on their own merits. Nevertheless, I also consider their (inter)relationships as a means of working toward an interrelational environmental justice heuristic that assembles their strengths.

FEMINIST MATERIALIST PERSPECTIVE

Feminist materialism (FMP), as a theory, views oppression as a basic reality of women's lives. It is born from the work of feminists who sought not to be considered as organizing under the branches of conventional Marxism or feminisms of difference in the 1970s. Feminism alone presented a problematic embrace of women's essential identity whereas Marxist feminism was insufficient because Marxism had to be augmented in order to articulate the sexual division of labor (Hennessy 1993). It is important to note that the use of *materialism* originates from Marx's notion of historical materialism. Yet, the work of feminists such as Christine Delphy, Stevie Jackson, Rosemary Hennessey, Nicole-Claude Mathieu, Maria Mies, and Iris Marion Young placed patriarchy rather than capitalism as their objects-of-inquiry. These scholars understood historical materialism as a means of understanding relations between the sexes as social in nature and not natural. This view of the world articulates that capitalism is a product of patriarchy rather than an effect. Thus, materialist feminist analyses focus on the social practices of capitalism as they affect women, people of color, and political minorities in the pursuit of social change. In fact, it should be regarded coterminously as a theory and political strategy that confronts social institutions and demands their demolition and the adoption of more equitable and just practices. The landscape of feminist material realities in the present requires us to consider what does it mean to have bodies that are entangled with the physical world. Specifically, how do gender, class, and environmental relations reinforce each other through the logics of domination? To answer this question, we have to adopt a frame of analysis that joins social and ecological relations, as it is social relations that lead to environmental crisis due to "patterns of unsustainability" (Mellor 2000, 108).

This frame addresses how we have culturally come to understand nature and its relations to society. For quite some time it has been convenient to deflect blame for emerging environmental problems upon unexplainable natural forces rather than a product of human material culture. We should look no further than discussions regarding how best to deal with global climate change. Of particular concern regards: who is directly responsible? Who is most affected? And whether science-driven policy is the most effective means of redress? These are moments where communicators need theory, but oftentimes discussions regarding environmental policy become contentious because we lack the language necessary for moving forward. This crisis in language reflects an initial concern of some materialist feminists and cultural materialists, who have insisted on focusing on language because it is a way of thinking and acting in the world that bears material consequences as well as affordances. From the standpoint of FMP, if we can understand and isolate the material constraints of a problem, then we can develop a solution. For example, Sandra Steingraber (2000), in discussing the discursive social production of cancer, highlights the fact that for such a long time we have continued to ignore our immediate environments as contributing to rises in cancer rates (23). Instead, we have chosen to focus exclusively upon individual lifestyles in isolation as contributing to risk. Her argument is that our approach to looking at disease (i.e., cancer) has always been to look at heredity and behavior rather than environment. Her suggestion that we must look at the environment is important because it asks that we interrogate how the material aspects of human culture (specifically created through technological innovation) place burdens upon the environment, which in turn affects our health. Looking at the environment as an actor rather than a scene forces us to consider the larger context under which environmental problems manifest.

FEMINIST POLITICAL ECOLOGY

Political ecology, as one of the dominant areas of research within human and cultural geography studies, joins the concerns of ecology with political economy (Blaikie and Brookfield 1987). This is the study of the continually-shifting dialectic between civil society groups, resource management, and world systems. Specifically, it addresses how global capitalism, conflict, and unequal power relations work to destabilize and transform human-environment interactions. Earlier work in political ecology featured "in-depth environmental histories and examinations of methods of environmental assessment that appear to owe much to established traditions of cultural ecology and ecological science"; however, there has been debate as to what extent ecology has been de-emphasized over the years (Walker 2005, 75). It is the writing of history that punctuates the "political" in political ecology. Stories of how marginalized groups experience environment revises the *we* and *our* that frame mainstream environmental historiography, as the benefits and shortcomings of resource management are unequally distributed (Hornborg 2007). In addition to offering critiques regarding the acceptance of a collective human environmental experience, political ecology raises the voices of marginalized others as it holds true to the idea that "there are very likely better, less coercive, less exploitative, and more sustainable ways of doing things" (Robbins 2012, 20). Political ecologies also engage in the de-liminalization and de-localization of sites by strongly contesting the borders we place around environmental problems. If place is always seen as local, then we can readily ignore how forces beyond the local affect how local conditions are experienced. Engaging in this politics of symmetry (or scale) is an argument that all places are local and are networked to each other, and it offers a foundation to critique power dynamics (Escobar 2001, 2008).

If political ecology is a focus on world systems and the environment, then feminist political ecology (FPE), as an articulation of political ecology, is a confluence of identity, economic, political, and environmental relations into a single framework. As an analytical approach, it does not simply "add gender to class, ethnicity, race, and other social variables as axes of power in investigating the politics of resource access and control and environmental decision-making" (Rocheleau, Thomas-Slayter, and Wangari 1996, 287). The idea here is that these systems are already entangled and mixed to the point where separation would create and legitimate unnecessary hierarchies of oppression. Thus, FPE seeks to redress de-gendered analyses that comprises much of political ecology's history. FPEs understand spaces as historically-gendered and raise the experiential knowledge of women as well as women's participation in political processes as integral to environmental decision-making. Most FPEs have focused on gendered rights to space and access to social power in mostly rural locations of the Global South (Bailey and Bryant 2005; Paulson, Gezon, and Watts 2003). This underscores that there is a need for analysis of gendered-spatial relations in urban locales. Nevertheless, the strength of FPE as an analytical approach is the insistence on breaking down both the urban-rural and post-industrial-developing economy divides that govern production systems as they illustrate connections between the local and the global.

ECOFEMINISM

Ecofeminism (EF) is a set of theories born from radical feminism, which claims that biological determinism, reproductive and maternal roles, the oppression of patriarchy, and the closer connection of women to nature allow women to care about environmental issues more than men. As a branch of feminism, EF has based itself on the phrase "it's all connected," the idea that all forms of oppression are equally linked to nature and patriarchal hegemony. Within the literature on women, environment, and development/women in development (WED/WID), EF ideas frame environmental and women's exploitation as the same since both are considered victims of development (Buckingham 2004; Jackson 1993; Williams and Millington 2004). EF is the most essentialist of any theoretical frame I offer in this chapter, as it strategically engages with western duality. Feminist theories have dedicated much time toward understanding the complex relationships that arise as gender, sexuality, and sex coalesce. They have chosen to organize their theoretical understanding of these relationships under the matrix of domination that they describe as masculine hegemony (or patriarchy). One basic premise in feminism is the assertion that hegemonic masculinity established and maintains many cultures around the globe for its own benefit. Commenting on hegemonic dominance, Richard Twine (2001) states, "[It] is formed through an alliance of position in which a dominant group secures the consent and complicity of others" (7). Masculine hegemony is ubiquitous in many societal power structures both in the United States and around the world. It acts to control nature and women through the establishment of dualism as a schema that drives the logics of western society. These dualistic forces are categories where, when placed against each other, one is privileged (masculinized), and the other is subordinated (feminized). Some examples of oppositional forces include but are not limited to: reason/nature, mind/body, culture/nature, male/female, heterosexual/homosexual and reason/emotion (Gaard 2010; Shiva 2010; Stein 2004).

In terms of dualism, EF asserts that any concept that is not comparable to reason is associated with nature and is subordinated. Western society for centuries has associated women with nature and men with reason; thus, women were subordinated to their male counterparts.

Twine (1997) notes that ecofeminists challenge the dualistic schema by claiming "if hegemonic constructions of masculinity have colonised [sic] what we have come to assume constitutes 'human nature' then a profeminist critique of hegemonic masculinity clearly also involves a simultaneous exposure of the arbitrary definitions of our dominant understanding of 'humanity'" (2). The task of ecofeminists as a whole is to destroy masculine hegemony by attacking the societal structures that enable such hegemony—that is to say—to attack the arbitrary definitions that comprise the foundation of the hegemony. This positions ecofeminism as seeking to craft an identity for men, which holds them as a part of nature in a post-patriarchal society where men are in concordance with nature instead of counter to it.

Under ecofeminism, dualism and essentialism coalesce as a strategy. Accordingly, strategic essentialism stems from postcolonial/decolonial theory and describes the strategies that minority groups adopt to present themselves as a unified group. Though there may be strong differences among members, temporary essentializing is necessary as a means of bringing about certain goals. The group always retains the power and the right to define their essential qualities rather than being defined. Furthermore, when ecofeminists deploy essentialism, they realize that to some extent their essential characteristics are socially constructed rather than natural, but their reliance upon certain dualistic categorical distinctions serves as a politically convenient tool.

Shiva discusses dualism in a way that links women with environment as a means of theorizing oppression. Her idea of the feminine principle produces an embodied and manifested ontology "charaterised [sic] by (a) creativity, activity, productivity; (b) diversity in form and aspect; (c) connectedness and inter-relationship of all beings, including man; (d) continuity between the human and natural; and (e) sanctity of life in nature" (Shiva 2010, 40). She defines the feminine principle as being equal to the non-dominant side of Cartesian dualism. A read of nature via the feminine principle forces us to reconsider how it has traditionally been valued. Cartesian dualism has given primacy to conceive nature as "environment" and "resource," which has produced a cultural relationship that allows people to see themselves as existing outside of nature and has allowed for the exploitation and destruction of natural systems. This opposition between (hu)man(ity) and nature positions nature as passive, separable, and inferior. Yet, Shiva's argument rests upon the questions: what would happen if we chose to value nature in a different way? And what would happen if we chose to see nature from the standpoint of the feminine principle?

ENVIRONMENTAL JUSTICE

Even Environmental Justice (EJ), which is understood as a grassroots community-driven response to threats against community health that disproportionately affect poor people of color, has jostled between those who espouse Dorceta Taylor's (2000, 2002) Environmental Justice Paradigm (EJP) and the Just Sustainability Paradigm espoused by Julian Agyeman, Robert Bullard, and Bob Evans in an edited collection and Michael Jacobs (Agyeman, Bullard, and Evans 2003; Jacobs 1999; see also Agyeman 2005). Despite the existence of different theoretical strands, focus on race and sustainability has always been consistent with organizers who use EJ as a frame. It is the meaningful involvement of people in the development, implementation, and regulation of environmental policy. This is governed by a set of 17 principles drafted and adopted in October 1991 at the First People of Color Environmental Leadership Summit in Washington, DC. At the core of EF is deep engagement and re-evaluation of how it is that we think of research and who is involved in research. It requires that research and inquiry be a collaborative endeavor that reaches across all communities and draws upon various situated knowledges.

We might refer to this as a "democratization of science," which, despite the increasing number of citizen science initiatives, still continues to trouble scientists and academic researchers alike (Bryant 1995). Assumptions regarding the inability of community groups to ascertain scientific processes or the belief that community members are too irrational and emotional have led many in the scientific community to argue for the maintenance of a status quo where policy decision-making is less community and more science-centered. Beyond disagreements concerning data collection or data analysis, the issue of causality is often contentious. What the community presents as a concern is often never a concern for scientists if there is no scientific evidence to support such a concern. Inability to prove causal relationship is a politically expedient way to impede political action. Those concerned with adopting an EJ frame should inquire as to (a) what level of risk is enough to support political action? (b) And to what extent can we build systems of trust between researchers and community members to the point where we can marshal all the available tools in order to address environmental crisis? Some Feminist EJ approaches have raised *intuition* as an answer.

Robert Verchick (2004) describes *intuition* as deriving from the local contexts in which grassroots environmental justice movements grow and thrive (66). This encompasses the everyday experiences situated in local contexts and the values that stem from experiences within those spaces. By situating the language of "space and place" in regards to the environment as being the places where people work and live, EJ causes us to realize that we are not apart from the environment—that the environment is everywhere we go. In traditional scientific research, *intuition* remains invisible during both the research process and the research write-up. Intuitive knowledge built from local contexts are probably the best primers and informants for inquiry as researchers are removed from contaminated/contested sites. In addition, intuition, as a form of contextual reasoning, when coupled with personal stories and empirical data, enables community members to link pollution and forms of discrimination (e.g., environmental racism).

A MORE CRITICAL FOCUS

In presenting these four heuristics, my goal is not to privilege one heuristic over another. In fact, the driving purpose is to offer a sense of the types of theories that could drive the work technical communicators do. It is worth noting that despite their affordances, many of these theories present a set of limitations. It might be more beneficial to martial the best aspects of each theory in order to build a more intersectional and inclusive heuristic. One way to build an intersectional methodology that attends to the emplaced and material aspects of social justice concerns is to bring certain aspects of each theory by focusing on space and place and agency (LaDuke 1999; Nagy Hesse-Biber 2013; Robbins 2012; Schlosberg 2009; Sprague 2005). By raising the idea of space and place, I am focusing the need to understand location in proportion to power within analysis. With respect to agency, I see the necessity to focus on power as it relates to marginalized groups. Are community members seen as having more power to change and affect their lives? Or are they more likely positioned as always exterior to political structures? The emphasis on power is important because it continues Grabill and Simmons's call to continually develop means by which we can push issues of power to the foreground in technical communication practice.

Space and Place

Power is seen more as an abstraction within EF. Since EF's primary focus is on the body and how power enacted upon the body via socially-constructed dualisms has become the very foundation of all present-day cultural relationships, it is more likely to see power as being dispersed everywhere. Therefore, its physical scope of analysis remains unbounded; thus, its interest in examining relationships never moves beyond the theoretical level. While FMP agrees with EF in that power is a lot more widely dispersed unevenly throughout all relationships, in terms of *space and place* it is more grounded in its analysis. FMP is concerned with the material dimensions of all relationships. This is what results in a more localized analysis of power relations within situated locales—the body. FMP differs from EF because it does not rely upon gender, essentialism, and Cartesian dualism as a means of understanding power relations. Specifically, FMP moves away from gender as an organizing principle because it regards gender as just one of many elements that need to be taken into account in order to understand the material effects of uneven power relations (Agarwal 2001; Fowlkes 1997; Jackson 2001; Korovkin 2003; Mellor 2000; Steingraber 2000). We can see FMP as being like EF in that it retains a theoretical quality but has a broader understanding of relationships as it discusses the physical reality of situations (Cornwall, Harrison, and Whitehead 2007; Jackson 1993; Romberger 2007; Shiva 2010). For example, women are not oppressed because they share a natural affinity with the environment. Instead, power enacted upon women's bodies in culture possibly results from their occupying cultural positions (socially-constructed and mediated) that place them as intermediaries between nature and man.

FPE is the more material of the two theoretical apparatuses. Its reliance upon the ecology metaphor forces it to bind its analysis to a problem in a local setting and then determine all of the causal relationships that create the problem. FPE relates to FMP's centering on the material, but it is less of a theoretical construct as it highlights issues on the local level (physical locales) and then tries to make arguments on a global-level based upon its local analysis of situations (Hovorka 2006; Rocheleau 2005a, 2005b; Schroeder 1993; Walker 2005). While FPE does not always focus on specific environmental issues, when it does involve the environment, environment serves more as a backdrop and is not regarded as important as political policy issues. Peter Walker (2005) highlighted that a primary characteristic of many political ecology pieces is that many do not engage in discussions of "biophysical ecology or environmental change in more than a glancing manner" (76). This is a fundamental problem found in Richard Schroeder's (1993) political ecology of agroforestry in Gambia where environmental problems are largely discussed as historical rather than how present-day local politics continue to shape the environmental problems that he historicizes. *Space and place* is always local and always material in EJ movements. This focus is always driven by the needs, demands and goals of citizen groups working to solve problems within their communities.

Agency

We should see *agency* as both a tool deployed by the four movements and/or the ways in which each theory positions oppressed groups in relation to hegemonic power. EF defines *agency* through the use of *strategic essentialism* as both a tool for feminist organization and as a point of reference against hegemonic culture. In terms of oppressed groups, EF offers limited agency to marginalized people as they work against hegemonic power. They are always oppressed and cannot work outside of socially-constructed western dualistic forces. FMP sees itself as a tool for reorganizing the dualistic forces that EF strategically embraces and highlights as the source of oppression. From the standpoint of FMP, if we can understand and isolate what causes a problem then we can possibly develop a solution. Again, Steingarber's work on the social production of cancer raises questions that dramatically change the way in which we understand the material effects of culturally-driven scientific development on both the environment and (more important) our bodies.

In terms of *agency*, FPE is less of a tool than the other movements, because it is more of an analysis of how groups have solved problems and changed material circumstances. Its focus on groups, however, provides more agency for them to manipulate systems and enact change. I draw this understanding of power via Diane Rocheleau (2005a) who argued that political ecologies are more pragmatic in their view of power/agency as they "consider more entangled and embedded workings of *power alongside, power under, power in spite of,* and *power between*" in their post-analysis of material situations (84; original emphasis). This works outside of certain feminist circles that embrace essentialism and falls more in line with feminist poststructuralist views of feminine agency within cultural circles. In terms of EJ, groups on the ground always have agency because their movements are driven by their own goals and experiences. The only limits placed upon their agency stems from unequal power relations between themselves and either corporate entities and/or government agencies.

BRINGING IT ALL TOGETHER

Thinking about the strengths and weaknesses of each theory, I want to push for a heuristic that takes on the FMP, EF, and EJ focus emphasis on the local but also continues to oscillate between the local and sites beyond the local as a means of understanding and solving environmental problems, much like FPE. Moreover, this is a heuristic that is not simply concerned with agency with respect to how communities are disempowered, but a generative activity that identifies possibilities whereby people can be empowered. This is an attempt to move outside of the analytical space of each theory in favor of action. Here I am asking that we trace associations and interrogate the ways people, places, and things are linked as a means of understanding place. Specifically, I am talking about the importance of context. As Kristen Moore (in this collection) articulates, context matters in that it creates the conditions to act in response to social situations, like the creation of genres (254). This means that rather than having predetermined notions of place driven by theory, it is the conditions of sites that form the basis of our theory and action. This requires that we map complexity in a way that engages local and global concerns by continuously asking how issues of gender, race, class, sexuality, and ability shape the way environments look and relate to other spaces and even how spaces often are designed in ways to produce specific subject positions.

With the latter I am thinking of Langdon Winner's classic study of the architect Robert Moses's designs. Winner posited that it was not so much the *what* regarding structures but the *where*. Specifically, when attending to the idea that artifacts carry and conduct politics, "what matters is not technology itself, but the social or economic systems in which it is embedded" (Winner 1989, 25). Moses was responsible for constructing many public works, which included parks, roads, and bridges, from the 1920s to the 1970s. This also included the construction of overpasses in Long Island, New York, that presented exceptionally-low clearance (e.g., as low as nine-feet in some places) for twelve-foot buses. This resulted in the restriction of buses and more importantly people from being able to move freely about the city. The design was purposeful in that it reflected Moses's socio-economic biases against the working-poor and ethnic minorities. Now Moses was neither the first nor the only person to ever to design racist and classist infrastructure. Many urban laws about parking cars overnight, food-sharing limits, or the construction of parking benches with handrails are anti-homeless disciplinary technologies, although their architects might not freely admit such aims (Lockton, Harrison, and Stanton 2008). Moses died in 1981; however, he left behind an enduring legacy in that his structures continue to practice his ideology into the present.

I want to take some time and space and shift to a pedagogical moment that should ground my idea a little further. In Fall 2012, I taught a course entitled, "Nature, Environmental, and Travel Writing." The course is housed in Michigan State University's (MSU) Professional Writing program, but it is also cross-listed as an intensive reading/

writing experience for Fisheries and Wildlife students. Although I was a graduate student, I was given free rein to completely revamp the course. I wanted to re-design the course so that my students and I could trouble what environmental, nature, and travel writing are as genres, but also interrogate the often unexamined cultural positions that these writers rely upon to create a view of nature. Earlier constructions of the course focused on traditional nature writers like Aldo Leopold, Charles Darwin, and John Muir. I did not want to make a course where we simply stylistically analyzed these writers' texts. Instead, I wanted to design a course that would allow students to interrogate how writers like these are designing an experience for their audiences that is reliant upon distinct cultural positions, be they tacit or acknowledged. This worked doubly because it also forced students to consider the systems that they use while writing and even how they understood environment. Whether we read non-fiction, travel guides, or environmental impact statements, I consistently repeated that professional writing (especially in relation to environmental issues) is about designing space and that there are affordances and consequences that come with our spatial practices as writers. Aside from the course being a survey of genres, my vision was to encourage students to realize that as professional writers they are designing experiences for audiences. As Elise Verzosa Hurley writes in this collection, "When translated to classroom settings where students all too often perceive technical communication as a neutral and objective practice, a singular, closed view of space limits the possibility for students to understand how the work they do in our classrooms are tied to knowledge work in a multiplicity of other spaces beyond the university" (94). How people (physically/cognitively) move within and between spaces can reveal how they relate to spaces, other people, and things. Here I posit that engagement with spaces and the stories that populate them present opportunities for environmental rhetorics to make critical interventions in both cultural and political realities.

For one unit, I took my students to an exhibit in the MSU Museum's Heritage Gallery titled, "Echoes of *Silent Spring*: 50 Years of Environmental Awareness." We had just read a couple chapters from Carson's book, as well as writing from her critics, and the work of others who had analyzed the entire controversy surrounding the book. The purpose of the exhibit was to examine the larger ecology of *Silent Spring* by situating it within the context it was written and its effects subsequent to publication. Adding another layer of complexity to the exhibit is the connection between the book's content and the university. Her most famous case study examined how *dichlorodiphenyltrichloroethane* (DDT), which was used on campus to fight the spread of Dutch elm disease, indirectly caused the deaths of birds, particularly robins. This was primarily facilitated by the efforts of MSU ornithologist George Wallace and his students, who collected, documented, and tested dead and dying birds for the insecticide.

For my students, there was a rich layering of stories and histories in the exhibit. Within the confines of this space, multiple actors and their worlds emerged to connect and either bolster particular realities or lay waste to others. This presented a great opportunity for the students to reflect upon the relationship between the spaces of texts to the physical world. Literally, how is it that the exhibit's designers make bridges to create meaning and facilitate understanding, especially as they must navigate complexity that frames the controversy surrounding Carson's work? The ultimate goal was twofold: (1) to recognize that we can critique constructions of place at multiple levels of abstraction and that such analytical moves are a necessity; and most important, (2) to make inferences into how writing can allow us to remake place and offer possibilities for action. Here I believe that if the act of questioning space and arrangement is the analysis of the dynamics of culture through the understanding of power via race, gender, economic, and ableist privilege, then attempts to reorder relations through rhetoric and writing become the work of justice—environmental justice.

ENVIRONMENT, JUSTICE, AND TECHNICAL COMMUNICATION

As Nancy Blyler (1998) so eloquently articulated, "Our discipline has acknowledged that professional communication is closely allied with science, technology, and business" (289). Training in technical and professional communication studies has always been geared toward helping students best prepare for the professional, personal, and contextual communicative situations that arise with a diverse group of stakeholders. The question is whether students will continue to have the right tools to address wicked problems. For this reason, it makes sense to continually introduce theoretical perspectives that might shape the work they do as part of environmental organizations working in communities like 48217, as empowered citizens working for change within their communities, or even change-agents working within institutions.

Regardless of the location, rhetoric remains a means (tactics/tools) whereby people come together to solve localized problems in movement that frequently oscillates between local and global foci. Within human and cultural geography studies there has been an increased focus toward looking at theories of human/environment interactions and their usefulness in addressing social change in relation to cultural politics and deliberative mechanisms. How people deliberate and who has access to these tools are important locations of inquiry as we both try to understand how culture means and resolves civic problems. Many of the heuristics presented above are geared toward achieving cognitive justice in the pursuit of environmental justice. Just policy requires more local participation and knowledge. There is an increasing role for technical communicators to serve as advocates or community organizers. They might even serve in the capacity of creating technical interventions that bridge spatial differences across many divides. For example, one might co-design community asset maps that might facilitate better collaboration within the community or between the community and researchers. Yet, this type of transformative work cannot take place unless we continue to provide students with a broad range of heuristics.

References

Agarwal, Bina. 2001. "Participatory Exclusions, Community Forestry, and Gender: An Analysis for South Asia and a Conceptual Framework." *World Development* 29 (10): 1623–1648. https://doi.org/10.1016/S0305-750X(01)00066-3.

Agboka, Godwin. 2013. "Participatory Localization: A Social Justice Approach to Navigating Unenfranchised/Disenfranchised Cultural Sites." *Technical Communication Quarterly* 22 (1): 28–49. https://doi.org/10.1080/10572252.2013.730966.

Agyeman, Julian. 2005. *Sustainable Communities and the Challenge of Environmental Justice*. New York: NYU Press.

Agyeman, Julian, Robert Bullard, and Bob Evans, eds. 2003. *Just Sustainabilities: Development in an Unequal World*. Cambridge, MA: MIT Press.

Bailey, Sinead, and Raymond Bryant. 2005. *Third World Political Ecology: An Introduction*. New York: Routledge.

Blaikie, Piers, and Harold Brookfield. 1987. *Land Degradation and Society*. London, UK: Methuen.

Blyler, Nancy. 1998. "Taking a Political Turn: The Critical Perspective and Research in Professional Communication." *Technical Communication Quarterly* 7 (1): 33–52. https://doi.org/10.1080/10572259809364616.

Bryant, Bunyan. 1995. "Issues and Potential Policies and Solutions for Environmental Justice: An Overview." In *Environmental Justice: Issues, Policies, and Solutions*, ed. Bunyan Bryant, Roger Bezdek, Deeohn Ferris, and Jamal Kadri, 8–34. Washington, DC: Island Press.

Buckingham, Susan. 2004. "Ecofeminism in the Twenty-First Century." *Geographical Journal* 170 (2): 146–154. https://doi.org/10.1111/j.0016-7398.2004.00116.x.

Cornwall, Andrea, Elizabeth Harrison, and Ann Whitehead. 2007. "Gender Myths and Feminist Fables: The Struggle for Interpretive Power in Gender Development." *Development and Change* 38 (1): 1–20. https://doi.org/10.1111/j.1467-7660.2007.00400.x.

Escobar, Arturo. 2001. "Culture Sits in Places: Reflections on Globalism and Subaltern Strategies of Localization." *Political Geography* 20 (2): 139–174. https://doi.org/10.1016/S0962-6298(00)00064-0.

Escobar, Arturo. 2008. *Territories of Difference: Place, Movements, Life, Redes (New Ecologies for the Twenty-First Century)*. London: Duke University Press. https://doi.org/10.1215/9780822389439.

Fowlkes, Diane L. 1997. "Moving from Feminist Identity Politics to Coalition Politics through a Feminist Materialist Standpoint of Intersubjectivity in Gloria Anzaldua's *Borderlands/La Frontera: The New Mestiza*." *Hypatia* 12 (2): 105–24.

Gaard, Greta. 2010. "New Directions for Ecofeminism: Toward a More Feminist Ecocriticism." *Interdisciplinary Studies in Literature and Environment* 17 (4): 643–665. https://doi.org/10.1093/isle/isq108.

Grabill, Jeffery T. 2005. "Globalization and the Internationalization of Technical Communication Programs: Issues for Program Design." *International Professional Communication Conference.* IPCC 2005. Proceedings. International, 373–78. https://doi.org/10.1109/IPCC.2005.1494200.

Grabill, Jeffery T., and Michelle W. Simmons. 1998. "Toward a Critical Rhetoric of Risk Communication: Producing Citizens and the Role of Technical Communicators." *Technical Communication Quarterly* 7 (4): 415–441. https://doi.org/10.1080/10572259809364640.

Haas, Angela. 2012. "Race, Rhetoric, and Technology: A Case Study of Decolonial Technical Communication Theory, Methodology, and Pedagogy." *Journal of Business and Technical Communication* 26 (3): 277–310. https://doi.org/10.1177/1050651912439539.

Hennessy, Rosemary. 1993. *Materialist Feminism and the Politics of Discourse*. New York: Routledge.

Hornborg, Alf. 2007. "Introduction: Environmental History as Political Ecology." In *Rethinking Environmental History: World-System History and Global Environmental Change*, ed. Alf Hornborg, J. R. McNeill, Joan Martinez-Alier, 1–26. Plymouth, UK: Altamira.

Hovorka, Alice J. 2006. "The No. 1 Ladies' Poultry Farm: A Feminist Political Ecology of Urban Agriculture in Botswana." *Gender, Place and Culture* 13 (3): 207–225. https://doi.org/10.1080/09663690600700956.

Jackson, Cecile. 1993. "Doing What Comes Naturally? Women and Environment in Development." *World Development* 21 (12): 1947–1963. https://doi.org/10.1016/0305-750X(93)90068-K.

Jackson, Stevi. 2001. "Why a Materialist Feminism is (Still) Possible—and Necessary." *Women's Studies International Forum* 24 (3-4): 283–293. https://doi.org/10.1016/S0277-5395(01)00187-X.

Jacobs, M. 1999. "Sustainable Development as a Contested Concept." In *Fairness and Futurity: Essays on Environmental Sustainability and Social Justice*, ed. Andrew Dobson, 21–45. Cambridge: Oxford University Press. https://doi.org/10.1093/0198294891.003.0002.

Keller, Evelyn Fox. 2001. "Gender and Science: An Update." In *Women, Science and Technology: A Reader in Feminist Science Studies*, ed. Mary Wyer, Mary Barbercheck, Donna Geisman, Hatice Öztürk, and Marta Wayne, 132–42. New York: Routledge.

Kelso, Louis O., and Mortimer J. Adler. 2011. *The Capitalist Manifesto*. 3rd ed. Whitefish, MT: Literary Licensing.

Korovkin, Tanya. 2003. "Cut-Flower Exports, Female Labor, and Community Participation in Highland Ecuador." *Latin American Perspectives* 30 (4): 18–42. https://doi.org/10.1177/0094582X03030004005.

LaDuke, Winona. 1999. *All Our Relations: Native Struggles for Land and Life*. Cambridge, MA: South End Press.

Lamb, Jonathan, and Katie Koster. 2012. "Severstal Timeline of Non Compliance Issues." [Memorandum] Michigan Department of Environmental Quality. August 14, 2012.

Lockton, Dan, David Harrison, and Neville Stanton. 2008. "Design with Intent: Persuasive Technology in a Wider Context." Persuasive Technology: Third

International Conference, PERSUASIVE 2008, Oulu, Finland, June 4–6, 2008, ed. Harri Oinas-Kukkonen, Per Hasle, Marja Harjumaa, Katarina Segerståhl, and Peter Øhrstrøm, 274–78. Berlin: Springer-Verlag. https://doi.org/10.1007/978-3-540-68504-3_30.

McArdle, John. 2011. "Health Worries Stalk Neighborhoods in Detroit's 'Sacrifice Zone.'" *The New York Times*, September 12, 2011.

Mellor, Mary. 2000. "Feminism and Environmental Ethics: A Materialist Perspective." *Ethics and the Environment* 5 (1): 107–123. https://doi.org/10.1016/S1085-6633(99)00026-1.

Nagy Hesse-Biber, Sharlene, ed. 2013. *Feminist Research Practice*. Thousand Oaks, CA: Sage Publications.

Paulson, Susan, Lisa L. Gezon, and Michael Watts. 2003. "Locating the Political in Political Ecology: An Introduction." *Human Organization* 62 (3): 205–217. https://doi.org/10.17730/humo.62.3.e5xcjnd6y8v09n6b.

Robbins, Paul. 2012. *Political Ecology: A Critical Introduction.* 2nd ed. Sussex, UK: John Wiley and Sons.

Rocheleau, Diane. 2005a. "Maps as Power Tools: Locating Communities in Space or Situating People and Ecologies in Place?" In *Communities and Conservation: Histories and Politics of Community-based Natural Resource Management*, ed. Peter J. Brosius, Anna Lowenhaupt Tsing, and Charles Zerner, 327–62. Oxford: Altamira Press.

Rocheleau, Diane. 2005b. "Political Landscapes and Ecologies of Zambrana-Chacuey: The Legacy of Mama Tingo." In *Women and the Politics of Place*, ed. Wendy Harcourt and Arturo Escobar, 72–85. Bloomfield, CT: Kumarian Press.

Rocheleau, Diane, Barbara Thomas-Slayter, and Esther Wangari 1996. "Feminist Political Ecology: Crosscutting Themes, Theoretical Insights, Policy Implications." In *Feminist Political Ecology: Global Issues and Local Experiences*, ed. Diane Rocheleau, Barbara Thomas Slayter, and Esther Wangari, 287–307. New York: Routledge.

Romberger, Julia. 2007. "An Ecofeminist Methodology: Studying the Ecological Dimensions of the Digital Environments." In *Digital Writing Research: Technologies, Methodologies, and Ethical Issues*, ed. Heidi A. McKee and Danielle Nicole DeVoss, 249–67. Cresskill, NJ: Hampton Press.

Savage, Gerald, and Kyle Mattson. 2011. "Perceptions of Racial and Ethnic Diversity in Technical Communication Programs." *Programmatic Perspectives* 3:5–57. http://www.cptsc.org/pp/vol3-1/Savage&Mattson_3-1.pdf.

Schlosberg, David. 2009. *Defining Environmental Justice: Theories, Movements, and Nature.* New York: Oxford University Press.

Schroeder, Richard A. 1993. "Shady Practice: Gender and the Political Ecology of Resource Stabilization in Gambian Garden/Orchards." *Economic Geography* 69 (4): 349–365. https://doi.org/10.2307/143594.

Shiva, Vandana. 2010. *Staying Alive: Women, Ecology, and Development.* Cambridge, MA: South End Press.

Sprague, Joey. 2005. *Feminist Methodologies for Critical Researchers: Bridging Differences.* Oxford: AltaMira Press.

Stein, Rachel. 2004. "Introduction." In *New Perspectives on Environmental Justice: Gender, Sexuality, and Activism*, ed. Rachel Stein, 1–20. New Brunswick, NJ: Rutgers University Press.

Steingraber, Sandra. 2000. "The Social Production of Cancer: A Walk Upstream." In *Reclaiming the Environmental Debate: The Politics of Health in a Toxic Culture*, ed. Richard Hofrichter, 19–38. Cambridge, MA: MIT Press.

Tanner-White, Kristi, and Tina Lam. 2010. "Database: Toxic Zip Code Rankings." *Detroit Free Press*, June 20, 2010.

Taylor, Dorceta. 2000. "The Rise of the Environmental Justice Paradigm." *American Behavioral Scientist* 43:508–80.

Taylor, Dorceta. 2002. *Race, Class, Gender, and American Environmentalism.* Gen. Tech Rep. PNW-GTR-534. Portland, OR: US Department of Agriculture, Forest Service, Pacific Northwest Research Station. https://doi.org/10.2737/PNW-GTR-534.

Twine, Richard. 1997. "Masculinity, Nature, Ecofeminism." *Ecofeminism e-journal.*

Twine, Richard. 2001. "Ecofeminism in Process." *Ecofeminism e-journal.*

Verchick, Robert. 2004. "Feminist Theory and Environmental Justice." In *New Perspectives on Environmental Justice: Gender, Sexuality and Activism,* ed. Rachel Stein, 63–77. New Jersey: Rutgers University Press.

Visvanathan, Shiv. 2007. "Knowledge, Justice, and Democracy." In *Science and Citizens: Globalization and the Challenge of Engagement,* ed. Melissa Leach, Ian Scoones, and Brian Wynne, 83–94. New York: Zed Books.

Walker, Peter. 2005. "Political Ecology: Where is the Ecology?" *Progress in Human Geography* 29 (1): 73–82. https://doi.org/10.1191/0309132505ph530pr.

Williams, Collin C., and Andrew C. Millington. 2004. "The Diverse and Contested Meanings of Sustainable Development." *Geographical Journal* 170 (2): 99–104. https://doi.org/10.1111/j.0016-7398.2004.00111.x.

Winner, Langdon. 1989. *The Whale and the Reactor: A Search for Limits in an Age of High Technology.* Chicago: Chicago University Press.

Young, Iris Marion. 1990. *Justice and the Politics of Difference.* Princeton, NJ: Princeton University Press.

Young, Iris Marion. 1999. "Residential Segregation and Differentiated Citizenship." *Citizenship Studies* 3 (2): 237–252. https://doi.org/10.1080/13621029908420712.

PART III

Interfacing Public and Community Rhetorics with Technical Communication Discourses

7
STAYIN' ON OUR GRIND
What Hiphop Pedagogies Offer to Technical Writing

Marcos Del Hierro

Every year during Black Entertainment Network's (BET) Hip Hop Awards, one of the most popular segments is "The Cipher," where old school legends, prolific rappers, and young up-and-comers participate in rap circles, commonly known as "ciphers," where they show off their talents and skills. Although these segments are glamorized for television, they pay tribute to one of hiphop culture's most important practices. Most of these ciphers are filmed in locations meant to depict old warehouses, parks, and other public spaces where many hopeful rappers show up to challenge the best on the block, to refine their skills, and grow their reputations. As an art form, the cipher shows what it takes for rappers to become experts in their craft, but this practice also creates a workspace. As a nontraditional workspace, it offers a different perspective of how organizational cultures conduct communication, writing, and rhetoric. It also offers an opportunity to see how factors like race, ethnicity, and culture influence workspace communication.

This essay argues for hiphop practices as valuable contributors to the field of technical communication, specifically in the ways hiphop challenges the field to imagine technical workspaces and communication beyond dominant conceptions. First, I will situate this essay among scholars who have been challenging the disciplines of technical communication and rhetoric and composition to recognize why culture, social justice, and decoloniality matter. Next, this essay will look at the role of culture in technical workspaces and how hiphop complicates what counts as both technical and a workspace. We will look particularly at hiphop ciphers and what I call "hiphop digital booklets" as examples of technical communication in hiphop. Finally, in joining the call for radical pedagogies in technical communication, we will look at how hiphop can be applied to the classroom. Hiphop culture grew as a response to systemic racial and colonial oppression by young people of color in postindustrial New York City. While many scholars talk about the artistic contributions hiphop made to speaking out and fighting racism, hiphop can do and does more. This essay offers scholars interested in decolonial approaches in general, and hiphop approaches in particular, accessible pedagogies that recognize why power, culture, and social justice matter in technical communication.

DECOLONIAL SCHOLARSHIP IN TECHNICAL COMMUNICATION

The value of decolonial critique and other theories rooted in social justice matter to the field and our students because they challenge the field to broaden its understanding of how technical communication happens according to cultural context. Technical communicative practices coming out of non-Western traditions, such as the hiphop examples and methods I will present later in this chapter, are seen as irrelevant to dominant understandings of the field, and that is a major problem for scholars and students coming from non-Western communities. Scholarship on rhetoric and technology often ignores culture and difference as a focus because Western discourses about technology inadequately address the role ideological agendas play in the design, use, and dissemination of technology. The myth that scientific critical distance affords technology the privilege of objectivity argues that it is impossible for issues such as race and gender to influence how technologies are designed, used, and produced. The dominant history of technical communication argues it emerged during the late nineteenth and early twentieth centuries out of the desire to move the field of engineering out of the apprenticeship system and into university classrooms coupled with communication needs of the US military during World Wars I and II (Kynell and Tebeaux 2009, 112). For instance, Kynell and Tebeaux offer a history of the founders of both technical communication as a discipline and its most powerful professional association. They describe early technical writing teachers and scholars as honest people who were invested in objectivity, practicality, and pedagogy. They cite John Harris, the first president of Association of Teachers of Technical Writing (ATTW), who describes the founders as "old hands" who "were diligently apolitical" when it came to their work in the field (112). Harris's portrayal makes a rhetorical move to categorize these individuals as objective, impartial, and scientific. Harris, and many in the field, see no place for politics, although that stance is a kind of political stance. Kynell and Tebeaux further establish the character of these early technical rhetoricians when they write, "These were people who believed that a reality that required descriptions of mechanisms, operations and instructions manuals, and proposals to launch projects did exist and could be captured in writing" (Kynell and Tebeaux 2009, 123–24). This statement makes the assumption that technical writing rhetorics and traditions capable of communicating technical information did not exist prior to the founders of ATTW and that technical communication needed to be "captured" and theorized by scholars in order to legitimize the field.

This act of using Western methods to "capture" the rhetorics of technology follows in the Western, colonial tradition of civilizing what is considered wild and/or untamed, often through narratives of discovery. Technical communication scholar Angela Haas (2008), through her analysis of John Gast's 1872 painting, *American Progress,* provides a persuasive example of how manifest destiny informs Western discourses about technology and technological literacies by demonstrating how all non-Western people and their technologies are relegated outside of civilization until they are reimagined through the West. In the painting, an angel symbolizing progress flies westward while illuminating land for white settlers. The angel's path extends the frontier by pushing Native Americans, animals, and darkness farther west. Behind the angel and settlers are signs of Western civilization made visible through symbols of technology, such as trains, ships, and housing. The angel also holds electric wire connected to a network of telegraph poles. Haas states, "Thus, God blesses only the invaders of the Americas with the cutting edge of technology and all others are left in the darkness of technological illiteracy and primitivity" (52). The land is not considered "civilized" until marked by Western technology.

The rhetorics sponsored by the work of Kynell, Tebeaux, and Gast are representative of how colonialism pervades through our society in general and technical communication in particular. Traditional curricula development and textbook options, as well as hegemonic histories of the history of the discipline and allied fields, contribute to a narrative of Western progress that moves linearly, tossing behind what it considers outdated. As this locomotive of technological history moves across the academic landscape, those traditions standing in its way are run over and/or abandoned. Whether bodies are forgotten, ignored, and erased does not matter as long as the train runs to the future.

Activist scholars such as Angela Haas, Adam Banks, and others challenge technical communication fields to decenter the West. Ignoring non-Western contributions to technical communication reinforce systemic colonialism, racism, and white supremacy in both the field and the classroom. Haas makes a call for decolonizing technical communication, particularly through Native rhetorics. She argues that "contemporary dig/viz rhetoric scholarship is currently positioning the field to make an interesting decolonial move" (Haas 2008, 84). Haas states that instead of defining dig/viz rhetoric through one rhetorical history, we must make room for listening to many traditions. Haas, through her analysis of indigenous digital rhetorics, dispels notions that the technologies and rhetorics employed in digital spaces are new. Rather, they are rhetorically placed in a line of progression that is inherent in Western frames of thought and history. Haas makes an important call toward a more inclusive approach to digital rhetorics that, "resist[s] the dominant notions of what it means to be technologically 'literate' or 'advanced' (with roots in manifest destiny), and to critically reflect on struggles for and engage with discussion on dig/viz rhetorical sovereignty . . ." (Haas 2008, 109).

She points out that traditional digital and visual rhetoric studies (dig/viz) continue privileging alphabetic texts while the fields claim to center digital and visual mediums. This preoccupation with the written word reinforces colonial legacies established by Europeans who felt that alphabetic texts represented ways to establish, maintain, and follow ideas considered to be static and eternal. The supposed preservation of history and ideas through the technology of books also justified the rendering of nonalphabetic communities and cultures as primitive. Rather than understand rhetorical practices and technologies as fluid and dynamic in accordance to cultural context, Western discourses often establish the need to standardize communication by disciplining non-Western rhetorics and bodies to fit the needs of the colonizer. For all these reasons, non-white subjects stand in a marginalized position when it comes to discourses about, and implementation of, technology as it is understood in the academy and in broader society.

Haas and other decolonial scholars in both rhetoric and composition and technical communication studies are influenced by the work of Latin American Studies scholars Elizabeth Hill Boone and Walter Mignolo, who discuss how Western epistemologies replaced and limited non-Western communication and rhetorical traditions as part of colonization. According to Mignolo, European colonizers never understood the value of nonalphabetic communication because the written word is fetishized in the west. He writes that the spread of Western rhetorics in the Americas "was also a massive operation in which the materiality and ideology of Amerindian semiotic interactions were intermingled with or replaced by the materiality and ideology of Western reading and writing cultures" (Boone and Mignolo 1994, 76). Elizabeth Hill

Boone talks about the use of signs as important and efficient communication that can make rhetorical moves that alphabetic texts could not. In *Writing without Words,* Boone defines writing as "the communication of relatively specific ideas in a conventional manner by means of permanent, visible marks" (Boone and Mignolo 1994, 15). Her definition functions as a move recognizing Western writing theories as colonial because they narrowly frame histories of writing as the evolution from the use of images to alphabetic texts.

Decolonizing technical communication matters because it has material consequences for marginalized peoples. African American scholar Adam Banks argues that technology should exist at the center of social justice issues, rhetoric and composition, and technical communication for black folks because "technologies are the spaces and processes that determine whether any group of people is able to tell its own stories on its own terms" (Banks 2008, 10). He argues technology is a pervasive factor across an exhaustive list of social, political, and economic issues such as unemployment, the prison industrial complex, inequalities in the education system, and the digital divide. In a decolonial and anti-racist move, more scholars need to center technology as it relates to culture if we are to continue making spaces that honor non-Western rhetorical traditions and ways of knowing.

Adam Banks echoes Haas's call when he states:

> Any attempt to foster meaningful access to communication technologies or to a working education system must include theoretical frameworks or conceptual models that build from the traditions and truths of a people and assume their agency and ability. Black people must see themselves in the digital story. (Banks 2011, 5)

Banks refers to the ways that black folks have been marginalized by digital technologies and rhetorics because dominant understandings about both inherently ignore them. I would extend this argument to Latinx and Native communities. He also makes a rhetorical move that emphasizes agency as an important factor in influencing discourses about technology. Haas and Banks both call for people of color to "see themselves in the digital story," which, under Haas's decolonial move to culturally situate what counts as digital and technological, listens to traditions continued and established through hiphop digital rhetorics. Later, I will demonstrate how young people coming from underrepresented communities can "see themselves" in technical communication through hiphop methods. In order to fully appreciate the social justice impact hiphop can make on students and the field, we must first take a moment to understand how culture produces and affects workspaces.

CULTURE AND WORKSPACE

The time is right for technical communication to revisit our interest and investment in better understanding the relationships between culture and workspaces. In fact, the lack of scholarship paying attention to culture is doing a disservice to our students. Jim Henry (2012) argues that technical communicators need the intellectual flexibility to understand and adapt to different organizational cultures, especially since workforce turnover is much higher than it has been in other eras. Henry urges technical communicators to see themselves as "cultural analysts" (76). Henry believes his heuristic offers a "bottom-up" method of analyzing organizational culture that is more effective than using "top-down" approaches. His heuristic involves composing field notes, interviewing organization members, collecting and analyzing artifacts, and using all this information when writing. Henry advises creating memos. He states, "Think of these memos as a *working paper* to which you can add as you go and that will serve as an invaluable documentation of your own enculturation at your place of work" (88; emphasis in original). Henry calls to mind that organizations are not uniform and neither are their individual cultures. Factors like turnover ensure that an organization's culture will always be in flux, which means technical communicators must be ready and equipped to adapt to those changes.

Making the shift of paying more attention to culture invites more theory into the conversation, particularly theories paying attention to power, privilege, and difference. This may cause anxiety for some and add tension to the relationship between industry and scholarship. However, former presidents of the Society of Technical Communicators and the Association of Teachers of Technical Writing urge technical communication to put theory and practice into conversation to bridge the historical chasms between the professions and the discipline toward our shared core competencies—both of which have implications for workplace cultures. For example, Hayhoe (2003) calls for academics and practitioners to set aside their differences and preconceived notions in order to fill each other's gaps and create grounds for innovation. He offers the perspective of someone with extensive experience as both a professor and a technical writer, and he knows the stereotypes of "academic snobs" and "industrial rednecks" that each side imagines the other to be (Hayhoe 2003, 102). Hayhoe proposes ten steps that lead to mutual economic, professional, and social benefits. His first step, "Don't Be Afraid of Contamination" underpins the rest, and it suggests that we should not be afraid of cross-cultural contamination, cross-gender contamination, cross-class contamination, cross-disciplinary contamination, and more.

Bill Hart-Davidson (2001) positions technical communicators as the most qualified members of an organization or business who best negotiate experimentation, collaboration, abstraction, and system thinking, but argues that we need theory in order to do so more fully. Hart-Davidson uses Jacques Derrida's theories on writing to show how technical communicators are rhetoricians who understand how sign technologies undergird the use of other technologies. In looking at experimentation, collaboration, abstraction, and system thinking, Hart-Davidson writes, "Each requires an attention to work practices such that day-to-day and highly situated activities are reflected on and are represented so that they can be improved and reemployed in future situations" (151). "Who should look out for those flexible strategies that can be noticed, recorded, refined, and redeployed to make work practices—or products meant to enhance work practices—better?" (151). Hart-Davidson's answer: technical communicators. In striving to find the balance between developing communication and technologies that account for the fluidity of identities, practicality, and context, he challenges us to think beyond the concept of practicality as always striving to serve particular majorities.

This argument is powerful in challenging technical communication to recognize diversity because practicality does not have to mean white. It does not have to mean male. And it does not have to mean privileged. Thus, we must be mindful of how we describe the academy and industry and who gets to participate in those roles. The snobby academic is a stereotype, and it is one that is often perceived as a privileged white man, while the "industrial rednecks" are also white men. Where does everyone else fit into this conversation?

Hart-Davidson draws upon Bonnie Nardi and Vicki O'Day's concept of "gardeners" (Hart-Davidson 2001, 54). Hart-Davidson explains, "Gardeners translate ideas and processes to make continuous improvements to workplace practice," and he argues their methods fall right in line with technical communicators' skills. He concludes that because they are trained to write and communicate effectively, "technical communicators are likely to be the ones who help a design team make the most of its own diversity, in terms of domain-specific expertise, by enabling cross-functionality and process efficiency whenever possible" (Hart-Davidson 2001, 154).

If we can recognize that culture matters, both existing in an organization and those held by individual members, then technical communication must recognize that workspaces are populated by different kinds of bodies and subjectivities that complicate what a workspace is, how it functions, and its particular technologies. This recognition of culture also means valuing social justice principles, such as inclusivity, and opening the field to what and who counts as technical communication and communicators on both the academic and industry ends. This means that the faces in conference presentations, classrooms, organizations, and internships need to diversify to reflect broader understandings of technical communication. Finally, if technical communicators are going to work as gardeners, what they learn in the classroom matters tremendously. They will foster better organizations if they are better equipped to understand changes in the field as well as theories, methods, and practices that move beyond the traditional base of the field.

One way to do this is through hiphop.

WHY HIPHOP MATTERS TO TECHNICAL WRITING

There is plenty of fertile ground for scholars interested in the relationship between hiphop and technical communication because hiphop's birth largely depended on technology. One only needs to hear Afrika Bambaataa's "Looking for the Perfect Beat" to hear the futuristic, robotic sounds that early hiphop explored. DJ techniques like scratching and sampling show the ways that technology was essential toward creating hiphop music. DJs adapted record players into musical instruments as both an innovation and a response to the lack of music programs and musical instruments in poor communities that also allowed for an expressive vehicle that shared all kinds of information. These rhetorics have had a profound societal impact, including in how people practice rhetoric. When Jim Ridolfo and Dánielle Nicole DeVoss argue that, "Remix is perhaps the premier contemporary composing practice" (Ridolfo and DeVoss 2009, n.p.), I would say hiphop played a major part. Their concept of "rhetorical velocity," or how people compose with an understanding and anticipation of recomposition by others, shows how digital communicative practices are profoundly impacted by hiphop rhetorics.

Hiphop scenes during the 1970s and 1980s developed communication methods reflective of their social and material realities, which included having to survive the impact of systematic racism and colonialism. Just because they were not working in traditional settings did not mean they had no use for technical communication. DJs and rappers needed to attract audiences to the next performance, whether it was in a park, someone's basement, or a dance club. Gang culture taught posses and crews to adopt mutual methods of communication to establish territorial lines and protocols for beefs and battles. Graffiti artists built art networks by making do with existing transportation grids to create moving canvases. This happened in the context of the post-industrial city, where the hiphop generation was being criminalized and discriminated against by the government, the police, and other citizens. Extending from hiphop culture's ability to make do, communication methods and practices developed that not only gave voice to a youth subculture and resistance movement, but it also demanded an understanding of all aspects of communication and meaning making that affected their goals, practices, and daily lives. Again, just because they were not in traditional workspace settings, did not mean they did not need technical communication and, in fact, makes the argument for hiphop communication as both social justice and technical communication.

Hiphop culture has revolutionized the way that music is made, the ways people communicate, and contemporary popular culture, and yet, it is often not recognized as a culture that retooled and repurposed discarded technologies while combining cultural traditions from Africa, the Caribbean, and the Americas. Tricia Rose warns in *Black Noise* that understanding hiphop as purely derived from Afro-Diasporic traditions is a mistake. She argues that technologies and economic conditions played essential parts in hiphop's birth. Rose writes that many of hiphop's founders worked as technicians and repairpersons that "were trained to repair and maintain new technologies for the privileged but have instead used these technologies as primary tools for alternative cultural expression" (Rose 1994, 63). Instead of only using these technologies for their intended purposes, these technicians recycled, repurposed, and "significantly revised in ways that are in keeping with long-standing black cultural priorities, particularly regarding approaches to sound organization" (1994, 63). Youth movements around the world have adopted hiphop rhetorics and practices because hiphop teaches how to make do by combining what is readily available to create resistance and agency.

HIPHOP DIGITAL BOOKLETS

I define hiphop digital booklets as brief, concise texts that combine rapping and storytelling to relay important technical knowledges. These texts may exist in print, but more often exist in other rhetorical forms of writing, such as rapping, deejaying, and graffiti art. Hiphop digital booklets are effective because they are ways hiphop communities can share knowledges in portable and accessible ways.

The art of rapping shares roots in West African griot traditions, specifically how community storytellers are responsible for sharing important community histories, knowledges, and art. Rappers, as griots, relate a variety of knowledges to their communities, including technical ones. Rapping as a method of communication often uses an inclusive approach that muddies the lines drawn by academic disciplines. We often forget that those lines are constructions and not naturally occurring. Digital booklets are one example of how rappers provide a form of technical communication that may not immediately reflect what the field imagines but does the same job. Sometimes better.

I offer the Notorious B.I.G.'s (1997) "The Ten Crack Commandments" as an example of a digital booklet. Biggie portrays himself as an expert in a volatile business and presents a "step-by-step booklet" sharing his secrets. He establishes ten basic principles for conducting business on the streets with the constant warning that the losers pay with their lives. Biggie's commandments are no different from free market, hyper-capitalist business practices employed by legitimate businesses. The first rule cautions against flaunting wealth because it attracts jealous and aggressive competitors. The other rules include keeping one's moves hidden, trusting no one, and the risks of lending and borrowing resources and money. The song prescribes the speaker's formula for success, but it also carries a constant and implied critique of drug dealing as an aggressive, dangerous line of work that creates jealousy, aggression, anger, paranoia, and violence. Again, these are not business practices that only happen in the drug game.

This song offers business survival practices and tactics through music. Biggie creates an auditory booklet and quick reference guide for people working in the drug business, but as stated earlier, this booklet is applicable beyond selling drugs. The aggressive nature of capitalism as practiced in the United States depends on establishing and maintaining power and competitive advantages through cutthroat business practices. It also means that businesspersons and businesses must always be on alert for possible threats. This digital booklet—concise, clear, and compelling—is the kind of educational tool that can be carried and distributed much easier than textbooks or actual booklets. Legal scholar Imani Perry (2004) lists Biggie's music as a reason she and her fellow Harvard law school colleagues survived final exams. After her friend remarks, "I wouldn't have been able to get through it without Biggie," Perry responds, "[t]he generated energy, the adrenaline rush, and the rhythm of the Biggie Smalls music he listened to while writing his exam all motivated him as he expressed his knowledge and skills of argumentation in text" (1). This law student uses music as a survival technology that powers him. The combination of compelling lyrics and beats shows how hiphop can transmit technical knowledges, from person to person, in nontraditional ways.

While Biggie offers business advice, Jay Z's song "99 Problems" (Jay Z 2003) focuses on the law. The second verse of the song describes an incident where the police stopped Jay Z for driving while black. The conversation between he and the cop provides an example of sharing legal knowledge. Jay Z raps:

> Well, my glove compartment is locked, so is the trunk in the back
> And I know my rights, so you gon' need a warrant for that

The conversation happening between Jay Z and the cop offers a moment for people to observe how a black man can challenge authority by using the laws in place. Rhetorically, he wins the battle with the cop, but he also offers direct and usable techniques to his audience. The most important lesson is to know one's rights and the rule of law so that police understand that they shouldn't attempt anything unethical or illegal.

In his memoir, *Decoded,* Jay Z (2010) shares how the business world applies the lessons he teaches in his lyrics. He states, "My friend Steve Stoute, who spends a lot of time in the corporate world, tells me about young execs he knows who say they discovered their own philosophies of business and life in my lyrics. It's crazy" (293). Jay Z's music often tells stories about drug dealing, including business practices. People ignore how technique is being taught to audiences who often find more credibility in Jay Z than traditional sources. For young people coming from historically marginalized communities, being able to access knowledge through hiphop music can have a profound impact on their lives.

Skeptics may look at digital booklets as a novelty, and it is understandable given the knack for instructors to depend on gimmicks and novelties to teach students. To dismiss digital booklets, and hiphop technical communication in general, is a disservice to students and teachers because it ignores how culture, technology, and rhetoric produce technical communication. Even traditional technical communication, as understood by the field, is a result of someone's culture, technology, and rhetoric. I believe skeptics can be dismissive out of the discomfort of trying something new and unfamiliar.

Pedagogies situated in radical and critical theories can help familiarize us with teaching underrepresented technical communication, such as hiphop technical communication.

THE CALL FOR RADICAL PEDAGOGIES IN TECHNICAL COMMUNICATION

The education system across all levels shares a sad relationship with the history of colonization and racial oppression in the Americas. The idea that the educational process converts young people from wild and misbehaved children into educated and refined citizens has dangerous consequences for non-white students. The history of genocide, erasure, and assimilation of people of color in the United States attests to their marginalization in the classroom. I believe students feel that marginalization, whether they realize it or not. Gloria Anzaldúa's (2007) theorization of borderlands when applied to classrooms helps illuminate high political stakes for nonwhite students. She states:

> Borders are set up to define the places that are safe and unsafe, to distinguish *us* from *them* . . . The prohibited and forbidden are its inhabitants. *Los atravesados* live here: the squint-eyed, the perverse, the queer, the troublesome, the mongrel, the mulato, the half-breed, the half dead; in short, those who cross over, pass over, or go through the confines of the "normal." Gringos in the US Southwest consider the inhabitants of the borderlands transgressors, aliens—whether they possess documents or not, whether they're Chicanos, Indians or Blacks. Do not enter, trespassers will be raped, maimed, strangled, gassed, shot. The only "legitimate" inhabitants are those in power, the whites and those who align themselves with whites. (25–26; emphasis in original)

Anzaldúa speaks to how heteronormativity organizes the boundaries for acceptance as part of the continued colonization of the Mexico-US border. Her theorization of borderlands, especially the concept of "los atravesados," can be used to understand how these same boundaries and identities function in classroom spaces. Students of color, working class students, queer students, female students, and students with disabilities all must confront not being white, affluent, and male. These same students are often also first-generation and navigating unfamiliar places, which emphasizes their status as atravesados. When students struggle to find their place in the university, the self-doubt they feel feeds into the model of hyper-individualism that also characterizes the college experience and is a larger part of US mythologies about self-reliance, bootstraps mentalities, and success.

Ana Louise Keating (2007) argues that Enlightenment philosophies about the individual from thinkers like Locke, Hobbes, and Hume ground ideas about the "self-made man" in US society (25). The emphasis on the individual creates a hierarchal relationship "between the self and other, where the individual and society occupy mutually exclusive poles" (26). This relationship "presumes and reinforces a model of domination, scarcity, and separation in which intense competition leads to aggressiveness and fear" (26). Education becomes a process of competition and dominance rather than creating the possibilities for community-building and social change. Education based on dominance is especially damaging for students on the margins because they must compete on an already unfair playing field. Students with power and privilege dominate classroom discussions, expect to make the highest grades, and feel no obligation to interrogate their power and privilege. Keating writes that this model makes failing students believe that "[t]hose people who do not succeed have only themselves to blame, and their failure has absolutely no impact on anyone but themselves . . . Just pull yourself up by your own bootstraps!" (27) In *Borderlands*, Anzaldúa (2007) explains that Chicanx engage in self-hate, self-blame, and self-terrorism that mostly " . . . goes on unconsciously; we only know that we are hurting, we suspect that there is something 'wrong' with us, something fundamentally 'wrong'" (67). When forces seek to discredit students, and when they lack a community willing to listen, explain, and bear their burdens alongside others, what choices do students really have?

These issues are no different in technical communication classrooms where homogenous students and curricula still dominate. Students of color know they are atravesados who must figure out how to "fit" into the class and the broader academy, which often means assimilating into the hierarchy as best they can. Students practicing this model of success based on hyper-individualism and dominance continue applying these ideas as they participate in the workplace, their home communities, in their families, and in their roles as citizens. The possibilities for community change and social justice work—not to mention transformative technical communication—ultimately suffer.

HIPHOP PEDAGOGIES AS RADICAL PEDAGOGIES

Scholars theorizing and writing about hiphop pedagogies argue that hiphop culture promotes social justice through its recognition of how oppression works against people of color and its response through the five main elements (rapping, deejaying, graffiti art, break dancing, and doing the knowledge). The implementation of pedagogies based on the elements offers exciting possibilities for changing how knowledge is taught and the methodologies employed in producing theories, ideas, and discourses. Hiphop pedagogies run a broad spectrum of theories, tactics, and practices that are meant to engage young people, interrogate power structures, build communities, and grow hiphop scholarship directly from grassroots, bottom-up efforts.

One of the most visible examples of how hiphop culture has influenced US society is through the classroom because students bring that influence with them through the ways they dress, speak, and act. A. A. Akom (2009) argues that classrooms—starting in the 1990s—have been affected by the ways hiphop aesthetics merged with youth cultures, yet critics of hiphop culture characterize it as encouraging anti-intellectualism and detrimental lifestyles (53). Akom locates the relationship between hiphop culture and critical pedagogies in the ways that hiphop culture is rooted in black struggles for freedom and social justice. Excluding the ways hiphop culture can influence and produce pedagogies, including those in technical communication, ignores a space of exciting intellectual possibilities.

Akom introduces the concept of "Critical Hip Hop Pedagogy" (CHHP) as a framework that builds on previous student-centered pedagogies that promote social justice (54). Students are seen as active knowledge producers who interrogate how knowledges affect their lives and provide practical application to problems and issues in their communities. Akom suggests that the most important aspect of CHHP is that it "challenges the role that schools play in reproducing social inequality" by "the creation of pedagogic spaces where marginalized youth are enabled to gain a consciousness of how their own experiences have been shaped by larger social institutions" (63).

While we want to offer students a better classroom experience, we should also be careful to not rehearse the same actions that have students believing that they will leave the academy as reformed savages. Akom's use of hiphop as a critical tool offers marginalized students a familiar space for intellectual engagement. Kermit E. Campbell (2005) writes, "And after ten years of teaching this class, I gotta say that the oral-literate art of hip hop is one of the few things that inspire me to teach writing" (149). Campbell refers to a composition course based on the art of rap he created as a way of trying to engage with African American students. His course has helped him think of the ways that black students engage with how academic writing is defined. He concludes that these experiences unexpectedly helped him think of his own pedagogical practices and how they work with and against the academy. He explains, "Along with playing [Tupac Shakur] records (and many others, of course) in a writing class, I would also want the class to bust a few moves, to get students completely out of traditional classroom mode" (149). He presents a rap he wrote called "Hip-hopology 101" that he performed for students as a way of practicing what hiphop pedagogies look like beyond using traditional classroom practices. After having performed this rap for one of his classes, he writes, "Maybe if we treated writing more like rapping, taught writing like it was something you felt and got a groove to, then students like E[1] could really show us how much game they got" (152). What do technical communication classrooms look like if, like Campbell's example, they are designed to engage with African American students and other students of color?

Hiphop pedagogies require a holistic approach to teaching that asks more from teachers and students. In order to create the kind of space where students of color are validated, the instructor must enact the kinds of pedagogical practices that decenter the teacher as the most important person in the classroom. They must also design a curriculum that is inclusive and aware of how power and privilege affect the classroom. Teachers and scholars, especially those of us who are people of color, must recognize that we are often professionals who struggle with what we want to get out of a classroom, and we too struggle with rhetoric and writing. Those of us who achieve MAs and PhDs followed through the tracks set by the academy, and we should not lose sight of how our bodies have been disciplined by the academy, whether we resisted or not. Even the most radical scholars in the academy are still working in the academy. We must face this reality not as an indictment or reason to give up, but rather something that influences our work. Being aware of that influence can help us make the kinds of strides of which we dream.

Students must push themselves to engage with their own subjectivities as members of a community of learners to listen to all voices as deeply as possible before making a judgment or forming an opinion. Paulo Freire (2000) suggests, "The radical committed to human liberation, does not become the prisoner of a 'circle of certainty' within which reality is also imprisoned" (12). Embracing discomfort and seeing what is produced is one of the most powerful behaviors a teacher can model as well as an invaluable experience for students. To create the conditions for engaging classroom experiences where students can offer, create, test, and interrogate knowledge, they must feel willing to take intellectual risks with each other. The teacher must also feel this way. Failure, mistakes, and errors should be valued as much as agreements, successes, and strong arguments. Everyone in the room must be committed to making the space a safe, fun experience. Aya de Leon (2013, n.p.), who has worked primarily with young people in the Oakland Bay Area states, "Hip hop culture, with its emphasis on personal expression and noisy, high-energy interaction, requires a different level of chaos-tolerance on the part of the adults in charge." In a technical communication classroom, this means bringing the noise and the funk to the wack, quiet classrooms where all the bodies in the room must sit quietly as the teacher lectures. We must find ways of breaking the norms and makin' noise all through the halls of the academy—if merely to remind ourselves that we are entire human beings in the room and not just floating heads consuming knowledge.

THE BENEFITS OF RADICAL PEDAGOGIES IN THE TECHNICAL COMMUNICATION CLASSROOM

In *Teaching to Transgress*, bell hooks (1994) discusses her childhood, pre-integration classroom as a space where teachers and students understood that education was a tool for social justice. She remembers, "We learned early that our devotion to learning, to a life of the mind, was a counter-hegemonic act, a fundamental way to resist every strategy of white racist colonization" (3). Her teachers were mostly women who made intellectual inquiry a fun experience. Teachers built relationships with students' families out of the communal recognition that colonialism, racism, and white supremacy affected their lives. Her experiences in integrated classrooms starkly differed. She describes, "The classroom was no longer a place of pleasure or ecstasy. School was still a political place, since we were always having to counter white racist assumptions that we were genetically inferior, never as capable as white peers, even unable to learn" (4). Although much has changed to promote better relationships between different students and teachers, and diversity initiatives are popular across college campuses, the stakes remain high and difficult for students of color. Departments, classrooms, and campuses still struggle and maintain power structures that marginalize students of color for the ways they look, speak, and think. The assumption remains that students of color are to enter the academy to painfully shed aspects of their cultures and ethnicities in order to adopt the vestments of upward mobility.

Radical pedagogies that seek to build community and create alliances transform classroom projects, spaces, and relationships from reifying the status quo to interrogating it, deconstructing it, and imagining new possibilities. According to hooks (1994), inclusive and fun teaching deeply transgress the dominant educational model, and "any radical pedagogy must insist that everyone's presence is acknowledged. That insistence cannot be simply stated. It has to be demonstrated through pedagogical practices" (8). Teaching students to value learning as opportunities to enter conversations, challenge ideas, build knowledge, and find pleasure in the process through a community-based approach demands more out of students, teachers, and the academy. It also chips away at the idea that education should be an elitist practice built on hierarchies that privilege white, male, heterosexual, wealthy, and able-bodied students.

Radical pedagogies are sorely needed in technical communication classrooms. If we acknowledge how histories of technology in the West erase, exclude, and barbarize nonwhite communities, then we must also acknowledge that the ways technologies are designed, implemented, and used are also a part of that history. The field must acknowledge that collaboration and communication involve learning how to adapt, listen, and communicate in inclusive ways. Using radical pedagogies in technical communication impacts who becomes a technical communicator and how they practice their art.

Hiphop culture has offered people the ability to reclaim spaces in ways that create community, confidence, and more possibilities for advancement. Hiphop pedagogies acknowledge that the education process fails marginalized communities, especially black, brown, and red communities. Education in the United States often follows the conversion narrative of civilizing nonwhite bodies through the classroom. Mary Louise Pratt's (2007) conceptualization of "contact zones" as "the space[s] in which peoples geographically and historically separated come into contact with each other and establish ongoing relations, usually involving conditions of coercion, radical inequality, and intractable conflict" (6) helps us understand how spaces we like to think of as safe are oftentimes dangerous, threatening, and debilitating for underprivileged students. The academy brings students from different backgrounds and perspectives into the room and asks them to interact with each other, the instructor, and knowledge while often ignoring that power relationships heavily influence what voices, ideas, and perspectives will dominate the learning process.

HOW TO USE CIPHERS AND DIGITAL BOOKLETS IN THE CLASSROOM

Digital booklets and ciphers are two ways teachers may use hiphop, and therefore social justice based, pedagogies in the technical communication classroom. These two activities are easy to do and bring the kind of radical fun mentioned earlier. Before beginning either one, it is important to culturally situate these rhetorical activities so that students gain an understanding of how culture impacts the ways people communicate. This can be as simple as assigning students readings about African American rhetorics, such as Adam Banks's *Digital Griots*, and/or assigning reading on hiphop history. Jeff Chang's *Can't Stop, Won't Stop* offers a basic history of the culture and its roots. I teach Jay Z's *Decoded* often because most people have heard of Jay Z, and the book features close readings of his songs. This step is crucial because it makes students more aware of the relationship between hiphop, culture, and struggles for social justice. Students also become better prepared to see diverse communities in the story of technical communication.

An extremely useful pedagogical tactic that builds community is to call on any hiphop heads in the classroom to help. In my experience, every classroom has one student who would consider her/himself a hiphop head, or aficionado, and is usually eager to share her/his knowledge. Teachers unfamiliar with hiphop will find that these students will fill many knowledge gaps. Even teachers who know plenty about hiphop will benefit from collaborating with hiphop heads. Many times they are aspiring rappers, DJs, and other kinds of artists. When they learn that they have been practicing forms of technical communication and/or have the potential to do so, their investment in your course will grow.

THE DIGITAL BOOKLET ASSIGNMENT

Asking students to bring a song for analysis to class is nothing new to composition and literature classrooms. Many rookie instructors have found this exercise as one of their first successful ones because students have an easy entry point toward engagement with the object of study. What this kind of activity often lacks is sufficient cultural situating by the instructor and students. In the case of the digital booklet exercise, the instructor has already introduced background reading as suggested at the beginning of this section.

For the digital booklet assignment, technical communication students should look at the effectiveness of this technology. For example, the earlier discussion over "The Ten Crack Commandments" shows the potential for several discussions. Does the digital booklet provide important technical knowledge? Is it practical? Students could also analyze the organizational culture. In the context of the song, why would music do a better job of teaching business tactics over a college textbook? One could also take a different approach and try to identify how a digital booklet could influence traditional technical communication genres.

Students could look for examples, such as Jay Z's anecdote about business school students adapting hiphop knowledges to their professional careers.

The instructor could ask students to write an analysis paper based on a song they feel would fit the criteria of a digital booklet. This exercise could be adapted across genres, but this would also demand a discussion of how the concept of a digital booklet changes in different contexts. For example, a hiphop song is going to draw from traditions and theories different from a country song based on factors like location, race, class, and ethnicity.

Finally, students can be assigned their own digital booklet project. Instead of writing traditional instruction manuals, memos, and resumes, they could figure out how a digital booklet could accomplish a similar goal. Digital booklets are also easily adaptable for projects and internships involving community engagement since students could offer an additional tool for solving a wide range of technical communication challenges. Students may find advantages and disadvantages to technical communication through this different method and reflect on how they could use it in their future careers.

THE CIPHER

Ciphers are rap circles where people engage in freestylin'. The mythology of ciphers is that the best rappers improvise everything, although many pre-prepare rhymes. Ciphers are spaces where rappers establish their ethos as master wordsmiths. The best establish their reputations, while the rest refine their skills. The point of the cipher is to keep coming back. The subject matter of a particular freestyle session varies, but community, social, and technical knowledges are often shared. Because ciphers always carry the potential for inclusivity and change, any instructor can combine this activity with another methodology. For example, what would a cipher look like if it applied a Feminist Disabilities or Ecocriticism perspective? What if the instructor challenged students to combine more than one? I teach using my own version of the cipher as a safe space where we all show up to share, critique, and build knowledge. Students are not required to rhyme, but they are encouraged to keep the cipher movin' by feeding off each other's comments and ideas. The cipher is useful in introductory technical communication classrooms because it helps build community through an organizational culture. It also gives students a hands-on approach to thinking about technical communication in a social justice context. In thinking about how they need to shift and adjust to a different way of communicating, they can theorize and think about what will happen when they enter different communities, spaces, and organizational cultures.

The cipher can also serve a very practical purpose: waking students up. Every college instructor knows that moment in the semester when students appear tired, sleepy, and stressed. I have been known to stop class, ask everyone to get in the cipher, and play a funky song. I often choose "The Big Payback," by James Brown, not only because it is so damn funky but because it is heavily sampled by hiphop artists. I'm always the first to dance, and I'll usually get a few students to join in the fun. It always gets a laugh out of students and it often leads to discussion of how we should be aware of our entire bodies, and not just our brains, when we are learning. It also builds rapport with students because it reinforces the idea that learning should be fun.

CONCLUSION

Hiphop methods of practicing technical communication can shift the field in radical ways. First, it changes the definition of technical communication by being more inclusive of nontraditional methods and methodologies. Second, this inclusion creates opportunities for nonwhite people and communities to participate in the shape and scope of the field. Finally, placing these theories into practice offer solutions across a variety of fields. Hiphop digital booklets are easily transmitted ways of communicating technical knowledges across time and space. They also offer marginalized peoples a subversive communication method. Much like the Notorious B.I.G. uses "The Ten Crack Commandments" to teach business practices, other rappers have used their songs to teach other important techniques. "Be Healthy," by rap duo Dead Prez offers dietary advice to people living in food deserts. Their hope is not only offering ways of healthier cooking and eating but fostering social justice through nutrition. The GZA of the Wu-Tang Clan raps about science as a way of making scientific knowledges available to young hiphop heads stuck in inadequate school systems. Health advocates in New York and Houston are using hiphop to promote health and lifestyle messages to their communities, like the partnership between Hip Hop Health New York and legendary rapper/producer Doug E Fresh.

In 1999, scholar Victor Villanueva asked people in rhetoric and composition to "break precedent" with the traditional conceptualization of the discipline (Villanueva 1999, 659), and I believe we should do the same in technical communication. Breaking precedent has an effect on material conditions and real bodies by making room for perspectives that have been historically ignored as a result of systemic oppression. Engaging in hiphop rhetorics can influence the inclusion of more people of color in STEM fields. It can mean creating better methods of access for the marginalized and oppressed. It also means giving the ignored and erased tools for combating obstacles and surpassing barriers.

Note

1. "E" is a student Campbell uses as an example of students who engage intellectually in ways that are not recognized by the academy, and thus, label him as at-risk and unproductive.

References

Akom, A. A. 2009. "Critical Hip Hop Pedagogy as a Form of Liberatory Praxis." *Equity & Excellence in Education* 42 (1): 52–66. https://doi.org/10.1080/10665680802612519.

Anzaldúa, Gloria. 2007. *Borderlands/La Frontera: The New Mestiza*. 3rd ed. San Francisco: Aunt Lute.

Banks, Adam J. 2008. *Race, Rhetoric, and Technology: Searching for Higher Ground*. New Jersey: Lawrence Erlbaum Associates.

Banks, Adam J. 2011. *Digital Griots: African American Rhetoric in a Multimedia Age*. Carbondale: Southern Illinois University Press.

Boone, Elizabeth Hill, and Walter D. Mignolo, eds. 1994. *Writing without Words: Alternative Literacies in Mesoamerica and the Andes*. Durham, NC: Duke University Press.

Campbell, Kermit E. 2005. *"Gettin' Our Groove On": Rhetoric, Language, and Literacy for the Hip Hop Generation*. Detroit: Wayne State University Press.

De Leon, Aya. 2013. "Hip Hop Curriculum: A Valuable Element for Today's Afterschool Programs." Accessed December 12, 2013. Hiphoparchive.org.

Freire, Paulo. 2000. *Pedagogy of the Oppressed: 30th Anniversary Edition*. New York: Bloomsbury Academic.

Haas, Angela. 2008. "A Rhetoric of Alliance: What American Indians Can Tell Us about Digital and Visual Rhetoric." PhD diss., Michigan State University, East Lansing.

Hart-Davidson, Bill. 2001. "On Writing, Technical Communication, and Information Technology: The Core Competencies of Technical Communication." *Technical Communication (Washington)* 48 (2): 145–55.

Hayhoe, George F. 2003. "Inside Out/Outside In: Transcending the Boundaries that Divide the Academy and Industry." In *Baywood's Technical Communications Series: Power and Legitimacy in Technical Communication, Volume I: The Historical and Contemporary Struggle for Professional Status*, ed. Teresa Kynell-Hunt, 101–14. Amityville, NY: Baywood Publishing Company. https://doi.org/10.2190/PL1C5.

Henry, Jim. 2012. "How Can Technical Communicators Fit into Contemporary Organizations?" In *Solving Problems in Technical Communication*, ed. Johndan Johnson-Eilola and Stuart A. Selber, 75–97. Chicago: University of Chicago Press.

hooks, bell. 1994. *Teaching to Transgress*. New York: Routledge.

Jay Z. 2003. *"99 Problems." The Black Album*. Roc-A-Fella Records. Compact Disk.

Jay Z. 2010. *Decoded*. New York: Spiegel & Grau.

Keating, AnaLouise. 2007. *Teaching Transformation: Transcultural Classroom Dialogues*. New York: Palgrave Macmillan. https://doi.org/10.1057/9780230604988.

Kynell, Teresa, and Elizabeth Tebeaux. 2009. "The Association of Teachers of Technical Writing: The Emergence of Professional Identity." *Technical Communication Quarterly* 18 (2): 107–141. https://doi.org/10.1080/10572250802688000.
Notorious B.I.G. 1997. *"Ten Crack Commandments." Life After Death.* Bad Boy. Compact Disk.
Perry, Imani. 2004. *Prophets of the Hood.* Durham, NC: Duke University Press. https://doi.org/10.1215/9780822386155.
Pratt, Mary Louise. 2007. *Imperial Eyes: Travel Writing and Transculturation.* New York: Routledge.
Ridolfo, Jim, and Dánielle Nicole DeVoss. 2009. "Composing for Recomposition: Rhetorical Velocity and Delivery." *Kairos* 13 (2) Accessed May 1, 2011. http://kairos.technorhetoric.net/13.2/topoi/ridolfo_devoss/intro.html.
Rose, Tricia. 1994. *Black Noise.* Hanover, MD: Wesleyan University Press.
Villanueva, Victor. 1999. "On the Rhetoric and Precedents of Racism." *College Composition and Communication* 50 (4): 645–61.

8
BLACK FEMINIST EPISTEMOLOGY AS A FRAMEWORK FOR COMMUNITY-BASED TEACHING

Kristen R. Moore

At the 2014 Conference on College Composition and Communication, Angela Davis (2014) advised, "We need new epistemological horizons." Advocating for dialogic alliances between feminism and abolitionism, Davis provoked activist scholars to consider the ways their own work might be making a change. As I sat in a row of other technical communication scholars [three of whom are represented in this volume], it became obvious that Davis's talk was relevant not only for composition scholars but for technical communicators as well. Indeed, her call for new epistemological horizons echoed and reiterated calls by Miriam Williams and Octavio Pimentel (Williams and Pimentel 2012), Angela Haas (2012), and Gerald Savage (2013), among others, who have exposed the limits of technical communication's theoretical frames. Her call also reflected my own struggle as a researcher in St. Louis, Missouri, where I spent two years learning from and with Vortex Communications, a company of public engagement specialists.[1] Having been quickly named the resident white girl[2] among the eight black women who ran and worked for the company and having spent even just a few days in their office and in the field, I was faced with my own need for new epistemological and methodological horizons if I were to effectively [and ethically] conduct research with these women. My go-to theoretical and methodological lenses [like Bruno Latour's Actor Network Theory and Michel Foucault's theory of power] failed to help me account for and do justice to the activist approaches to public engagement and technical communication I encountered at my site of research. Drawing on this experience, this chapter demonstrates the ways new theories benefited me and can benefit technical communicators who work in the public sphere and responds to scholars in technical communication like Haas and Eble in this collection as well as Savage (2013) and Williams and Pimentel

(2012, 2014), who see new theoretical frames as a step toward diversifying the field and pursuing social justice.

In this chapter I highlight one particular theoretical approach, Black Feminist Theory, explaining both the ways it augmented my abilities to ethically and effectively study and write about the public engagement[3] firm I studied. Black Feminist Theory (and its corollary theories of Womanism, Africana Feminism, and African Feminism)[4] has traditionally not been used or acknowledged in technical communication. This omission is not surprising since only in the last two decades has the field begun to expand its theoretical frames to include cultural theories and questions of race, ethnicity, and identity. I follow technical communication scholars in seeing a need for expanding our theoretical approaches but further argue that Black Feminist Theories aid technical communicators in developing the skills necessary to both enact social justice and successfully and ethically engage communities in the public sphere.

BLACK FEMINIST THEORY, ROUGHLY DEFINED

Black Feminist Theory is a theoretical perspective that "makes room" for black women's experiences against a backdrop of generally exclusive white feminisms or, as some scholars call it, feminism.[5] While certainly the omission of black women's experiences from feminism is not an across-the-board truth, black women have noted, quite rightly, that feminism has long represented the inequities, concerns, and lives of white middle class women while ignoring those of women of color. One need only read Sojourner Truth (2005) or Audre Lorde (1984) among many others to find that, indeed, white feminism as a movement and theoretical lens has historically treated the categorical oppression of "gender" as sacrosanct while ignoring other forms of oppression like race, class, and sexuality. White feminism has likewise taken the experience of white middle class women as representative of all women. Notably, as black women joined white women in the suffrage fight, white women failed to advocate (in turn) for equal rights for black women. In the late twentieth century, black women scholars like bell hooks (1991), Barbara Smith (1978), and Clenora Ogunyemi (2001) noted the ways that white feminism excluded black women from its considerations and its activism. Ogunyemi (2001) describes, for example, that white feminist authors have the luxury of writing about patriarchy in ways that black women writers do not. Indeed, as Crenshaw (1991) articulates, black women's experience of sexism counters white women's in that black women experience intersecting, multiple oppressions at the same time—a difference seldom acknowledged by white feminists.

Such critique is applicable to technical communicators who attempt to understand and study women or people of color without accounting for intersectional oppression, but it is also applicable to those who see feminist theory as sufficient for understanding cultural positions that don't benefit from the privilege of whiteness, straightness, wealth, etc. Crenshaw explains this particularly well: "Because the intersectional experience is greater than the sum of racism and sexism, any analysis that does not take intersectionality into account cannot sufficiently address the particular manner in which black women are subordinated" (Crenshaw 1989, 140). White feminists have historically theorized oppression along only one axis, making their theories exclusionary of women of color and inapplicable to women of color's lived experiences. Recently, white feminists[6] (like myself) have accepted the limits of feminism's earlier frames. According to Mama, in more recent years, "[W]estern feminists have agreed with much of what [African feminists] have told them about different women being oppressed differently and the importance of class and race and culture in configuring gender relations" (Mama 2001, 63). She suggests that "[t]he constant tirades against white feminists do not have the same strategic relevance as they might have had 20 years ago" (63). And she may be right. Yet in my mind (and, I think the mind of Hill Collins, whose Black Feminism I draw on most heavily), the call to acknowledge rather than suppress black women's experiences, theories, and knowledges is still relevant, particularly in fields like technical communication that rely on empirical frameworks. These historical roots have clear consequences for current feminist practices, which often (in technical communication, rhetoric, and writing) do not acknowledge Black Feminism or the differences in black women's intersectional experiences of oppression. Indeed, while the importance of feminism has been acknowledged by a handful of scholars, particularly women scholars (Brasseur 1993; Durack 1997; Flynn 1997; Gurak and Bayer 1994; Koerber 2000; Lay 1991; Ross 1994; Sauer 1994; Sullivan and Porter 1997), feminism in technical communication most often refers to white feminism, privileging the importance of gender-based oppression over other forms of oppression, including race, class, and sexuality. This white feminist scholarship potentially contributes to the suppression of Black Feminism because it sidesteps the intersectionality of oppression in favor of a focus on gender.

Despite the neatly-told origin story[7] I articulate above, Black Feminism is not a monolith. Many scholars have contributed to the body of knowledge that might be included in Black Feminism: Johnetta Cole (1986), Kimberle Crenshaw (1989, 1991), Ula Taylor (1998a, 1998b), Barbara Smith (1978), among many others. Rather than prioritize the particular orientation of theory that privileges Western male philosophy, Black Feminist Theory prioritizes action, experiences, and epistemological frameworks beyond the theoretical. Further, Black Feminism grows out of black women's experiences, often the oppressive and exclusionary experiences of living in a world dominated by white privilege (Cole and Guy-Sheftall 2003; Combahee River Collective 1977; Davis 1999; Hill Collins 1996, 2008; King 1988; Smith 1983; Springer 2002).

The need for including Black Feminism as a legitimate and important intellectual contribution to technical communication grows from what I see as the longstanding suppression of black women's thought and roles in technical communication. These suppressive patterns are extensions of what Hill Collins describes as systemic and ideological patterns that dominate the United States at large. Despite calls for social justice and diversity (Haas 2012; Savage and Mattson 2011; Williams and Pimentel 2012) and responses by scholars like Flourice Richardson (2014), Godwin Agboka (2013), and Natasha Jones (2014), students and scholars in the field of technical communication have little exposure to women of color's scholarship, practices, or theories. These patterns of suppression support the oppression of black women [and all women of color], and "the supposedly seamless web of economy, polity, and ideology function as a highly effective form of social control designed to keep African-American women in an assigned, subordinate place. This larger system of oppression works to *suppress* the ideas of Black women intellectuals . . ." (Hill Collins 2008, 7; emphasis added). The patterns of suppression that limit black women's prominence in dominant theories and practices (particularly in the academy) include:

- silencing black women's work in order to protect white male interests and world views
- employing white women's feminisms as a representative of gender inequality
- calling for diversity, but maintaining one's own practices
- appropriating black women's work and thereby developing a form of symbolic inclusion

These forms of suppression are not necessarily found whole in the field of technical communication. But critical examination of published work and curricula demonstrate that women of color's work is traditionally unacknowledged at best and suppressed at worst. Clark's (2004) article, for example, which discusses his collaborative work with a Coalition of African American women activists signals a (perhaps unintentional) silencing of black women's work. In his study, he and students worked with a local black women's advocacy group, and his article reports on their efforts to help the group. But the article does little to address either the advocacy as a mission to help black women in the community or to acknowledge the role of the women in the projects he worked on. One need only scan journals in technical communication to note the very few examples of black women technical communicators or studies that use (or even acknowledge) Black Feminist Theories.

As one key figure in Black Feminist Theory, Patricia Hill Collins stands out both as a prolific writer and a field researcher whose work straddles the line of activist and academic. A sociologist by trade, Hill Collins's research has worked to develop a black women's standpoint and to unearth the critical frames black women's experiences draw from and create.[8] In *Black Feminist Thought*,[9] Hill Collins explores black women's public intellectualism, community work, and activism, noting the collocation of intellectualism with community and social activism. Importantly, Hill Collins reminds scholars that public intellectualism exists outside of the academy as much as within in it. She joins other black women scholars, like Barbara Christian, who note that intellectualism for black women is not located solely in the university or in the academic practices endorsed by the elite patriarchy that helped build academic and scientific standards. Christian (1988) reminds us that "people of color have always theorized—but in forms quite different from the Western form of abstract logic . . . [O]ur theory," she explains, "is often in narrative forms, in the stories we create, in riddles and proverbs, in the play with language, because dynamic rather than fixed ideas seem more to our liking" (68). So Black Feminist Theory is built by literary theorists, activists, teachers, public intellectuals, etc., and it prioritizes action, experiences, and epistemological frameworks beyond the theoretical. As technical communicators move beyond the theoretical and into the public sphere, then, Black Feminist Theory, like Del Hierro's Hip Hop approach to technical communication, provides an opportunity for combining theory and experience, action and critical thought.

My own research, too, moves outside the academy to study technical communication as it occurs in the field, including cityscapes like St. Louis, Missouri, smaller Midwestern cities like Urbana-Champaign, Illinois, and Springfield, Illinois. My primary participants—nine women from the St. Louis area—worked for a for-profit company that sought to improve the community through their public engagement projects. For Vortex consultants like Rose and Jocelyn, their roles as black women *and* their work with communities drew attention to (1) the importance of acknowledging intersecting oppression in public decision making, (2) the strength [and necessity] of listening to citizens' stories and experiences, and (3) the benefits of building knowledge through shared dialogue and empathetic relational work. As I learned from the women at Vortex, organizing public decision-making in technical contexts like transportation planning requires diverse forms of knowledge-making, particularly in the diverse communities that Vortex worked in. Black Feminist Theory allowed me to see beyond the genres and media the consultants worked with or technical presentations they did in the public sphere. Instead, I analyzed the ways the women built relationships, dialogue, and knowledge from positions of difference and empathy. Thus, I adopted Black Feminist Theory both because I believe in moving outside the white patriarchal canon for knowledge-work and also because it refocuses technical communication in the public sphere on building relationships—a key concept for the Vortex firm.

TERMS AND LIMITS OF USE

There is some risk in adopting Black Feminist Theory as a framework. Anti-racist scholars, womanists, and black feminists have rightly been skeptical about white women's use of black women's work and theories, in part because of the historic patterns of co-opting black women's energy and expertise without reciprocity. I tread lightly, then, working not to appropriate but to incorporate, ameliorate, and privilege the work of the black women I studied, and my study was designed with many stages of reciprocity and—in the spirit of Black Feminism—opportunities for my participants to challenge, dialogue about, and retract my claims. My own methodology leans toward coalition-building and Womanist[10] approaches that draw together activists, scholars, and citizens who seek equity across all oppressive structures. In her more recent talks, Hill Collins (2012) advocates for coalition-building, a move toward working across social justice objectives to repudiate claims that racism is over and address white supremacy. In offering a framework for new forms of institutionalized (rather than more overt, individual) racism, she called on scholars to work together to resist the new more insipid and insidious forms of racism that dominate both our policies and workplaces. My own scholarship responds to this call.

An additional impetus for my inclusion of Black Feminist Theories in my research, however, was the realization that the methodological and theoretical milieu of technical communication was not only insufficient for my research in diverse public spheres but also inappropriate and unethical, particularly given my own investment in redressing the inequities associated with white supremacy, including the privileging of white male theories over other traditions.[11] In designing an ethnographic study of a black women's firm, I resisted pressure to frame the work of black women through traditional theoretical approaches in Technical Communication—including white male and feminist lenses. Samantha Blackmon, a leading scholar in race and game studies (see Blackmon, 2002, 2004, 2014; Layne and Blackmon 2013) and a member of my dissertation committee, cautioned me against "laying my white theory on Black bodies." Blackmon's guidance prompted me to explore new epistemological, methodological, and theoretical approaches to my work.

My own subjectivity as a white woman researcher does, in some ways, complicate my use of Womanism and Black Feminism. I have been asked after presentations, for example, whether or not it's appropriate for me to use Black Feminism or, even, to use the term black rather than African American. However, the need to *privilege* and *acknowledge* black women's work in technical communication supersedes the risks—indeed, the calls for inclusivity in technical communication indicate a need for offering narratives of black women as technical communicators and Black Feminist frameworks for technical communication. I follow Blackmon (2014) in the critical frame that insists work that represents those unlike us requires mindfulness, critical reflection, and care—she suggests that Black Feminism and studying the rhetorical practices of black women is not "off limits" to anyone. Blackmon notes, "What it all boils down to is that we have all got to do research in order to represent anyone well" (Blackmon 2014, para. 1). And so my use of Black Feminism is tricky, of course, but also I see it as due diligence necessary in representing those unlike me: women whose experiences and local activisms are not appropriately understood through technical communication's "go to" theories. In advocating for Black Feminism, I also advocate for a critical approach to teaching and scholarship, wherein the use of Black Feminist Theories is coupled with critical reflexivity (see Sullivan and Porter 1997).

PATRICIA HILL COLLINS'S BLACK FEMINIST EPISTEMOLOGY

In chapter 11 of *Black Feminist Thought,* Patricia Hill Collins asserts that black women's experiences provide a foundation for wisdom, knowledge, and the assessment of knowledge claims. In so doing, she resists the knowledge frameworks that subjugate and exclude black women and other people of color. Hill Collins articulates it this way: "Because elite white men control Western structures of knowledge validation, their interests pervade the themes, paradigms and epistemologies of traditional scholarship. As a result, US black women's experiences as well as those women of African descent transnationally have been routinely distorted within or excluded from what counts as knowledge" (Hill Collins 2008, 269). In articulating ways of knowing that align with black women's experiences, Hill Collins develops four tenets of Black Feminist Epistemology, ways knowledge is made and assessed among black women.

  I. Lived Experience as a Criterion of Meaning, locating wisdom in knowledge differentiates between education and experience and reveres experience as a way of understanding the world (275–79)

  II. The Use of Dialogue in Assessing Knowledge Claims, valuing the antiphonal back-and-forth that generates knowledge through discussion rather than adversarial debate (279–81)

  III. The Ethics of Caring, emphasizing the expressiveness, emotions, and empathy required to validate and build knowledge (281–84)

  IV. The Ethic of Personal Accountability, reflecting on the relationship between the knowledge claim an individual makes and assuming responsibility for it vis-à-vis its connection to an individual's character, values, and ethics (284–85)

Taken together, these tenets produce "standards that are consistent with Black women's criteria for substantiated knowledge and with [their] criteria for methodological adequacy" (275).

Methodologically, I also follow Patricia Sullivan and James Porter in their belief that the theoretical and methodological positions we adopt ought to adequately respond to the local situations we study (Sullivan and Porter 1997). I did use Black Feminist and Womanist Theory because, as I discuss above, it seemed appropriate to use theories developed by black women to understand black women's work. But more than that, Hill Collins's Black Feminist Theory lends itself particularly to work in communities, where so often the experiences and wisdom of citizens is subjugated to the expertise of the engineers and public officials who do the planning.

Many scholars in technical communication, including a number of the authors in this collection, provide frameworks, methodologies, and strategies for working within communities. Michele Simmons, for example, provides a blueprint for developing public decision-making processes that are equitable and just; Jeffrey Grabill's participatory action research, too, presents a methodological approach to working as both a researcher and citizen advocate. Similarly, other scholars across the field have investigated the civic responsibility and potential of technical communication courses, often (though not always) through the discussion of service learning and client-based projects (Dubinsky 2001; Flower 2008; Long 2008; Ornatowski and Bekins 2004; Scott 2004). This scholarship reflects a broad interest in the ways technical communicators can engage more directly in community knowledge-making. Beyond a mere interest in engagement, Simmons (2007), Grabill (2001), Blythe, Grabill, and Riley (2008), and Williams (2017) illustrate the need for technical communicators to question the ethics of both their research methodologies and the institutional infrastructures that dictate work in the public sphere.

I, too, am a supporter of participatory design, participatory action research, and the many approaches to teaching students to ethically engage with communities, but these approaches fail to adequately integrate theories that raise awareness of the suppression and oppression of particular groups of people in the community or the academy. As a supplement, I offer a localized discussion of theories and practices that advocate for social justice advocacy. In this chapter I push the field to consider the need for resisting intersecting oppressive structures, arguing that this consideration is necessary if we want to successfully engage diverse publics in an equitable and thoughtful way. Because Black Feminist Theory was born from the need to resist oppressive structures, I suggest that it can aid in our strategies for resisting the top-down decision-making and design structures and further assert that it can profitably contribute to our approaches to working as advocates with community and public engagement projects.

My research with Vortex Communications aimed to understand the ways the company developed and implemented public engagement in local communities. In order to fully understand their public engagement processes, I analyzed their work using a Black Feminist Epistemology—valuing not merely the end product of their engagement but their process of developing and implementing the public engagement approach. While it's beyond the scope of this chapter to fully explain my study, I use this research to elucidate the tenets of Black Feminist Epistemology and its usefulness for repositioning the advocacy work of technical communicators in local communities. My findings suggest that Vortex Communications draws on some of the same epistemological frames that Hill Collins has articulated as Black Feminist and that these might serve as a framework for other scholars whose work focuses on technical communication in the public sphere—even if that work is not produced by black women or focused on black women. What follows, then, is my uptake of this framework and brief examples of the ways its criteria were useful in my analysis of Vortex's approach to technical communication.

Lived Experience as Meaning. Written into many of the daily practices of Vortex is a respect and reverence for daily, lived experience—and for allowing the "everyday" to inform the business practices. In my field notes, I wrote, "During the brainstorming session, Aleia and Rochelle [two participants in my study] offered up examples of their mothers, sisters, and aunts in order to develop strategies for the Great Rivers Greenway[12] project. No one had a computer or data; the brainstorming drew from personal stories." As a participant in this brainstorming project, I was struck by the ways they chose to apply their experiences to decisions—and by the respect they offered those experiences. Aleia, for example, began by saying, "Well, when I think of this campaign, I think of my mom. Y'know . . . what would she want? Why would she get out and do exercise." Rather than shut down this kind of discussion, the other participants added to these narratives. This one example is not isolated or an anomaly. This valuing of experiential knowledge is woven into their public engagement development and enactment.

Using Dialogue to Assess Knowledge Claims. Relatedly, the Vortex consultants' tie experiential knowledge-making to dialogue as a relevant approach to assessing claims. In order to develop a community—or a family, as Leah describes the company—the women adopt the practice of Cheers and Sharing during their bi-weekly staff meetings. Cheers and Sharing prompts consultants to share stories of joy and struggle with one another; these stories range from personal accounts of their children and parents to experiences during projects and other professional endeavors. I see the inclusion of Cheers and Sharing in business meetings as (1) an acknowledgment that the personal lives and experiences of the employees at Vortex Communications matter for the health of the company, (2) an opening to draw connections between home life and work, and (3) an indication that emotions, stories, and personal expression are epistemologically valuable.

Much of the background information I learned about a key transportation planning case study was articulated during a Cheers and Sharing session. Rose had just returned from some of her first meetings with citizens in Springfield, Illinois. Her Cheers and Sharing narrative focused on this meeting, wherein she had to defend the importance of race for the citizens; she was both angered and indignant at the move to initially dismiss the racial problems in the city. But she also shared that she was overwhelmed by the warm welcome the citizens offered her. "And y'know, they *hugged* me and said, 'We are SO glad that you're here.'" This sharing quickly developed into a strategic plan, however, for how to deal with the project. Jocelyn and Lana both raised questions about the project, and the dialogue about Rose's experience became central to the project. In other words, Rose's Cheers and Sharing prompted a dialogue that refined and focused the Springfield Project.

An Ethic of Caring. Hill Collins suggests that an ethic of caring includes three interrelated components: (1) personal expressiveness, (2) emotions, and (3) empathy. These are the criteria through which individuals ought to be evaluated. This ethic of caring moves through the way Vortex Communications consultants interact both with one another and with clients and citizens. The consultants engage with their own and others' emotions empathetically, and their expressions of emotion are interwoven into the workplace. In addition to Cheers and Sharing, I observed countless other examples of the women responding to one another's ideas, thoughts, and feelings with good humor and with empathy. Further, during interviews I asked the women to describe what Vortex Communications brings to the community, and Rose said, without hesitation, "Empathy." She explained, "If you don't have empathy, then you won't be successful—that's PR, not PE." Their ethic of caring should not indicate a lack of professionalism; to the contrary, it arms them as they enter into public and community spaces, and it is accompanied by the realization that as black women, they have to be twice—or if you ask co-founder Lana, three times—as good as their competitors.

Black Feminist Theory provides a framework for seeing professionalization in tandem with—not in opposition to—an ethic of care. Too often, technical communicators teach professionalism to the detriment of training and knowledge that develop from an ethic of caring. Hill Collins, of course, is not the first to suggest an ethic of caring as a principle for doing knowledge work. Carol Gilligan (1982), for example, has offered an ethic of care as an important theoretical move that conflates the binary between reason and emotion. Hill Collins does nod to Gilligan in places, and she acknowledges the assonance between African-influenced and feminist principles: "White women may have access to women's experiences that encourage emotion and expressiveness, but few White-controlled U. S. social institutions except the family validate this way of knowing" (Hill Collins 2008, 283). Hill Collins points to the black church as one site that validates empathy, emotion, and expressiveness. My research with Vortex Communications suggests that these principles, as they constitute an ethic of caring, can also be an important part of technical communication knowledge-work.

An Ethic of Personal Accountability. Hill Collins first discusses the Ethic of Personal Accountability in her chapter on black women and motherhood, specifically as she describes the role of community othermothers. Community othermothers are often foundational for political community activism and grow from "experiences both of being nurtured as children and being held responsible for siblings and fictive kin within kin networks" (Hill Collins 2008, 205). Within an ethic of personal accountability, black women—particularly educated and well-nurtured ones—often engage in community work that can lead to social activism (205–206). Vortex Communications' mission has roots in social activism, and the company is devoted to community work, as evidenced not only by their policies but by the consultants' individual community participation. Vortex Communications requires all employees to commit eight of their work hours per month doing community service. One of the consultants, for example, spends a week each year volunteering for a children's camp through her enlistment in the National Guard. Other consultants connect professional capabilities with community service requirements. In 2010, for example, Aleia facilitated a strategic plan seminar for the National Association of Black Business Women. The institutional expectation of community work follows the ethic of personal responsibility as articulated through Hill Collins's discussion of community othermothers. In addition, every member of the Vortex Communications sits on nonprofit boards or participates in community leadership in a variety of ways. The consultants worked through an ethic of personal accountability to increase the effectiveness of their public technical communication.

APPLICATIONS OF BLACK FEMINIST THEORY

These four themes provide a basis for further developing technical communication pedagogy and methodologies. Ellen Cushman (1996) suggests that reciprocity and an acknowledgment of the ivory tower position provides one way to ensure our research is ethical. I want to go a bit further. Drawing on these (and other alternative) epistemological frames is an important step toward expanding the ways professional and technical communicators work toward activist and just ends. Perhaps more important, it is also a step toward including rather than suppressing black women and their theories in our academic frameworks of understanding. Black women, along with women of all ethnicities, African-Americans, Latino/as, Native Americans, Arab- and

Asian-Americans, have not been written into the history of professional and technical communication (see Williams and Pimentel 2012). This is an oversight, an erasure that, as I discussed earlier, results from the prejudices inherent in the academic system, STEM fields, and the field of technical and professional writing. The inclusion of Black Feminist Epistemologies in both our research and our teaching demonstrate an acknowledgment that black women and their theories have much to offer as we develop new mechanisms for change, both in the academic sphere and public sphere.

What do we gain from Black Feminist Theory as an addition to the technical communication classroom?

- New frameworks for understanding traditional problems, like the expert-novice relationship
- A more inclusive pedagogy that can accommodate students from diverse traditions and experiences
- Theoretical frames for understanding and valuing emotion, empathy, and personal expression in the classroom
- An approach to seeing community members as knowledgeable, even if that knowledge looks different from academic practices

For community-based projects, these become particularly important tools for helping researchers and students alike to conscientiously enter communities where they will encounter difference in various forms.

The strength of the women who worked at Vortex Communications was, perhaps, their savvy ability to write with multimedia for wide ranges of audiences. That's one conclusion to draw from their successes. However, I locate their broad range of strategies for building knowledge about sites, projects, and stakeholder needs and for enacting social justice in their local communities. These foundations are not exclusively found in Black Feminist Theory, but Black Feminist Theory crystallizes their work in ways that ameliorate typically undervalued work for technical communication in the community and public sphere.

USING BLACK FEMINIST THEORY IN THE CLASSROOM

This section concludes my chapter by suggesting ways we might integrate Black Feminist Theory into the technical communication classroom. Specifically, I offer a pedagogy that works from Black Feminist Theory to value lived experience and dialogues as the central activity of community-based projects in technical communication. Through this approach, activities focused on experience and dialogue are nested within the larger ethics of care and personal accountability that dominate Hill Collins's community activism and Vortex Communications' daily practices.

Alternative approaches to making knowledge can aid students in interpreting and engaging with diverse communities because it foregrounds learning from communities rather than providing knowledge to communities. Maori decolonial studies scholar and activist Linda Tuhiwai Smith (2012) suggests this as an approach to community research, and we can apply a similar framework to the classroom. By humbling ourselves to community members' wisdom, experience, and dialogues, we can expose and value (rather than subjugate or suppress) local knowledge. I use the four tenets of Hill Collins's Black Feminist Epistemology as guideposts for preparing students to engage with community-based projects.

Within the field of technical communication, a wide range of scholars have addressed the strength of community-based work in technical communication (i.e., Eble and Gaillet 2004; Scott 2009) and the struggles instructors encounter as they try to work ethically with those communities (i.e., Coogan 2002; Cushman 1996; and Grabill 2007). Both Del Hierro and Moeller in this section also ask us to think more broadly about the work technical communicators can do in the community through Hip Hop and Feminist Disability Studies approaches. In developing community-based projects through a Black Feminist framework, I extend efforts to approach communities ethically and to prepare students for this work through reflection and critical practices. Of course, detailing a pedagogical approach for community-based work is difficult: every community, classroom, and student is different. This framework approaches communities flexibly through a Black Feminist framework. For example, some instructors might organize a whole class around the tenets. In my own classroom, however, I typically assign one community-based project, using the first two tenets to guide students' in-community activity and the last two tenets as assessment principles. Thus, as I discuss the tenets below, I'll group the first two and the second two and provide activities that facilitate student learning.

APPLYING TENETS 1 AND 2: DIALOGUE AND LIVED EXPERIENCE

This approach to Community-Based Teaching takes dialogue and lived experiences (two tenets of Black Feminist Theory) as the central form of knowledge-making about local communities, projects, and the ways technical communication classes might enter into the community. As I discuss elsewhere (Moore 2013) and as a number of the authors in

Table 8.1. Enacting Tenets 1 and 2

Tenets #1 and #2 Integration: What Students Do in the Community	
Tenet #1	Students focus their community work on learning from citizens rather than merely helping citizens.
Tenet #1	Students draw on citizens' lived experiences to assess what they already know about a project, topic, place, or concern.
Tenet #2	Students use dialogue with citizens and with one another to determine the accuracy of their conclusion.
Tenet #2	Students engage in projects that create lived experiences they can use to make knowledge claims.

this collection note, storytelling, lived experiences, and dialogue about knowledge claims can be a cornerstone of understanding local communities—and I argue that entering communities requires a sensitivity and awareness of difference and diverse values. Inviting students to become learners about the local community rather than merely practitioners in the field revises the objectives and values of the classroom in ways that value Black Feminist Epistemology.[13] In addition, this pedagogy asks students to approach the community as a site of both learning and doing. What does this look like? One approach to valuing lived experience and dialogue is to ask students to engage with citizens through stories and storytelling. Such a move solicits knowledge about lived experiences vis-à-vis dialogue.[14] But the approach I advocate here moves beyond dialogue and lived experiences in that it insists that students assess their community work through an ethic of care and personal responsibility (see table 8.1).

Implementing these tenets can be tricky, but using both readings and classroom activities, we can prompt students to both value community members' knowledge and engage with dialogue and the lived experiences of others. Because the particulars of any community-based project should depend on local situations, here I focus on classroom activities that help students build knowledge using Black Feminist Epistemology. For example, I often use mapping to help my students engage with multiple ways of understanding the makeup of a project, community, or organizations. I've used the activity below (see figure 8.1) to prepare students to attune themselves to what they might not know about a community and to identify places where their own ideas can be contested and/or augmented by another perspective, by a dialogue, by an interaction in the community. Sullivan and Porter (1997) suggest that postmodern mapping and the repeated return to and revision of maps can keep us from overly certain portraits of our research sites. Drawing

> Sample Activity:
> Engaging Dialogue and Lived Experience to Assess Knowledge Claims
> BEFORE CLASS: Read Chapter 4 of Sullivan and Porter's *Opening Spaces* [Alternatively, instructors can model postmodern mapping ahead of time.]
> IN CLASS, BEFORE COMMUNITY ENGAGEMENT: Map What You Know—and What You Don't
> Students map what they know about the community they're working in, both the broader community and the local organizations or community members they'll be researching, working with, etc. Maps should include institutions, cultural knowledge, demographic information, and additional knowledge that students have. Instructors prompt students to identify what they don't know about the community and also what parts of the map might be contested either by other students or by citizens.
> IN THE COMMUNITY: Students meet with organization, conduct interviews or observations, and/or conduct a work session. [This is dependent on what kind of project students are doing.]
> POST-MAPPING: Students identify unknowns or contested ideas that they might be able to discover either through experiences in the community or talking with community members
> POST-COMMUNITY ENGAGEMENT: Students report back and add to or revise the map.

Figure 8.1. Sample Activity for Tenets 1 and 2

on this idea, this activity moves toward valuing the lived experience of both students and community members and prompts students to assess and revise their maps through dialogue.

APPLYING TENETS 3 AND 4: ETHICS OF CARE AND ACCOUNTABILITY

Black Feminist Theory provides a critical framework that legitimizes the importance of care in the classroom, and I suggest that this focus on ethics can become the frame for understanding the lived experiences and dialogues advocated above (see table 8.1). In other words, when we ask students to enter the community and value citizen stories and experiences, we are also asking them to care about their community partners and to spend emotional energy on working with others. Banks-Wallace (2005) divides the Ethic of Care into three parts: personal expression, emotion, and empathy. In this framework, students are assessed not only on *what* they've done for their community partners but also *how* they've done it. We can ask students to consider these kinds of questions:

- What kinds of empathetic moves have you made?
- How have you expressed your own emotion and personal relationship to these community members?
- Describe an activity in which you worked to be empathetic with your community partners.

Such prompts require reflection and an awareness of the demonstrable moves students make to listen humbly to citizens' stories, knowledge, and lived experiences. These criteria are difficult to assess. However, through an emphasis on personal accountability, students can assess themselves; these tenets serve as the foundation of student self-reflection assignments and/or a student self-assessment tool, depending upon the project. Through reflective reporting of their activities, students become accountable for their own actions and interactions and attuned to the attitude with which they engage with community-based work.

Some technical communicators will, of course, be tempted to ask: is this really technical communication? How does this align with the values of the field? My research with Vortex Communications, however, suggests that in the public sphere, especially when working with diverse communities, demonstrating care for the community is fundamental in building relationships and effectively communicating across cultural bounds. Scholars in civic and community research often sidestep, ignore, or overlook the important steps of developing working relationships with their community partners. There are, of course, notable exceptions, like Ellen Cushman and Miriam Williams and a number of the authors from this collection, but our students and the field at large can benefit from more pronounced treatments of relationship-building and the ethics of relational work.

Teaching students to engage with these last two tenets is more difficult, perhaps, than the first two tenets. Although the first two tenets are challenging to teach and enact well, fairly straightforward strategies and activities like mapping, interviewing, etc., allow students to stumble into

Table 8.2. Enacting Tenets 3 and 4

Tenets #3 and #4 Integration: How Students Enter the Community	
Tenet #3	Students develop strategies for listening empathetically to community partners.
Tenet #3	Students express their emotional responses to community work, including discomfort, incredulity, and joy.
Tenet #4	Students detail goals that meet criteria of an ethic of caring and engage in reflective discussion and writing that holds them personally accountable for those goals.

this new epistemological framework. Tenets 3 and 4 can also be enacted through activities, but instructors, community members, and students alike make themselves vulnerable in enacting these tenets. For example, in encouraging students to express their emotions, we risk inviting hostility and potentially complex feelings about themselves and the community. In one of my classes, for example, giving space for my students to express their emotions invited a discussion of the discrimination they endured as part of the projects. That said, the activity in figure 8.2 helps students (and instructors) become empathetic and expressive in and about their community engagements. This activity is most appropriate for undergraduate students—though it might also be used in graduate level classes—because it helps them practice an ethic of care through model and dialogue. In asking students to use an "improv" logic, instructors prompt students to listen and affirm rather than contest, critique, or resist. This activity does *not* make students care, and it doesn't force students to use this within the community, but students can use this model as a way to describe their efforts to empathize with community partners.

In working with graduate students on these two tenets, I assign readings (rather than activities) that help them analyze their experiences and their orientations to the community. These readings include *Black Feminist Thought* by Hill Collins (2008), *La Frontera* by Gloria Anzaldúa (2007), *Sister Outsider* by Audre Lorde (1984), *Decolonizing Methodologies* by Linda Tuhiwai Smith (2012), *Reading Chinese Fortune Cookie* by LuMing Mao (2006), among other shorter works. These readings provide students with a critical frame for considering difference through a range of lenses. As graduate students in one class entered community projects, they used the concepts in the readings to conceptualize their own subjectivity and the subjectivity of other community members. In addition to seminar-like discussions, I developed a "studio time" where we applied readings to the community projects they were undertaking, providing opportunities to map, discuss, and write about the ways the

> **IN-CLASS ACTIVITY: ENGAGING EMPATHETICALLY AND EXPRESSIVELY: AN IMPROV DISCUSSION**
>
> The first rule of improv is that you always say yes—and then respond. The purpose of this discussion is to begin practicing empathetic engagement with others. Each pair of discussants will receive a controversial statement and have two conversations about the statement: first, we seek to contest; second we seek to understand.
>
> In the first discussion Student A expresses his or her thoughts about a topic, and Student B attempts to contest Student A's statements.
>
> In the second discussion, Student B expresses his or her thoughts about a topic, and Student A uses improv logic: to always agree and then try to learn more and/or try to move the discussion along.
>
> Writing Prompt: What did you learn about Student A during the Contestable Discussion? What can you remember of your partner's views?

Figure 8.2. Activity for Enacting Tenets 3 and 4

readings and their lived experiences of the community project converge, diverge, and/or complicate one another.

INSTRUCTORS ASSESS AND PLAN WITH BLACK FEMINIST THEORY

A Black Feminist Theoretical orientation toward classroom projects resists the tendency to solely value effectiveness of documents, the clarity of ideas, and the efficiency of the project they're undertaking. Rather, students are encouraged to value the new knowledge they've gathered with citizens. The effectiveness of this approach relies on instructors also valuing dialogue and lived experience. In figure 8.3, I visualize the way this works for instructors. Instructors both engage with students' lived experiences and use dialogue to develop claims (tenets 1&2), and they do so through an ethic of caring for which they are personally accountable (tenets 3&4). Thus, an instructor invested in this pedagogical approach will value students' experiences and engage with dialogue about them as part of the project evaluation and assessment. Traditional notions of

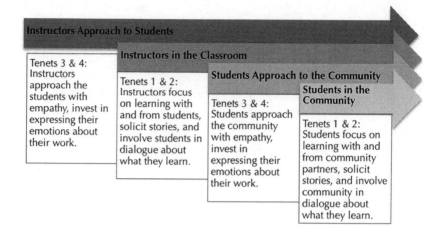

Figure 8.3. Black Feminist Approach to the Classroom

assessment get revised to include students' experiences and dialogues with students about their work. Rather than assign grades from the top down, for example, instructors can invite students to discuss their criteria (tenets 3&4 as well as others) and progress and can use dialogue to assess claims about their course grade. Using dialogue with students to assess their work is tricky, but it demonstrates at an infrastructural level that the ways of making knowledge in the classroom—as in the community—are diverse and not merely created by an "expert." Rather, instructors seek wisdom from the students with the same humility that students are asked to enter the community. This approach to assessment echoes Cruz and Medina's discussion of grade contracts in this collection in that it, too, seeks to "avoid re-articulating systems of exclusion and oppression" (47).

Finally, instructors using Black Feminist Theory apply the same kinds of ethics of care and personal accountability that students are expected to demonstrate. While I support the social contract that asks students to respect the authority of the instructor, a Black Feminist Theory approach to community work asks both students and instructors to be sensitive, empathetic, and emotionally engaged with one another. This may not be comfortable for all instructors, and the appropriate levels of empathetic engagement vary from context-to-context. However, students working in the community experience vulnerability, uncertainty, and—in some cases—rejection and discrimination. Instructors asking students to do this kind of emotional work, too, must hold themselves accountable for engaging in dialogues that extend beyond the content-based knowledge often valued in the classroom.

LOGISTICS AND IMPLEMENTATION

Even without a focus on Black Feminist Epistemology, community-based projects pose particular problems for technical communication instructors. Problems arise in part because they are difficult to control and because their implementation requires securing community sites and projects. I want to briefly address these problems with the pedagogy I lay out here. Instructors who adopt a Black Feminist approach *and* teach through community-based projects should first devote time to understanding and listening to the communities they work in. Community-based researchers, for example, may have already engaged with the community through dialogue and lived experience in their research, and they use these experiences to identify community struggles, sites for engagement, or projects appropriate for their students. Because engaging with communities is more than just serving up the special skills we have, instructors should resist the impulse to charge into a community with their document design skills and technical editing expertise. This impulse, while often benevolent, sidesteps the importance of valuing community dialogue and lived experiences.

If instructors want to employ Black Feminist Theory in community-based projects but do not conduct community-based research, the first step is to learn about the community. Class projects and students can help with this, and an effective first research project in the classroom can be to investigate the community. In a graduate level course on Contemporary Rhetorics, for example, I assigned students to engage with the community through volunteer work, participation in a community organization, or community research. The objective of the assignment was to learn about the local community, so the final deliverables were primarily discussions, maps, and informal oral reports of discussions with members of the local organizations they worked with. Such a project seemed strange to the students, many of whom wanted to "do community work." However, in order to fully prepare both me (new to the local community) and them (thoroughly entrenched in the academy) for understanding local community needs, this open-ended project guided them to see the local community apart from their own expertise and outside their own course-driven objectives. In an undergraduate class, this kind of approach has seemed inappropriate. Instead, I assign undergraduates to develop research skills by investigating the local community before developing proposals, strategies, or identifying community problems. These projects frustrate undergraduates, who want to jump to proposing solutions to problems, but, like Sackey's approach to environmental justice, it also re-frames the idea of problem-solving as always already within a rhetorical and cultural context.

In short, the logistics of securing a community-based project challenge instructors to interrogate their own motivation for assigning these projects, relegating our own desires to the backburner and holding ourselves personally accountable for our actions. If you're reading this chapter and want to work on a community-based project, you may need to begin with your own community-based research, relying on a Black Feminist Methodology. Or you can develop community-based research projects for your students that ask them to rely on storytelling, dialogue, and lived experiences with an ethic of caring and personal accountability.

CONCLUSION

Using a Black Feminist Epistemology in community-based research and classroom projects presents challenges, but it also presents opportunities for instructors to resist the elite patriarchal approaches to technical communication that dominate the field. Integrating this pedagogy whole or in parts helps to expose rather than suppress knowledge work that represents and extends from black women's work. I employ this framework differently in different classes, consistently asking students to engage with stories, lived experiences, and dialogue as legitimate and important ways of making knowledge. Because theory is not built in isolation, at levels of abstraction, or through reading alone, I maintain that Black Feminist Theory (and others) helps technical communicators build knowledge that is ultimately more effective, practical, and locally flexible. Demonstrating the value of black women's theoretical and experiential knowledge in the classroom is one way to encourage students of technical communication to re-see the way theory works in relationship to practice. In addition, integrating Black Feminism into technical communication broadens the traditional scope of Black Feminism, expanding the ways it has been used. In fact, when I spoke with Hill Collins at a symposium in 2012, I mentioned that I was using her work to understand public engagement in transportation planning, and she was quite surprised at the reach of her work. In drawing on Black Feminist Theory, then, we not only strengthen our abilities, but we also begin building an interdisciplinary effort toward social justice and equity that includes Black Feminists in the kinds of user-driven discussions highlighted in technical communication.

Much more work can and should be done to expand the theoretical and methodological foundations for technical communication. As our field faces the complexities of globalization, diversification, and ever-expanding student populations, more researchers will confront ethical problems like the one I faced at Vortex: do I try to apply traditional frames to understand this new context or problem? Methodological treatises like Sullivan and Porter's *Opening Spaces*, pedagogical cases like Haas's "Race, Rhetoric, and Technology," and the many discussions throughout this collection suggest good reasons to arm our field with new and more ethical theories and frames. Sullivan and Porter caution researchers against pulling methodologies off the shelf and applying them blindly to situations, asking instead that we expand our approaches because "to do less risks our larger aims . . . to help empower and liberate through the act and art of writing" (Sullivan and Porter 1997, 187). Published fifteen years later, Haas (2012) reports on her work with developing a course on "Race, Rhetoric, and Technology" using a decolonial, critical race framework. Part pedagogical case study, part review of the field, Haas's article demonstrates both the limits of the field's attentiveness to critical race and decolonial frameworks as well as the potential gain: that using a "decolonial, critical race framework in our research, pedagogy, and other technical communication workplace practices will rupture dominant notions of what it means to be—and who gets to be considered—technical or technologically advanced" (304). Not only is our research better when we diversify our pedagogies but so too can the sites and workplace practices we study become more equitable. More cases, theoretical discussions, field reports, and research projects are needed to effectively equip practitioners and scholars in technical communication with the expanded methodologies necessary to respond to the diverse contexts in which they work, learn, and research.

Notes

1. This study was approved by Purdue University's Institutional Research Board.
2. This naming occurred when the one white woman who works for the company was out of town and therefore not in the office. At the time of my research, nine women worked for Vortex Communication.
3. Vortex Communications is a firm that develops public engagement processes for large-scale public planning initiatives, including strategic planning for cities and nonprofits and transportation and urban planning for governmental organization. Here, public engagement refers to these large-scale, often-mandated projects.
4. Banks-Wallace (2005) sees a particular consonance of Womanist Theory, Womanism, and Black Feminist thought, which she claims has been used "interchangeably to designate [theoretical frameworks that articulate and interpret an African American woman's standpoint]" (316). Other scholars (Alexander-Floyd and Simien 2006; Hudson-Weems 1989; Phillips 2006) make a finer distinction. Hill Collins's (2008) *Black Feminist Thought*, unsurprisingly given the title, focuses on Black Feminism rather than Womanism. In this chapter I use the term Black Feminist Theory to reference specifically Hill Collins's naming of theory that draws on black women's experiences. But the corollary theories of Womanism Africana Womanism, Africana Feminism, and African Feminism are "sibling" extensions of black women's standpoints, according to Phillips (2006).

5. The point here is not to vilify feminists or feminism; rather, I hope to acknowledge the kinds of exclusionary practices that extend out of feminism and move forward with feminists to include women of color in both theory and practice.
6. My self-identification as a white woman complicates the notion of white feminism/Black Feminism. Can feminists who are white be anything other than a white feminist? How do we create opportunities for acknowledging the historical dismissal of women of color and then, as Mama (2001) has worked to do, rescue or recover the term feminist? I write this chapter using Black Feminism because I identify technical communication as a discipline that has yet to acknowledge the need for multiple feminisms, but I want to be clear that I do see a space to move forward *with* rather than *against* the term feminist.
7. The origin, of course, is not so neat as I've told it. I draw on a wide range of black feminist and womanist authors in cobbling together the origins of Black Feminism, but I want to be careful not to assert a truthiness to the story that exceeds my own expertise or experience.
8. Theory, as I describe it—and as Black Feminists describe it—does not occur separate from experiences; rather, it's through experiences that theory is built.
9. *Black Feminist Thought* was released in its second edition in 2008. The second edition transformed early perspectives on Black Feminism as exclusive to black women into a more inclusive social justice effort. Where the first edition featured clearly demarcated boundaries, Hill Collins's second edition draws together, perhaps in the spirit of Womanism, all efforts towards equality while still maintaining that Black Feminism draws from black women's experiences.
10. Again, the demarcation between Black Feminism and Womanism is blurry, but where I would not call myself a Black Feminist, I do call myself a Womanist. I follow Phillips (2006) who advises that anyone dedicated to social justice and equality can be a Womanist. The lines of inclusion for Black Feminism are more clearly tied to racial identity, and therefore less permeable.
11. In this way, I follow in the tradition of other critical race scholars who eschew the use of white male frames and traditions in order to privilege more inclusive traditions.
12. The Great Rivers Greenway organization is a long-standing client of Vortex Communications.
13. It's important to note that this revision doesn't completely rectify racial problems or oppressive regimes—and it doesn't necessarily require a focus on addressing racial difference. Rather, it chips away at expectations that academic classroom approaches ought to be in service of patriarchal, white supremacist expectations.
14. For more details, see Moore 2013.

References

Agboka, Godwin. 2013. "Participatory Localization: A Social Justice Approach to Navigating Unenfranchised/Disenfranchised Cultural Sites." *Technical Communication Quarterly* 22 (1): 28–49. https://doi.org/10.1080/10572252.2013.730966.

Alexander-Floyd, Nikol, and Evelyn Simien. 2006. "Revisiting 'What's in a Name?': Exploring the Contours of Africana Womanist Thought." *Frontiers* 27 (1): 67–89. https://doi.org/10.1353/fro.2006.0011.

Anzaldúa, Gloria. 2007. *Borderlands/La Frontera: The New Mestiza*. 3rd ed.. San Francisco: Aunt Lute Books.

Banks-Wallace, Joanna. 2005. "Womanist Ways of Knowing: Theoretical Considerations for Research with African American Women." In *The Womanist Reader*, ed. Layli Phillips, 313–326. New York: Routledge.

Blackmon, Samantha. 2002. "'But I'm *Just* White'; Or, How 'Other' Pedagogies Can Benefit All Students." In *Teaching Writing with Computers*, ed. Brian Huot and Pam Takayoshi, 92–102. New York: Houghton Mifflin.

Blackmon, Samantha. 2004. "Which Came First? On Minority Recruitment and Retention." *CPTSC Proceedings*: 1–3.

Blackmon, Samantha. 2014. "Who Gets to Represent the Other?: On Race and Difference." *Not Your Mama's Gamer*. February 7, 2014. http://www.samanthablackmon.net/notyourmamasgamer/?p=4572.

Blythe, Stuart, Jeffrey T. Grabill, and Kirk Riley. 2008. "Action Research and Wicked Environmental Problems: Exploring Appropriate Roles for Researchers in Professional Communication." *Journal of Business and Technical Communication* 22 (3): 272–298. https://doi.org/10.1177/1050651908315973.

Brasseur, Lee E. 1993. "Contesting the Objectivist Paradigm: Gender Issues in the Technical and Professional Communication Curriculum." *IEEE Transactions on Professional Communication* 36 (3): 114–123. https://doi.org/10.1109/47.238051.

Christian, Barbara. 1988. "The Race for Theory." *Feminist Studies* 14 (1): 67–79. https://doi.org/10.2307/3177999.

Clark, David. 2004. "Is Professional Writing Relevant? A Model for Action Research." *Technical Communication Quarterly* 13 (3): 307–323. https://doi.org/10.1207/s15427625tcq1303_5.

Cole, Johnetta. 1986. *All American Women: Lines That Divide, Ties That Bind*. New York: Free Press.

Cole, Johnetta, and Beverly Guy-Sheftall. 2003. *Gender Talk: The Struggle for Women's Equality in African American Communities*. New York: Ballentine Press.

Combahee River Collective. 1977. "The Combahee River Collective Statement." In *Home Girls, A Black Feminist Anthology*, ed. Barbara Smith, 264–274. New York: Women of Color Press.

Coogan, David. 2002. "Public Rhetoric and Public Safety at the Chicago Transit Authority: Three Approaches to Accident Analysis." *Journal of Business and Technical Communication* 16 (3): 277–308. https://doi.org/10.1177/1050651902016003002.

Crenshaw, Kimberlé. 1989. "Demarginalizing the Intersection of Race and Sex: A Black Feminist Critique of Antidiscrimination Doctrine, Feminist Theory, and Antiracist Politics." *University of Chicago Legal Forum* 1989 (1): 139–167.

Crenshaw, Kimberlé. 1991. "Mapping the Margins: Intersectionality, Identity Politics, and Violence against Women of Color." *Stanford Law Review* 43 (6): 1241–1299. https://doi.org/10.2307/1229039.

Cushman, Ellen. 1996. "The Rhetorician as an Agent of Social Change." *College Composition and Communication* 47 (1): 7–28. https://doi.org/10.2307/358271.

Davis, A. Y. 1999. *Blues Legacies and Black Feminism: Gertrude "Ma" Rainey, Bessie Smith, and Billie Holiday*. New York: Vintage Books.

Davis, Angela. 2014. Keynote Speech at the Conference on College Composition and Communication, Indianapolis, IN.

Dubinsky, James M. 2001. "Service-Learning and Civic Engagement: Bridging School and Community through Professional Writing Projects." Paper presented at the Annual Meeting of the Warwick Writing Programme, Department of English and Comparative Literary Studies, University of Warwick, 5th, Coventry, England, March 26–27, 2001.

Durack, K. T. 1997. "Gender, Technology, and the History of Technical Communication." *Technical Communication Quarterly* 6 (3): 249–260.

Eble, Michelle F., and Lynee Lewis Gaillet. 2004. "Educating Community Intellectuals: Rhetoric, Moral Philosophy, and Civic Engagement." *Technical Communication Quarterly* 13 (3): 341–354. https://doi.org/10.1207/s15427625tcq1303_7.

Flower, Linda. 2008. *Community Literacy and the Rhetoric of Public Engagement*. Carbondale: Southern Illinois University Press.

Flynn, Elizabeth. 1997. "Emergent Feminist Technical Communication." *Technical Communication Quarterly* 6 (3): 313–320. https://doi.org/10.1207/s15427625tcq0603_6.

Gilligan, Carol. 1982. *In a Different Voice*. Cambridge, MA: Harvard University Press.

Grabill, Jeffrey T. 2001. *Community Literacy Programs and the Politics of Change*. Albany: SUNY Press.

Grabill, Jeffrey T. 2007. *Writing Community Change: Designing Technologies for Citizen Action*. New York: Hampton Press.

Gurak, Laura J., and Nancy L. Bayer. 1994. "Making Gender Visible: Extending Feminist Critiques of Technology to Technical Communication." *Technical Communication Quarterly* 3 (3): 257–270. https://doi.org/10.1080/10572259409364571.

Haas, Angela M. 2012. "Race, Rhetoric, and Technology: A Case Study of Decolonial Theory, Methodology, and Pedagogy." *Journal of Business and Technical Communication* 26 (3): 277–310. https://doi.org/10.1177/1050651912439539.

Hill Collins, Patricia. 1996. "What's in a Name? Womanism, Black Feminism, and Beyond." *Black Scholar* 26 (1): 9–17. https://doi.org/10.1080/00064246.1996.11430765.

Hill Collins, Patricia. 2008. *Black Feminist Thought: Knowledge Consciousness, and the Politics of Empowerment*. 2nd ed. New York: Routledge.

Hill Collins, Patricia. 2012. Keynote Address. Purdue University, West Lafayette, IN.

hooks, bell. 1991. "Theory as Liberatory Practice." *Yale Journal of Law and Feminism* 4 (1): 1–12.

Hudson-Weems, Clenora. 1989. "Cultural and Agenda Conflicts in Academia: Critical Issues for Africana Women's Studies." In *Excerpted in The Womanist Reader*, ed. Layli Phillips, 37–43 New York: Routledge.

Jones, Natasha. 2014. "The Importance of Ethnographic Research in Activist Networks." In *Race, Ethnicity, and Identity in Technical Communication*, ed. Miriam Williams and Octavio Pimentel, 46–62. Amityville, NY: Baywood.

King, Deborah K. 1988. "Multiple Jeopardy, Multiple Consciousness: The Context of a Black Feminist Ideology." *Signs (Chicago, Ill.)* 14 (1): 42–72. https://doi.org/10.1086/494491.

Koerber, Amy. 2000. "Toward a Feminist Theory of Technology." *Journal of Business and Technical Communication* 14 (1): 58–73. https://doi.org/10.1177/105065190001400103.

Lay, M. M. 1991. "Feminist Theory and the Redefinition of Technical Communication." *Journal of Business and Technical Communication* 5 (4): 348–370.

Layne, Alex, and Samantha Blackmon. 2013. "Self-Saving Princess: Feminism and Post-Play Narrative Modding." *Ada: A Journal of Gender, New Media, and Technology* 2.

Long, Elenore. 2008. *Community Literacy and the Rhetoric of Local Publics*. West Lafayette: Parlor Press.

Lorde, Audre. 1984. *Sister Outsider: Essays and Speeches*. Freedom, CA: Crossing Press.

Mama, Amina. 2001. "Talking about Feminism in Africa." *African Feminisms* 50:58–63.

Mao, LuMing. 2006. *Reading Chinese Fortune Cookie*. Logan: Utah State University Press. https://doi.org/10.2307/j.ctt4cgqqt.

Moore, Kristen R. 2013. "Exposing the Hidden Relations: Storytelling, Pedagogy, and the Study of Policy." *Journal of Technical Writing and Communication* 43 (1): 63–78. https://doi.org/10.2190/TW.43.1.d.

Ogunyemi, Clenora O. (Original work published 1985) 2001. "Womanism: The Dynamics of the Contemporary Black Female Novel in English." In *Excerpted in The Womanist Reader*, ed. Layli Phillips, 21–36. New York: Routledge.

Ornatowski, Cezar M., and Linn K. Bekins. 2004. "What's Civic about Technical Communication? Technical Communication and the Rhetoric of 'Community.'" *Technical Communication Quarterly* 13 (3): 251–269. https://doi.org/10.1207/s15427625tcq1303_2.

Phillips, Layli. 2006. "Introduction." In *The Womanist Reader*, ed. Layli Phillips, xix–lv. New York: Routledge.

Richardson, Flourice. 2014. "The Eugenics Agenda: Deliberative Rhetoric and Therapeutic Discourse of Hate." In *Communicating Race, Ethnicity, and Identity in Technical Communication*, ed. Miriam Williams and Octavio Pimentel, 7–22. Amityville, NY: Baywood Publishing.

Ross, Susan Mallon. 1994. "A Feminist Perspective on Technical Communicative Action: Exploring How Alternative Worldviews Affect Environmental Remediation Efforts." *Technical Communication Quarterly* 3 (3): 325–342. https://doi.org/10.1080/10572259 409364575.

Sauer, Beverly A. 1994. "Sexual Dynamics of the Profession: Articulating the Ecriture Masculine of Science and Technology." *Technical Communication Quarterly* 3 (3): 309–323. https://doi.org/10.1080/10572259409364574.

Savage, Gerald. 2013. "Global Technical Communication: A Voice of Neo-colonialism or Social Justice?" Keynote Speech, Texas Tech University May Seminar, Lubbock, TX.

Savage, Gerald, and Kyle Mattson. 2011. "Perceptions of Racial and Ethnic Diversity in Technical Communication Programs." *Programmatic Perspectives* 3 (1): 5–57.

Scott, J. Blake. 2004. "Rearticulating Civic Engagement Through Cultural Studies and Service-Learning." *Technical Communication Quarterly* 13 (3): 289–306. https://doi.org/10.1207/s15427625tcq1303_4.

Scott, J. Blake. 2009. "Civic Engagement as Risk Management and Public Relations: What the Pharmaceutical Industry Can Teach Us about Service-Learning." *College Composition and Communication* 61 (2): 343–366.

Simmons, W. Michele. 2007. *Participation and Power: Civic Discourse in Environmental Policy Decisions*. Albany: SUNY Press.

Smith, Barbara. 1978. "Toward a Black Feminist Criticism." *Radical Teacher* 7:20–27.

Smith, Barbara, ed. 1983. *Home Girls: A Black Feminist Anthology*. New Brunswick, NJ: Rutgers University Press.

Springer, Kimberly. 2002. "Third Wave Black Feminism?" *Signs (Chicago, Ill.)* 27 (4): 1059–1082. https://doi.org/10.1086/339636.

Sullivan, Patricia, and James Porter. 1997. *Opening Spaces: Writing Technologies and Critical Research Practices*. Santa Barbara, CA: Praeger.

Taylor, Ula Y. 1998a. "Making Waves: The Theory and Practice of Black Feminism." *Black Scholar* 28 (2): 18–28. https://doi.org/10.1080/00064246.1998.11430912.

Taylor, Ula Y. 1998b. "The Historical Evolution of Black Feminist Theory and Praxis." *Journal of Black Studies* 29 (2): 234–253. https://doi.org/10.1177/002193479802900206.

Truth, Sojourner. (Original work published 1851) 2005. "Speech at the Women's Rights Convention in Akron, OH." In *Available Means: An Anthology of Women's Rhetoric(s)*, ed. Kate Ronald and Joy Ritchie, 144–146. Pittsburgh: University of Pittsburgh Press.

Tuhiwai Smith, Linda. 2012. *Decolonizing Methodologies: Research and Indigenous Peoples*. London: Zed Books.

Williams, Miriam F. 2017. *From Black Codes to Recodification: Removing the Veil from Regulatory Writing*. London: Routledge.

Williams, Miriam, and Octavio Pimentel. 2012. "Introduction: Race, Ethnicity, and Technical Communication." *Journal of Business and Technical Communication* 26 (3): 271–276. https://doi.org/10.1177/1050651912439535.

Williams, Miriam, and Octavio Pimentel. 2014. *Communicating Race, Ethnicity, and Identity in Technical Communication*. Amityville, NY: Baywood.

9
ADVOCACY ENGAGEMENT, MEDICAL RHETORIC, AND EXPEDIENCY
Teaching Technical Communication in the Age of Altruism

Marie E. Moeller

What does it mean to be an advocate, to be a scholar-teacher focused on advocacy as a practice and an area of study? As a scholar-teacher of technical communication, this question directs my practices both in my research and in the classroom—how can I teach and discuss technical communication as a scholar-teacher and human invested in notions of advocating and advocacy? What does that work look like? How does technical communication drive action, and how can technical communicators engage in a kind of critical advocacy work that both advocates and is culturally and socially attentive to their own advocacy practices? In this chapter I outline one specific way advocacy work and technical communication can pair to alter the lives of people written about and to. Thus, this chapter grounds advocacy work within technical communication—accomplishing the goals of both an advocate and a technical communicator/ technical communication scholar-teacher.

To do so, I focus on the role of technical communication in theorizing, critiquing, and enacting advocacy-based medical rhetoric dissemination. I focus on medical rhetoric and data dissemination because our current medicalized culture gives birth to the kind of information dissemination that in turn galvanizes such medicalization, and I expand on this later in the chapter. Employing the theoretical framework of Feminist Disability Studies (FDS), I show how students in a technical communication classroom can engage with medical technical communication and data from nonprofit advocacy locations to de-stabilize harmful, multi-layered, normative medical narratives about bodies and health. I use an example from the medical advocacy website of Susan G. Komen for the Cure (SGKftC) to illustrate how and why the field of technical communication must focus on social justice action and also on a constant critical checks-and-balances approach to nonprofit rhetoric

DOI: 10.7330/9781607327585.c009

and engagement in our current cultural milieu. Terming something advocacy work, in other words, often creates a kind of cultural space absence of critique but full of action. This chapter calls for and argues that both critique and construction are needed for cultural change, especially with regard to the technical communication employed by advocacy organization and locations.

To explain one such method of critical engagement with advocacy-based technical communication and technical communication advocacy, I first articulate a theory of Feminist Disability Studies (FDS) as situated in technical communication. FDS is an important theoretical apparatus for technical communication, and specifically for advocacy organizations and advocacy practices, because FDS calls attention to bodies that comprise the margins, to the disempowerment and disenfranchisement of various populations of individuals often found outside of or bastardized by the cultural "norm." This is important for the field of technical communication, as FDS can assist technical communicators and teachers of technical communication in highlighting the potential to both employ and resist harmful normative narratives about health and wellness, as well as call attention to the power inherent in writing for advocacy organizations that currently play a large role in shaping medicalized discourse and understandings about health and well-being. Medical advocacy groups are a location ripe to do such critical technical communication work, as the organizations work on the behalf of marginalized populations, but some organizations simultaneously shore up the very cultural narratives that have caused the marginalization of such populations. Through such attention to location and language-work in that location, technical communicators are attending to the work of technical communication as a vehicle of language-into-action; Katz (1992) argues that technical communication is based upon deliberative rhetoric, upon the process by which language is used to forward and encourage particular types of action. Using FDS as a theoretical framework, then, provides a way to see how advocacy organizations have the power, through their dissemination of technical medical data and rhetoric, to shift problematic cultural norms rather than shore them up.

After explaining FDS as a framework, I discuss how and why, in my classrooms, I foreground the importance of nonprofits, and the technical communication work of nonprofits, in our current historical moment. I ask students to think of advocacy work under this theoretical framework to call attention to the bodies for which medical advocacy nonprofits purport to advocate on the behalf of and the technical communication they use to do it with. Next, I illustrate how Mark Ward Sr.'s discussion of exigence in technical communication design allows us to understand why cultural metaphors employed by advocacy websites may influence the expedient nature of advocacy group information architecture. Finally, I rhetorically analyze components of the SGKftC via FDS to illustrate how such expedient framing can authorize articulations of data and language that rely on narratives of normalcy about particular populations in order to shore up our dominant belief systems—altruism without change. I do so to provide students with concepts that travel in technical communication so that they can both develop their own critical awareness of the cultural work and power of technical communication, and of the place of technical communication in forwarding the un/ethical work of nonprofits.

FEMINIST DISABILITY STUDIES: ENGAGING AND CRITIQUING NORMAL

> "Feminist Disability Studies can suggest an avenue of critique for reductive biological understandings of both gender and disability" (5).—Kim G. Hall, "Introduction"

In this section I lay out the tenants of Feminist Disability Studies (FDS) as one possible theoretical framework by which technical communication can engage, analyze, and critique nonprofit advocacy organizations. I employ this framework as a viable and useful frame for understanding the cultural work as FDS is a framework that calls attention to the multi-layered activities of cultural identity construction, marginality, and normalcy—of language, of bodies, of human and organizational practices. By nature, advocacy organizations are created to advocate for causes, people, and difficulties that are deemed in need of more attention than such issues receive (either governmentally, socially, culturally, or otherwise) typically, issues that operate more on the margins than is deemed functional or appropriate. Employing FDS in technical communication settings thus highlights connection with marginality and the importance of technical communication's attention to the margins by calling attention to the ways in which advocacy organizations use medical technical communication to forward normative cultural narratives about bodies while simultaneously shifting movement of the issue such advocacy organizations work on the behalf of—a contradictory kind of politics.

Scholars have discussed feminist disability theory in numerous ways in recent years (Dryden 2013; Erevelles 2011; Hall 2011; Hamraie 2013; Kafer 2013; Kafer and Jarman 2014; O'Donovan 2013; Piepmeier,

Cantrell, and Maggio 2014; Schalk 2013; Söder 2009; Thomas 2006; Titchkosky 2005, 2011; Tremain 2013). For this chapter I'll rely on germinal work from Rosemarie Garland Thomson and Kim Q. Hall. In her early work, "Integrating Disability, Transforming Feminist Theory," Garland Thomson (2004) argues that disability studies did a great deal of wheel inventing by not integrating with feminist studies and vice versa. The goal of an FDS approach, then, is to integrate the two forms of theoretical dispositions as a way "to augment the terms and confront the limits of the ways we understand human diversity, the materiality of the body, multiculturalism, and the social formations that interpret bodily differences" (Garland-Thomson 2004, 75). In other words, employing the tools of disability and multiple forms of feminist studies in conversation with each other will deepen, expand, and challenge such theories. Bringing such work to technical communication responds to calls from Kristen Moore's chapter in this section of the book to challenge monolithic notions of feminism in that "feminism in technical communication most often refers to white feminism, privileging the importance of gender-based oppression over other forms of oppression including race, class, and sexuality" (269). In this chapter, then, I discuss how FDS as a theoretical apparatus may illustrate the dangers of using technical communication to create hierarchies of oppression and normalizing them by obfuscating the lived experiences of the bodies and lives that technical communicators write to and about.

As such, an FDS approach allows for the consideration of issues such as (but not limited to) the status of the lived body, the politics of appearance, the medicalization of the body, and the privileges and pressures of normalcy. In addition to these foci, I argue that FDS can also include attention to how **language** shapes and influences the status of the lived body, the politics of appearance, the medicalization of the body, and our understandings of normalcy. For this chapter I specifically focus on technical medical communication stemming from medical advocacy organizations, to understand how such technical communication contributes to the aforementioned categories.

Thus, the role of FDS is to denaturalize the understanding of disability-as-wrong and woman-as-wrong, thus altering the prevailing understanding of disability-as-deficiency/woman-as-deficiency. One way to do so, according to Hall (2011), is to see FDS as "an avenue of critique for reductive biological understandings of both gender and disability" (5). The trajectory of some FDS work, however, rings not necessarily as always transformative for some FDS scholars. As Hall reminds us in her introduction to *Feminist Disability Studies,* in an effort to connect feminist studies and disability studies, the category of woman is connected by being labeled disabled to illustrate a loss of power. Hall claims that, within FDS:

> the suggestion that "woman" is disabled by compulsory heterosexuality and patriarchy is met with ambivalence . . . it also reflects (and risks perpetuating) dominant conceptions of disability as lack and deficiency, to the extent that it is accompanied by a desire to show that the association of women with disability is unjust to women . . . Within feminist disability studies, exploring conceptual and lived connections between gender and disability helps to make visible the historical and ongoing interrelationship between all forms of oppression. (3–4)

Thus, an FDS technical communication approach must be careful to attend simultaneously to the marginalization of all bodies and not employ the marginalization of one body as a metaphor to explain the location of or shore up another. FDS scholars articulate that critically aware scholars can accomplish such tasks by making "the body, bodily variety, and normalization central to analyses of all forms of oppression" (Hall 2011, 6). I argue technical communication scholars can accomplish such tasks by employing FDS to attend to, critique, and subvert how narratives of normalcy about culturally static and reductive biological understandings of gender and disability get written into technical medical communication about the body and thus authorized and solidified by advocacy medical organizations. In particular, this is why I advocate that technical communication scholars, teachers, and practitioners pay attention to the location of advocacy nonprofits. Much information dissemination is stemming from these locations, yet little critical attention is being paid to the way such advocacy organizations are employing technical data to oftentimes problematic ends. In addition, I call for us to attend to advocacy nonprofit locations because of the rise in their existence and the kinds of locations our student writers may find themselves in when they leave our institutions.

COMING INTO FOCUS: TECHNICAL COMMUNICATION'S CONNECTION TO NONPROFITS

Carolyn Rude, in her 2008 article "Introduction to the Special Issue on Business and Technical Communication in the Public Sphere: Learning to Have Impact," articulates that while the expertise of teachers and researchers in business/technical communication is primarily corporate and governmental, we have sought ways to put that knowledge to use in wider spheres and with regard to controversial public issues (Rude

2008, 267). As Rude states, "In both research and teaching, our field has staked some places in the public sphere" (267). Such a stake isn't surprising, insofar as the sheer amount of business and technical communication that takes place in public arenas is staggering. In fact, Faber (2002) attached the need for such public issue participation to issues of professionalization. Faber argues that

> If professional communication continues to support professionalization, professional communication students need to learn how to be public advocates for their causes and their occupations . . . As a result, professional writing pedagogy must take seriously various calls for more political and social activism. (303)

This kind of field attention to public rhetorics and advocacy/activist issues appear in a myriad of locations in our scholarship—for example, Propen and Lay Schuster (2010) consider the rhetorical work of victim impact statements, specifically looking at how the genre/s respond to the systems in which they operate and works to extend knowledge about advocacy genres within institutional systems. Lindeman (2013) considers rhetoric forwarded by a grassroots movement seeking an end to an insect eradication aerial spray program in California. Evia and Patriarca (2012) consider how technical communicators can intervene in situations where more advocacy and knowledge is necessary—specifically, the authors considered workplace safety and risk communication for Latino/a construction workers. As Rude (2008) asserts,

> Our research is often driven by personal commitment to types of issues, but it is also motivated by a sense of civic responsibility that is enhanced by awareness that the field's knowledge gives it the potential to contribute to social justice. The same values that motivate research into public issues motivate service learning courses or projects for nonprofit organizations. (267)

The focus, as Rude articulates, is on seeing teacher/scholar's interests in civic responsibility and social justice as a catalyst for their research, as well as considering how technical communication students might participate as change-agents with service learning or with projects for nonprofits. This practical, useful, and activist way to transfer research into practice in technical communication classrooms provides a location where by students can become more aware of the potential of technical communication to participate in social justice actions. However, it might be useful for students to see other possibilities for change and inquiry with regard to the multiple ways such inquiry might take place—for example, it might also be useful to make advocacy rhetorics a location, not just for tangible projects but also a location for inquiry and critique within the technical communication classroom. In other words, in addition to the direct action of service learning and project generation, there might always be ways to welcome inquiry of the technical communication used by advocacy groups by asking students to critically engage with the rhetoric of nonprofit technical communication writing. In that way, students learn not only to be social justice participants but how to be critically engaged citizens in our current nation-state, and how to do so ethically.

GAINING GROUND: WHY ATTEND TO NONPROFITS?

Nonprofit employment is growing, according to a 2014 Nonprofit Employment Practices Survey, with all sectors (International/Foreign Affairs; Health; Public/Societal Benefits; Arts, Culture and Humanities; Regional Related or Faith-based; Human Services; Education; Environmental and Animal Welfare; and Membership Society/Association) projecting continued growth. Such growth was projected in a 2011 *New York Times* article by Catherine Rampell, who articulates that in 2009 alone, "16 percent more young college graduates worked for the federal government than in the previous year and 11 percent more for nonprofit groups" (Rampell 2011). The renewed interest in public service is shown by the number of applications for service opportunities such as Americorps, which has seen applications increase from 91,399 in 2008 to 258,829 in 2010 (Rampell 2011). That number reached 582,000 applications in 2011, according to The Corporation for National and Community Service. Rampell attaches this change in nonprofit workforce to the 2008 recession; however, a 2012 study on what workers want in their employment articulates an alternate view—that 72 percent of new employees/students seek jobs where they can make a positive social impact. I will return to the second of these two rationales later in this article, but for now, regardless of the impetus, the fact remains that job growth in the nonprofit sector remains significant, and it appears new college graduates are, more and more, seeking employment in such locations.

Understanding the current statistics on student employment, we must also consider why such employment-based information matters to technical communication scholarship and pedagogues. This information matters for two reasons—employability concerns and engaged-citizen concerns. First, I'll discuss this information importance from the perspective of the importance of employability.

Quick (2012) complicates the notion of transitioning as solely the movement from academic writing to writing in the workplace by arguing that many students now return to university from the workplace; thus, the technical communication classroom is a location where we can recognize students not as "workplace tabula rasa" but as individuals who have had knowledge in varying workplace rhetorical situations. She holds, in the end, that one of technical communication's primary concerns is how we can best assist students in learning to negotiate the varying rhetorical waters of academic and workplace writing:

> One of the primary purposes of technical and professional writing (TPW) programs, as well as a central concern of TPW scholarship, is transitioning students from academic writing to writing in the workplace... Writers who transition successfully from academic to workplace writing are those who learn and adapt within the discourse community of the workplace. They must experience and absorb the workplace culture; they must understand how that culture works and how writing works within it. (230, 232)

Therefore, if, as the statistics about rising nonprofit employment suggest, nonprofits are locations wherein our graduates are continually seeking employment, as technical communication scholars and pedagogues, we have a responsibility to focus teaching and scholarship on such locations and organizations—to engage, for example, with nonprofit practitioners, information, artifacts, genres, and issues. As I articulate above, technical communication scholarship has addressed work in the public sphere, most specifically with *Journal of Business and Technical Communication (JBTC)*'s special issue on technical communication in the public sphere in 2008 (Blythe, Grabill, and Riley 2008; Knievel 2008; Koerber, Arnett, and Cumbie 2008; Propen and Lay Schuster 2008; Rude 2008), but not necessarily focused on nonprofits as a location for critical inquiry, nor (as I discuss later in this chapter) the problematics of expediency as a key rhetorical concept that drives communication emerging from that category of work.

While there are many, many different nonprofit organizations and many, many different advocacy organizations, for this chapter I use several examples from SGKftC[1] to illustrate one potential avenue for technical communication to engage with nonprofit-based technical communication in teaching practices—to encourage technical communication students to recognize the difficult rhetorical positions involved in writing for the nonprofit, advocacy-based sector, and to illustrate the importance of critically engaging with the altruistic frameworks under which such nonprofits often operate. This second of two aims is where I intend to focus most of this chapter, because if our students must experience and absorb a workplace culture to transition from academic to workplace writing, we must understand what it means when that workplace culture is a nonprofit, advocacy-based culture. We must ask ourselves how might we engage such cultural locations as teachers and practitioners of technical communication, to help ourselves and students continually critique all workplace cultures. In this chapter, for example, I illustrate how, using FDS, students can read, interpret, and analyze medical advocacy web-based technical medical communication to understand the power such technical communication may potentially hold for bodies advocacy organizations purport to support. This is, of course, only one approach to engaging such locations. Through such engagement, we might ask questions such as "What role might technical communication play in shifting narratives of normalcy often culturally embedded within nonprofit work?" I speak further on in this chapter about what I mean when I say "narratives of normalcy," but what I ask students to consider here is how such technical communication illustrates the kinds of cultural commitments of a time period, echoing work from Ward (2010).

To answer such questions, teachers of technical communication must first find ways to ask students to engage, critically, with technical communication that might seem to be "off-limits" for the turn to critique—how, for example, to critique advocacy organizations, to teach students to seek meaning in the messages being disseminated not only by organizations that historically lend themselves to critique (corporate, governmental, etc.), but also to organizations whose ethos provides, at times, a cultural buffer from critical attention. And perhaps even more important, why one should undertake such work.

ENGAGING NONPROFITS: A LOCATION RIPE FOR TECHNICAL COMMUNICATION ENGAGEMENT

Now, more than ever, considering the rise in nonprofit employment, the rise in students seeking such employment, and considering the kind of political work such nonprofits seemingly undertake, we must engage nonprofit organizations and communication in the technical communication classroom to call attention to larger systems that shape workplace cultures and seek critical awareness of how those systems predispose us to think of and enact our writing work, especially for advocacy-based organizations.

The second of my two rationales, then, is connected to this goal—to address the growing notion of engaged-citizen concerns. In addition to students seeking work outside of the for-profit sector, other aspects of engaged-citizenry are on the rise. Charitable giving, for example, has risen continually for the past three years—in 2012 alone, Americans gave more than 316 billion dollars to advocacy, charity, and nonprofit organizations. 2.3 million nonprofit organizations exist in the United

States. The nonprofit sector contributed 804.8 billion dollars to the US economy in 2010—in the same year public charities, the largest component of the nonprofit sector, reported 1.51 trillion in revenue, 1.45 trillion in expenses, and 2.71 trillion in assets. In 2011, 26.8 percent of adults in the United States volunteered with an organization—volunteers contributed 15.2 billion hours of labor, worth an estimated 296.2 billion dollars. While the cultural narrative might indeed be that Americans are self-centered, self-motivated, and not driven by the common good,[2] such numbers indicate that our need and want for interconnectivity, and a drive to support a common good, is on the rise.

However, this chapter situates advocacy within the nonprofit sector and, more keenly still, within a kind of civic engagement that nonprofit advocacy groups often promote. As such, this article also responds to the call made by Simmons and Zoetewey (2012) when they ask us to consider how we design and test websites for web usability and increased civic engagement. The authors state:

> As Web site designers, teachers of Web design, and users of civic Web sites, we have much at stake in how citizens solve problems in their communities and in professional writing and rhetoric's role in enabling that participation. Civic online spaces can provide citizens with effective resources to solve problems in their communities but only if the Web site development takes into account the literacy practices that citizens must adopt to make sense of information and what counts as useful information for these citizens. (251)

As Simmons and Zoetewey articulate, there is much at stake in how citizens solve problems and participate in their communities, in particular with relation to how professional writing and rhetoric might enable (or not enable) such participation in web spaces. While Simmons and Zoetewey focus on civic issues, there are many problems in communities that technical communication can and does address, in particular in online spaces. Thus, I'd like to reiterate Simmons and Zoetewey concern that, in addition to our investments in students transitioning in and out of workplaces, we also have investment in how technical communication participates in how citizens understand and solve problems in their communities and, I would add, beyond their communities. Such stakes go beyond civic websites, and also of course beyond the minutiae I am offering as an example for this article—specifically, that we must focus on nonprofit, advocacy information dissemination both locally and globally, including focus on locations where nonprofit advocacy groups disseminate information—primarily, their web presence. Thus, nonprofit, advocacy, and charity websites must be engaged, analyzed, and critiqued in technical communication classrooms—as part of their goals, much like the goals of the civic websites analyzed by Simmons and Zoetewey, are often to help various stakeholders understand information, participate in larger discussions about issues affecting themselves and others, and potentially take action to change varying situations and conditions. Below, then, I provide two such analyses to illustrate ways that technical communication scholars and students can engage such critical work with advocacy organizations.

RACING FOR THE CURE: METAPHORS, ADVOCACY, AND TEACHING TECHNICAL COMMUNICATION

This section—which distills the macro discussion of why the field of technical communication must attend to nonprofit rhetorics/the uptake of altruistic space to one micro consideration of how we might engage with such nonprofit rhetorics—illustrates how an FDS analysis of technical communication from a global advocacy/nonprofit organization (SGKftC^3) may assist technical communication students/faculty and the broader field of technical communication in being cognizant of how intersections/confluences of ideological beliefs, metaphoric rhetorical action/s, and technical medical information can draw public attention toward altruism/altruistic ends and away from harmful ideological means.[4] With aims of social justice, this example illustrates one way to take up holding in tension and holding responsible organizations with altruistic end goals through a necessary critical lens toward their means; the end goal of this analysis, then, is to both understand how such organizations use technical information and communication principles in an expedient, harmful way, and be judicious in our own engagement with nonprofit rhetorics from both consumer and producer standpoints.

While there are multiple approaches to employing an FDS framework in a technical communication classroom, I engage in a multi-pronged approach that asks students to engage with nonprofit rhetoric in a specific manner—first, by analyzing metaphors that frame the larger organizations; second, to analyze the technical medical rhetoric being disseminated by the organization; and third, to put those readings in conjunction with one another to understand how nonprofit organizations use technical medical language to shape human life in both beneficial and harmful ways.

Thus, I open such work in my technical communication classrooms by introducing students to FDS principles broadly by moving from macro to micro analysis through an analysis—first, of major metaphor/s of a nonprofit organization that frame their work, and then more specifically in technical medical information employed under that frame. As FDS calls for us to attend to how **language** shapes and influences the status of the lived body, the politics of appearance, the medicalization of the body, and our understandings of normalcy, it's important to attend to how language is used to articulate the larger principles of any organization. One specific way to do so is to consider what metaphors are broadly employed by advocacy organizations as a way to frame and disseminate their larger commitments. For a foundation such as SGKftC[5] the rhetoric of *racing for a cure* is used as a branding tool on a global and local level—for example, the races are branded by US location (i.e., Komen South Florida Race for the Cure; Komen Central MS Steel Magnolias Race for the Cure) and by international (from Dar Es Salaam, Tanzania, to Tbilisi, Georgia). I ask students the implications of naming these locations—students often articulate that by providing such information, Komen illustrates that breast cancer touches every corner of the world, and thus, so does Komen. As a group, we then discuss how the connection of such locale-focused work operates multi-directionally as a way to show support for breast cancer survivors around the world, to illustrate that breast cancer is a universal issue and, more practically, to increase fundraising potential by operating locally and globally. We then discuss how such location-based persuasion isn't necessarily harmful as a stand-alone rhetorical act. However, taken together with other aspects of Komen's messaging and technical medical rhetorical work—including the implementation of the metaphor of *racing*—and read through an FDS lens with attention to how such language shapes human lived experience, students begin to piece together how messaging operates as a nexus and how careful attention must be paid to the creation and dissemination of such messaging.

This use of metaphor-as-branding approach to medical advocacy work, one that focuses on expediency, competition, and goal-oriented action to explain and justify action, is at once culturally effective and yet problematic. Specifically, the metaphor of "racing for the cure" rhetorically operates as a frame to persuade audience members and participants to see medical improvement and disease prevention, as supported by technical communication about the body, as a linear, expedient, normalizing process by which a disease (and by extension the diseased body) is managed and conquered, while simultaneously causing the healthy body to accept—by reorienting the work of competition to be altruistic—be controlled, and maintained. While many bodily iterations can race and can participate in races, it isn't necessarily the act of racing that is problematic in this metaphor—we should, of course, strive for a cure for breast cancer as efficiently as possible, but simultaneously, such a metaphor could also contribute to an understanding of "winning at all costs" with regards to health—that a cure for breast cancer is so important (which, of course, is true) that any means of getting to the finish line first is authorized; in other words—the end justifies any means.

In other words, through such metaphors, foundations such as SGKftC create space for an expediency-based approach to medical technical communication. I make this assertion because the altruistic end goal of these organizations—to end cancer—seemingly authorizes a cultural "any means necessary" approach to the treatment of bodies. As I show in the next section, by providing evidence of the information architecture, design, citation practices, and the framing of medical information on the SGKftC website, I continue to introduce my students to the traveling nature of such metaphors in advocacy rhetorics. Often, I find that college students at first respond favorably to the metaphor of racing for the cure, not just because of cultural "pinkwashing," which has been discussed more generally in cultural work on nonprofits and breast cancer research (Ehrenreich 2001, 2009; King 2008; Sulik 2011) and gestured toward within the realm of rhetorical studies (Kopelson 2005, 2013; Selleck 2010), or through volunteerism/ altruism (see above section for research on nonprofit growth in the United States and work in technical communication and nonprofits), but also because such college students are particularly conditioned, culturally, educationally, and otherwise, by the linear notion of racing to win—our students race to get good grades, to graduate, to make their resumes strong to get good jobs or get into good graduate schools. And, of course, these goals don't stop with schooling; however, I locate this discussion with school because first, our students inhabit a particular place within the institution as technical communication students to be able to understand how information "schools" and manages people (Foucault 1979), and second, such systematic schooling (literally and figuratively) asks students to see speed and accomplishment as means in and of themselves. Thus, it is particularly apt in technical communication classrooms, where students learn to disseminate technical information to a variety of audiences, to ask students to simultaneously interrogate linear, speed-driven public discourses such as those put out and on by medical advocacy groups.

This focus on culturally-situated and human-outcome based analysis isn't new to the field of technical communication. Miller's (1979) germinal article "A Humanistic Rationale for Technical Writing" reminds us that our words have social and ethical repercussions. Steven Katz's (1992) article "The Ethic of Expediency: Classical Rhetoric, Technology, and the Holocaust" calls the field to pay attention to the work of deliberative rhetoric with an eye toward its potential to redirect cultural and social ethics in problematic ways; Katz warned that Western rhetoric relies upon "an ethic of expediency" and, in doing so, that commitment often ends with disastrous human outcomes. In the past twenty years, scholarship in the field of technical communication has often referenced Katz's germinal work and used that frame to discuss ethics in technical communication, much as I do in my own technical communication courses.

Building on Katz's work, Mark Ward Sr.'s article "The Ethic of Exigence: Information Design, Postmodern Ethics, and the Holocaust" provides further discussion on the issue of ethics in technical communication (Ward 2010). Citing various professional organizations under whose principles technical communicators often operate, Ward illustrates how technical communicators must exist simultaneously in two intellectual lands—(1) "presenting information so that it is understandable and easy to use: effective, efficient, and attractive," and (2) recognizing how that information must also be interrogated by "examining the cultural, institutional, and socially situated particularities" of any given historical, social, and cultural context (62–63). Especially within this call, Ward asks technical communicators to not only "consider text but to examine why a particular arrangement of textual and graphic elements has symbolic potency within a given institutional or organizational culture" (63). To that end, Ward asks that we understand that information design is a process of co-constructing meaning between designers and users. Using the example of a Nuremberg Law poster from 1935 Nazi Germany, Ward illustrates how such a poster, by design, contributed to a co-constructed cultural meaning that excluded "an entire class of human beings from their community" by a seemingly rational process (65).

Co-construction, as Klawiter (2008) illustrates, is necessary for a non-profit foundation such as the SGKftC. In order to be successful, such foundations adopt and promote values that represent the ideological foundations of organizations and individuals who have brought them funding. Often, these organizations have roots in the very institutions seeking to understand bodies and medicine in a normative, expedient way. And, if not roots, the supporters have been persuaded, through the rhetoric of "racing for the cure," to approach bodies and medical treatment in an expedient and efficient manner. Thus, SGKftC continues to perpetuate cultural ideologies of normalcy that assist the foundation in securing funding readily by relying on ethics of expediency and exigence to authorize an any-means-necessary approach to fundraising, with little regard to the collateral damage that lies in the wake of such a cultural imperative. In an effort to ease apprehension about such an approach, SGKftC simultaneously attaches their conservative, normalizing work to making sure that low-income, under-insured women, especially women of color, are recipients of numerous funds. Using this moral imperative, SGKftC is thus able to run moral interference and thereby continue their normalizing work virtually uncritiqued. As Klawiter states:

> Susan G. Komen For the Cure embraced mainstream, and in some respects conservative, norms of gender and sexuality, and these resonated with corporate America and with ordinary Americans from across the political spectrum. Susan G. Komen For the Cure also maintained a respectful, uncritical stance toward the medical and research establishments . . . The foundation, in other words, embraced a traditional division of labor and distribution of power between the lay populace and the institutions of science and medicine. But the foundation was also at the forefront of legislatively focused efforts to force the health insurance industry to cover mammographic screening . . . Susan G. Komen For the Cure helped transform the issues of mammograms for low-income, uninsured, and underserved women, especially women of color, into a moral imperative. (139)

It is this very moral imperative—that promoting and supporting mammograms for low-income, uninsured, and underserved women authorizes the simultaneous promotion of norms of gender and sexuality—supported by the idea that we must *race for a cure* no matter the costs has allowed SGKftC to historically promote heteronormative, ableist narratives of normalcy—for example, claiming that homosexuality is a choice and may cause harmful health effects, that women with disabilities are logic-deficient, and other dominant narratives that circulate in this current historic moment (more on this in the section below). However, because SGKftC then focused their efforts on other problematic narratives of oppression (specifically early detection for low-income, uninsured, and in particular women of color), thus illustrating their commitment to issues of difference, they challenge some of those very oppressive narratives while simultaneously shoring up others. In doing so, the technical communication about risk and disease prevention on such a website remains virtually uncritiqued because of the commitment to the greatest good ethic of expediency and exigence.[6]

These commitments are further complicated by a deeper reading of expediency and exigence through disability studies. Jay Dolmage (2008) has argued that we have adopted the logic of retrofitting when forced to confront non-normal bodies, in that our actions and language are often driven by an ethic of accommodation, one that legally forces what is necessary to include "non-normal" bodies in various venues (14–27). In that forced interaction, we move expediently to accommodate, but give little thought to the mode of inclusion or method of framing, such accommodation can actually work antithetically. Therefore, this trifecta of ethics merges with the ubiquity of medical charity work to create conditions whereby rhetorics of normalcy are perpetuated, unquestioned, via websites such as the SGKftC.

In other words, retrofitting holds a commitment to expediency, work that is done mostly quickly and efficiently as a means to an end to become code or regulation compliant or, in the case of the SGKftC work, using political and social work that shores up dominant borders of hegemonic culture in order to eradicate the disease of breast cancer. In larger terms, we work as quickly as possible to remove the problem (whether that problem be a term, an experience, a body, or whatever is being seen an as obstacle) so that we can quickly reach on to the end point, so that all the rest is just considered a means to that end.

As Dolmage (2008) alludes, this approach highlights our commitment to an ethic of accommodation as expediency, an ethic that doesn't truly or even necessarily value inclusivity or altruism on the behalf of all, but rather culturally encourages us to do as little as possible while still meeting legal requirements to get to whatever end point it seems we are seeking, which might include, or even be, inclusivity. The effects of this ethic are varied but, perhaps most important, such an ethic functions to keep non-normal bodies marginalized while providing the illusion that actually the opposite is true. This type of approach is most clearly illustrated by Ward's notion of ethic of exigence, which is currently, I argue, an ethic of retrofitted information design on the SGKftC website.

Before I delve into this analysis, I want to discuss first why I believe SGKftC is employing what I call an ethic of retrofitted information design. In his analysis of Nazi information design, Ward (2010) articulates that paying attention to information design ethics "requires us to go beyond textual analysis to ask why a given arrangement of text and graphics has resonance for a particular culture" (65). In other words, Ward calls for us to consider not only individual ethics but also collective ethics to understand how particular rhetorical choices (in this case about information design) become authorized and accepted.

To get at our sense of collective ethics, as technical communicators and consumers and producers of technical texts, we must ask ourselves, and our students—what kind of political and social work are such metaphors, is this writing, doing to material bodies, in particular women's bodies, bodies that have continually been marginalized, with information design enactment and technical information dissemination? How could these distillations further structure and marginalize marginalized populations? How do we, as writers of and creators of knowledge, ethically and responsibly respond to such work? And how do we contribute when we subscribe to the idea that we should continually "race for the cure"?

The next section discusses the analysis of technical medical information articulated in various iterations from 2009 to 2014 using the expedient framework of racing for cure and an FDS lens. Specifically, using these examples, I illustrate how we might ask students to analyze advocacy technical communication with a focus on understanding how, as Garland-Thomson and Hall ask of us, to attend to ways in which such rhetorical data dissemination may forward logics of normalcy through discussion of varying populations of bodies.

FEMINIST DISABILITY STUDIES AND KOMEN: A CASE FOR INFORMATION DESIGN INTERROGATION

In an attempt to foster the kinds of interactions between feminism and disability studies Garland-Thomson calls for, and to illustrate the ways in which technical communication, in this case the technical communication present in 2009 and 2014 on the SGKftC website, frames "non-normal" bodies in an effort to normalize them, I will now discuss an example of a particular analysis assignment for a technical communication course that embodies these two purposes.

For this particular classroom assignment, I first focus on a particular section of the SGKftC website from 2009 and 2014 entitled "Risk Factors and Prevention." I do so because of the very technical nature of this risk-based disease information and the ways in which this information is disseminated and discussed. In addition to this discussion, I include a discussion of the rhetoric of the images on these websites and how the images work with and against the rhetorical, normalizing function of the information provided on the website about bodies. For the purpose of this chapter, I'll be discussing, much in the way we discuss in my own classroom, the construction of this particular section and the implications for the ethical obligations of technical communicators and teachers of technical communication—in particular technical communication involving risk, disease, and the human body.

In 2009, when one first entered the SGKftC website, readers were inundated with images of people, almost exclusively women—save images such

Figure 9.1.

as for Garth Brooks whose album is contributing to the SGKftC fund and two women's hands clasped together, articulating "One promise, two sisters" in reference to Komen and Brinker (who founded the organization in promise to her sister, who died of breast cancer).

In addition to the images, there are also staggering statistics regarding how many women are stricken with breast cancer each year, and how many people have joined the "fight" against breast cancer. To get to the section on "Risk Factors and Prevention," a section designed to better understand breast cancer and what we can all do to become more proactive in our own health, one clicks in the upper left hand on the button entitled "About Breast Cancer." "Risk Factors and Prevention" is located on the left-hand side of the screen and is the third button down. Once you click the "Risk Factors and Prevention" button, that's when the analysis truly begins.

In this section about risk factors and prevention, there are two lists. For this assignment, I narrow in on the list on the left, which is a category of factors that increase breast cancer risk. At the very top of the list of risk factors is "being female" and "getting older." I stop here and ask students about the categories and information design they see—about what these categories tell us about our knowledge of risks for breast cancer, about who we might hazard to believe the users are for SGKftC's website considering the information and how it is presented, why we might assume such things, and potentially how our cultural assumptions about populations/bodies/health/medicine shape the information design on this website.

Risk Factors & Prevention

This section of About Breast Cancer describes the many known risk factors for breast cancer—from those that have a large effect on risk (such as getting older, having a strong family history and having carcinoma in situ) to those that have a small effect on risk (such as being tall and eating an unhealthy diet). Also included are tips for lowering risk, information on genetic mutations (including genetic testing) and options like tamoxifen for women at higher risk.

Factors that Increase Breast Cancer Risk
- Introduction
- Being female
- Getting older
- Inherited genetic mutations
- Carcinoma in situ
- Family history of breast, ovarian or prostate cancer
- High breast density on mammogram
- Radiation exposure in youth
- Benign breast conditions (benign breast disease, hyperplasia)
- High levels of estrogen in the blood
- Personal history of breast cancer
- Menopause at age 55 or older
- Not having children or having first child after age 35
- High bone density
- Overweight/ weight gain
- High socioeconomic status
- Ashkenazi Jewish heritage
- Drinking alcohol
- Lack of exercise
- Postmenopausal hormone use
- First period before age 12
- Current or recent use of birth control pills
- Being tall
- Not breastfeeding
- Risk factors summary table of relative risks

Other Issues Related to Breast Cancer Risk
- Understanding risk
- Breast cancer risk factors table
- Understanding breast cancer prevention
- Healthy behaviors
- Race and ethnicity
- Gene mutations and genetic testing
- Options for women at higher risk (tamoxifen and other choices)
- Emerging areas in estimating risk
- Factors under study
- Factors that do not increase risk
- Questions for your provider
- References

Figure 9.2.

Entering the section about the risk factor of being female, it is first explained that women have an almost a hundred times higher risk of developing breast cancer than men. The rest of the page, then, is devoted to a discussion of "Women who partner with women and Lesbians."

I first ask my students to consider the information architecture of this page—what is presented? How is it presented? What does this presentation articulate to us about how writers for SGKftC were interpreting the dominant exigence of this time frame? Toward what end(s) does it seem that this work is doing?

In analysis, we consider the construction of this information—the first paragraph focuses solely on the absence of children from the lives of lesbians and women who partner with women,[7] and how that makes lesbians and women who partner with women more susceptible to breast cancer. The way this paragraph is framed makes the absence of children sound almost exclusively as if it is a choice (there is no mention, for example, of the rising number of women who are actively choosing not to reproduce, regardless of their sexual orientation, or of women who cannot conceive children because of a variety of reasons). There is also no mention of the social, cultural, and economic difficulties many women who partner with women and lesbians may encounter in trying

Being Female

Being female is the most important risk factor for breast cancer. Although men can develop this disease, it is about 100 times more common among women [3].

Breast cancer risk in women who partner with women and Lesbians
Although the data are somewhat limited, women who partner with women are believed to have a higher risk of breast cancer than other women. The reason? Women who partner with women tend to have more risk factors for the disease. For example, as a group, they are less likely to bear children or, if they do, are less likely to have them early in life. They may also have higher rates of obesity and alcohol use, both of which can increase the risk of breast cancer [4,5,6].

In addition to having more risk factors for breast cancer, women who partner with women may also be less likely to get routine mammograms and clinical breast exams. The reasons for this are not yet clear but may be due to issues like lack of insurance, financial hardship and past experiences of discrimination and insensitivity from health care providers [5,7]. For more on this, see "Women who Partner with Women and Lesbians".

Figure 9.3.

to conceive children, or of the societal issues that compound those problems. The normalizing work being done here, then, is to argue that one puts themselves at risk by "choosing" a lifestyle in which she may not continue to procreate as "normal" individuals do, and that one should consider the ramifications of those "choices." Of course, the problematic notion of "choosing" one's sexuality arises from the enactment of such medical rhetoric, making it an option that one can then choose to prevent by normalizing oneself to the heteronormative. Consider, too, the certainty with which the first paragraph of that section is written, the section of course that places the responsibility of conformity, of guilt and shame and responsibility, with the individual, and how the second paragraph, a paragraph that loosely implicates societal complications and oppressions, is much more tentative, stating that there may be a connection, perhaps, between doctors discriminating against women because of their sexuality.

At this point, I often direct my students to search for issues of disability, or to consider issues of access, both to the information and a general discussion of women with disabilities (as it seems SGKftC is willing to categorize women further as "normal" and "non-normal"). As students find, there is no direct section that discusses risk factors disabled women face in diagnosis and treatment of breast cancer, despite other identity markers (race, gender, socioeconomic status) having their own sections. In fact, the one pamphlet I did find on the SGKftC website that dealt specifically with women with disabilities was something I had to search for myself, and none of the information provided had alternate formats of the information. In addition to access issues, the pamphlet provides discussion of the "barriers to early detection"—the Komen mantra of early detection, and similar to the category of "Risk Factors and Prevention."

As shown, the pamphlet articulates that "Some disabled women believe that they are less likely to have breast cancer than other women, since they are already coping with one disability. They may believe that 'lightning doesn't strike twice.'" So, not only are women with disabilities ignored in the section on risk factors, in the pamphlet they are framed as irrational and logic-deficient. Disability, in this case, is also framed solely as a physical "problem" and thus does not take into account other disabilities that may affect a woman's ability to seek medical attention. As teachers of technical communication and potential disseminators of nonprofit, medical information, we must ask students to not just pay attention to what is present but what is absent, as well, and what and how that absence says to real, material bodies.

> **Barriers to screening**
>
> Research has shown there are several reasons why women with disabilities may not receive breast cancer screening:
> 1. It is hard to access the place where the screening is offered:
>   - Women may have a hard time making and keeping medical appointments. For example, a woman who is deaf may not be able to easily call a clinic that does not have a TDD text telephone. Her doctor's office may not have a sign language interpreter that can be present at the appointment.
>   - Facilities for breast cancer screening are not always accessible to some women, such as those who use a wheelchair. For example, there may be no ramp or dressing room that is large enough to accommodate her wheelchair.
>   - The equipment used is not always accessible to some women, such as those who have trouble walking or standing still in one position. For example, mobile mammography vans are not always wheelchair accessible. Mammography equipment may not adjust enough to allow some women to easily position themselves or sit while being screened.
> 2. Some disabled women believe that they are less likely to have breast cancer than other women, since they are already coping with one disability. They may believe that "lightning doesn't strike twice."

Figure 9.4.

The next images show similar, and problematic, technical communication. Many of the solutions and preventions that the SGKftC website provides, in fact, center around having children, being thin, and creating the feeling that one's body has betrayed—there is nothing one can do to stop from being female, or from getting older, or from loving who one loves, or perhaps from having children or not having children, from breastfeeding or not, yet those are risk factors that are shown in 2009. Consider, for example, the section on birth control and the connections being made between birth control and breast cancer. Consider, then, as well, the high focus again on the importance of having children to preventing breast cancer. Also included in that risk is not breastfeeding. So, not only do you need to have children, but you must be able to breastfeed them as well, both for your own benefit and for the benefit of the child. Invoking the rhetoric of "for the child" foreground's the importance of

Lesbians and women who partner with women

Lesbians and women who partner with women have a greater risk of breast cancer than other women, but this is not because of their sexual orientation. Rather, it is linked to risk factors for breast cancer that tend to be more common in these women such as never having children or having them later in life [55-57]. Lesbian women also tend to have higher rates of obesity and alcohol use, both of which can increase breast cancer risk [55-57].

Lesbian women may also be less likely to get routine mammograms and clinical breast exams [58-59]. The reasons for this are not yet clear. However, lack of insurance, a perceived low level of breast cancer risk and not seeing a health care provider on a regular basis may all play a role [58-59]. For many women, reproductive health issues are their main link to the health care system (for example, during pregnancy). Even when seeing a provider about reproductive health, other health issues are often addressed, including having clinical breast exams or mammograms. But because fewer lesbians have children and therefore, may not seek routine health care, they may have fewer opportunities to have routine breast cancer screening. As a result, breast cancer may not be found at an early as stage, when it is most treatable. One step lesbians can take for breast health is to find a provider who is sensitive to their health issues, and to see that provider regularly—especially for clinical breast exams and mammograms.

Transgender people

At this time, data on breast cancer among transgender men and women are too limited to comment on any increased or decreased risk in these populations. If you are transgender, talk to your health care provider about your breast cancer risk. Your provider can assess your situation.

Figure 9.5.

heteronormative action and that the mother should first be concerned about her children and then her own health. Indeed, SGKftC's research writers state that having two children is even better than having one, and being able to breastfeed all your children is the best option. Considering this information, we must ask our students: what kind of political, social, and cultural work is this technical communication doing to material bodies, in particular women's bodies, bodies that have continually been marginalized and normalized in the name of medicine and health?

This research, done back in 2009, has subsequently been retrofitted by SGKftC writers with specific information design choices to illustrate both how we are to co-construct knowledge about women, in particular the lesbian body, and to understand information retrofitting. This ethic of information retrofitting is illustrated by this image.

By adding the phrase "but this is not because of their sexual orientation," SGKftC has made accommodation for what would be considered non-normal bodies in hegemonic culture, yet has left all other problematic information and discussion, thus encouraging readers to still conceptualize particular bodies in a non-normal/othering way. Even in the segmentation of the information design, non-normal bodies are set off as a category, thus using the status of altruism to encourage readers to continue to authorize marginalization and violence against such bodies. An addendum to this information came in 2013, wherein to address such problematic practices, SGKftC highlighted a piece by Eric Brinker, Nancy Brinker's son and Susan G. Komen's nephew who is gay, to address the critiques the foundation had incurred as part of this information campaign. Further rhetorical analysis could be conducted of this decision, including having a gay man respond to critiques of the treatment of lesbian populations by this particular foundation. This conversation, then, seems to be a continual source of complication with regard to the dissemination of technical medical communication.

Finally, important to note are the citation practices that are at use on the website. Some of the citations are dated back to the 1980s, and often only one citation is provided for each claim. Yet, using that one or two, typically dated citation, the writers were able to make such sweeping generalizations about a whole population of women. Technical communication students often come to our classes knowing to triangulate medical information from 2001 with more current data for use in 2013, and yet SGKftC has not done that kind of due diligence. In fact, some citation practices have research that directly contradicts one another—in looking to update this version in 2014, SGKftC used an updated 2010 article (Zaritsky and Dibble 2010) that articulates in their study that the lesbian population they studied did not have a higher rate of alcohol use than their heterosexual counterparts. Yet, Komen continues to rely on this narrative to illustrate that the lesbian population is inherently in danger. This retrofitted, expediency-based, goal-oriented, get the work out without much thought on who or how particular bodies are being harmed rhetoric is tied, deeply, to the metaphor of racing for the cure. This rhetoric of racing for the cure creates conditions whereby such harmful, heteronormative, normalizing narratives, and harmful structures of information design are ignored, obfuscated, or otherwise made palatable because of the ultimate goal of eradicating breast cancer.

Having just finished this analysis project, in which students not only analyze and critique the information dissemination process of this medical writing but then are charged with revising a section of the website to include in their final portfolio and present to the class, one student approached me with their project, handed it to me, and laughing said, "I want you to know I kept asking myself your favorite question: how is this framing real, material bodies?" And, while the student was making a joke about how many times I ask that question in class, it is my hope that we can ask all technical communicators, ourselves included, the same kinds of questions. What kind of impact might technical communication have on bodies who need this information to assist them or to help them to understand their body in a way that works to resist objectification and harmful dominant narratives? How can I ethically respond to research that is contradictory? How can I bring to light issues with social systems rather than foregrounding the problematics of the body? Technical writers, as James Porter (1998) effectively persuades in his chapter "The Exercise of Critical Rhetorical Ethics," have an ethical obligation to consider the impacts we have on users, intended or not. Included in this call for an ethical approach to considering the impact of technical communication is the role of the teacher: we must, as educators, challenge our students to be critical, both of information being disseminated by others, and especially of technical medical communication being written by, co-constructed, and disseminated to them. Employing an FDS lens, with a critical eye toward expediency and retrofitting practices, will provide a way in which students can further and deepen their critical engagement with cultural texts, and thus with their own lives, with an awareness of their potential, expedient participation in varying altruistic locations that operate simultaneously as locations of oppression.

Notes

1. Some might find the use of SGKftC here suspect, considering the kinds of press and decisions that this advocacy organization has made, organizationally, especially in the last few years. However, dismissal of a site of inquiry doesn't negate that that site of inquiry still exists and that as critical eyes move away from such locations, SGKftC, for example, still earns in excess of 35 million dollars in Cause Marketing (brand-attachment to the cause) as well as hosting the largest breast cancer fundraiser in the United States—Komen Race for the Cure. Ubiquity necessitates constant, critical attention.
2. A study in *Psychological Science* in 2013, for example, found that trying to get Americans to be more interdependent—to think and act as such—failed and may have decreased motivation for tasks.
3. SGKftC has found themselves in some complicated public relation locations as a result of varying organizational decisions, in particular their participation in the pinkification of breast cancer—through partnering with complicated corporate sponsors like American Express who in their "Change for the Cure" sponsorship gave a single penny per charge, regardless of amount of purchase, to partnering with oil field service company Baker Hughes who produced 1,000 pink drill bits to raise awareness regarding breast cancer, while fracking (for which such drills are used) involves chemicals that are endocrine disrupters and are shown to increase the risk of cancer (Jagger 2014). For further information about the gaps between ideology and practice (or as I'll articulate the formula in this article, a privileging of expediency in pursuit of an altruistic goal) in the Komen Foundation and other breast cancer foundations, see King (2008) *Pink Ribbon Inc.* and Sulik (2011) *Pink Ribbon Blues*.

4. As a woman who has had friends/family members suffer and die from breast cancer, SGKftC is an organization with which I have particular pathos- and ethic-based attachments. I do not, with this work, critique the end goal of Komen's organization. I do, however, wish to critique the ways in which SGKftC authorizes achievement of that end goal, or to attach narratives that don't belong to achieving that end goal to the telos of the organization. Additionally, and perhaps even more important, in recognizing the various stakeholders and locations of this medical advocacy foundation, I am cognizant of the prominent and important location of advocacy groups and advocacy work, especially within our Western, capitalistic frame of reference. In many ways it seems problematic to analyze and critique the rhetoric of an organization whose articulated goal is to end suffering from a particular disease. And, in many of those ways, I agree—this research makes me, as a privileged Western woman, especially one who has in the past contributed to the Susan G. Komen for the Cure foundation, fairly uncomfortable for a myriad of reasons, just as it may make many readers uncomfortable. However, it is that very discomfort that also hails us to recognize the importance of such cultural work.
5. This metaphor is also present in breast cancer organizations such as Avon Foundation's Breast Cancer Crusade.
6. The politics of The Susan G. Komen Foundation have been critiqued in widespread fashion, especially their post–2012 debacle involving Planned Parenthood and grant funding (see Watt 2012). I am less concerned, in this article, with the overt political nature of Susan G. Komen for the Cure's work, and more concerned with the subversive nature of the construction of technical information and structure within the Komen website, and how that information is disseminated to varying audiences online.
7. As a class, we found it interesting the way in which this information is categorized as "lesbians and women who partner with women"—we are uncertain if that is SGKftC being sensitive to the labeling of groups, to create a sense that any homosexual activity is detrimental to one's health, or a mixture of both, the latter of which would be keeping with Komen's multiple rhetorical trajectory.

References

Blythe, S., J. T. Grabill, and K. Riley. 2008. "Action Research and Wicked Environmental Problems: Exploring Appropriate Roles for Researchers in Professional Communication." *Journal of Business and Technical Communication* 22: 272–298.

Dolmage, Jay. 2008. "Mapping Composition: Inviting Disability in the Front Door." In *Disability and the Teaching of Writing: A Critical Sourcebook*, ed. Cynthia Lewiecki-Wilson and Brenda J. Brueggemann, 14–27. Boston, MA: Bedford/St. Martin's.

Dryden, Jane. 2013. "Hegel, Feminist Philosophy, and Disability: Rereading our History." *Disability Studies Quarterly* 33 (4). http://dsq-sds.org/article/view/3868.

Ehrenreich, B. 2001. *Nickled and Dimed: On Getting by in America*. New York: Henry Holt.

Ehrenreich, B. 2009. *Bright-Sided: How Positive Thinking is Undermining America*. New York: Metropolitan Books.

Erevelles, Nirmala. 2011. *Disability and Difference in Global Contexts*. Basingstoke: Palgrave Macmillian.

Evia, Carlos, and Ashley Patriarca. 2012. "Beyond Compliance: Participatory Translation of Safety Communication for Latino Construction Workers." *Journal of Business and Technical Communication* 26 (3): 340–367. https://doi.org/10.1177/1050651912439697.

Faber, Brenton. 2002. "Professional Identities: What Is Professional about Professional Communication?" *Journal of Business and Technical Communication* 16 (3): 306–337. https://doi.org/10.1177/105065190201600303.

Foucault, M. 1979. *Discipline and Punish: The Birth of the Prison.* New York: Vintage Books.

Garland-Thomson, Rosemarie. 2004. "Integrating Disability, Transforming Feminist Theory." In *Gendering Disability*, ed. Bonnie G. Smith and Beth Hutchinson, 73–103. Piscataway, NJ: Rutgers University Press.

Hall, Kim Q. 2011. *Feminist Disability Studies.* Bloomington: Indiana University Press.

Hamraie, Aimi. 2013. "Designing Collective Access: A Feminist Disability Theory of Universal Design." *Disability Studies Quarterly* 33 (4). http://dsq-sds.org/article/view/3871/3411.

Jagger, K. 2014. "Komen Is Supposed to be Curing Breast Cancer: So Why Is Its Pink Ribbon on so Many Carcinogenic Products?" *Washington Post*, October 21, 2014. https://www.washingtonpost.com/posteverything/wp/2014/10/21/komen-is-supposed-to-be-curing-breast-cancer-so-why-is-its-pink-ribbon-on-so-many-carcinogenic-products/?noredirect=on&utm_term=.6ca27e10b01d.

Kafer, Alison. 2013. *Feminist, Queer, Crip.* Bloomington: Indiana University Press.

Kafer, Alison, and Michelle Jarman. 2014. "Growing Disability Studies: Politics of Access, Politics of Collaboration." *Disability Studies Quarterly* 34 (2). http://dsq-sds.org/article/view/4286.

Katz, Steven. 1992. "The Ethic of Expediency: Classical Rhetoric, Technology, and the Holocaust." *College English* 54 (3): 255–275. https://doi.org/10.2307/378062.

King, S. 2008. *Pink Ribbons, Inc.: Breast Cancer and the Politics of Philanthropy.* Minneapolis: University of Minnesota Press.

Klawiter, Maren. 2008. *The Biopolitics of Breast Cancer: Changing Cultures of Disease and Activism.* Minneapolis: University of Minnesota Press.

Knievel, M. 2008. "Rupturing Context, Resituating Genre: A Study of Use-of-Force Policy in the Wake of a Controversial Shooting." *Journal of Business and Technical Communication* 22 (3): 330–363.

Koerber, A. E., J. Arnett, and T. Cumbie. 2008. "Distortion and the Politics of Pain Relief: A Habermasian Analysis of Medicine in the Media." *Journal of Business and Technical Communication* 22 (3): 364–391.

Kopelson, K. 2005. "Tripping Over Our Tropes: Of 'Passing' and Postmodern Subjectivity—What's in a Metaphor?" *JAC* 25 (3): 435–468.

Kopelson, K. 2013. "Risky Appeals: Recruiting to the Environmental Breast Cancer Movement in the Age of 'Pink Fatigue.'" *Rhetoric Society Quarterly* 43 (2): 107–133.

Lindeman, Neil. 2013. "Subjectivized Knowledge and Grassroots Advocacy: An Analysis of an Environmental Controversy in Northern California." *Journal of Business and Technical Communication* 27 (1): 62–90. https://doi.org/10.1177/1050651912448871.

Miller, Carolyn. 1979. "A Humanistic Rationale for Technical Writing." *College English* 40 (6): 610–617. https://doi.org/10.2307/375964.

O'Donovan, Maeve M. 2013. "Feminism, Disability, and Evolutionary Psychology: What's Missing?" *Disability Studies Quarterly* 33 (4). http://dsq-sds.org/article/view/3872/3403.

Piepmeier, Alison, Amber Cantrell, and Ashley Maggio. 2014. "Disability Is a Feminist Issue: Bringing Together Women's and Gender Studies and Disability Studies." *Disability Studies Quarterly* 34 (2). http://dsq-sds.org/article/view/4252.

Porter, J. 1998. *Rhetorical Ethics and Internetworked Writing.* Santa Barbara, CA: Greenwood Publishing Group.

Propen, A., and M. Lay Schuster. 2008. "Making Academic Work Advocacy Work: Technologies of Power in the Public Arena." *Journal of Business and Technical Communication* 22 (3): 299–329.

Propen, Amy D., and Mary Lay Schuster. 2010. "Understanding Genre through the Lens of Advocacy: The Rhetorical Work of the Victim Impact Statement." *Written Communication* 27 (1): 3–35. https://doi.org/10.1177/0741088309351479.

Quick, Catherine. 2012. "From the Workplace to Academia: Nontraditional Students and the Relevance of Workplace Experience in Technical Writing Pedagogy." *Technical Communication Quarterly* 21 (3): 230–250. https://doi.org/10.1080/10572252.2012.666639.

Rampell, Catherine. 2011. "More College Graduates Take NonProfit Jobs." *New York Times*, March 1, 2011.

Rude, Carolyn. 2008. "Introduction to the Special Issue on Business and Technical Communication in the Public Sphere: Learning to Have Impact." *Journal of Business and Technical Communication* 22 (3): 267–271. https://doi.org/10.1177/1050651908315949.

Schalk, Sami. 2013. "Metaphorically Speaking: Ableist Metaphors in Feminist Writing." *Disability Studies Quarterly* 33 (4). http://dsq-sds.org/article/view/3874/3410.

Selleck, L. G. 2010. "Pretty in Pink: The Susan G. Komen Network and the Branding of the Breast Cancer Cause." *Nordic Journal of English Studies* 9 (3): 119–138.

Simmons, W. Michele, and Meredith W. Zoetewey. 2012. "Productive Usability: Fostering Civic Engagement and Creating More Useful Online Spaces for Public Deliberation." *Technical Communication Quarterly* 21 (3): 251–276. https://doi.org/10.1080/10572252.2012.673953.

Söder, Mårten. 2009. "Tensions, Perspectives, and Themes in Disability Studies." *Scandinavian Journal of Disability Research* 11 (2): 67–81.

Sulik, G. A. 2011. *Pink Ribbon Blues*. London: Oxford University Press.

Thomas, Carol. 2006. "Disability and Gender: Reflections on Theory and Research." *Scandinavian Journal of Disability Research* 8 (2–3): 177–185.

Titchkosky, Tanya. 2005. "Clenched Subjectivity: Disability, Women, and Medical Discourse." *Disability Studies Quarterly* 25 (3). http://dsq-sds.org/article/view/589/766.

Titchkosky, Tanya. 2011. *The Question of Access*. Toronto: University of Toronto Press.

Tremain, Shelley. 2013. "Introducing Feminist Philosophy of Disability." *Disability Studies Quarterly* 33 (4). http://dsq-sds.org/article/view/3877/3402.

Ward, Mark, Sr. 2010. "The Ethic of Exigence: Information Design, Postmodern Ethics, and the Holocaust." *Journal of Business and Technical Communication* 24 (1): 60–90. https://doi.org/10.1177/1050651909346932.

Watt, S. S. 2012. "A Postfeminist Apologia: Susan G. Komen for the Cure's Evolving Response to the Planned Parenthood Controversy." *Journal of Contemporary Rhetoric* 2 (3–4): 65–79.

Zaritsky, E., and S. L. Dibble. 2010. "Risk Factors for Reproductive and Breast Cancers among Older Lesbians." *Women's Health* 19 (1): 125–131.

PART IV

Accommodating Different Discourses of Diversity

10
USING NARRATIVES TO FOSTER CRITICAL THINKING ABOUT DIVERSITY AND SOCIAL JUSTICE

Natasha N. Jones and Rebecca Walton

INTRODUCTION: RELEVANCE OF DIVERSITY AND SOCIAL JUSTICE
Extending the work of critical and cultural studies requires an intentional focus on social justice, diversity, and inclusion. Scott, Longo, and Wills (2006) asserted that technical communication needs "approaches that historicize technical communication's roles in hegemonic power relations—approaches that are openly critical of nonegalitarian, unethical practices and subject positions, that promote values other than conformity, efficiency, and effectiveness, and that account for technical communication's broader cultural conditions, circulation and effects" (1). They called for a critical-cultural studies approach to technical communication research and pedagogy that foregrounds a subjective, reflexive, and critical way of conceptualizing what technical communication is, what technical communication does, and why technical communication matters. Connections between critical-cultural studies and technical communication have since been explored, defended, and legitimized by many scholars (e.g., Haas 2012; Hunsinger 2006; Longo 1998; Palmeri 2006; Grabill and Simmons 1998).

Nearly a decade later, technical communication instructors and scholars still grapple with developing an understanding of what to do with this critical-cultural framework. The question is, "Now what?" We argue that extending the work of critical and cultural studies requires (1) a specific and intentional focus on social justice, diversity, and inclusion and (2) an action-oriented approach to addressing concerns of inequity. As Scott, Longo, and Wills (2006) assert, "Perhaps the most important function of cultural studies is to translate critique into ethical civic action" (15). Focusing on social justice, as we advocate in this piece, examines the hermeneutics of "lived realities" and asks not "what does it mean" but "what can we do" (Saukko 2005, 343).

We recognize that social justice can be defined in a number of ways by different disciplines and scholars, so we find it necessary to define social justice as it relates to technical communication. Extending Frey et al.'s (1996) understanding of social justice as advocating for people who are under-resourced, we propose the following definition for our field: **social justice research** in technical communication investigates how communication, broadly defined, can amplify the agency of oppressed people—those who are materially, socially, politically, and/or economically under-resourced. Key to this definition is a collaborative, respectful approach that moves past description and exploration of social justice issues to taking action to redress inequities. As such, our definition of social justice is broad and encompasses action-oriented research and pedagogy that can inform and integrate civic engagement, participatory research and action research, and minority studies (e.g., feminist, queer, critical race, etc.).

In recent years, scholars in technical communication have taken up the complicated work of research about social justice explicitly named (Agboka 2013; Haas 2012; Savage and Mattson 2001), as well as civic engagement (Bowdon 2004; Rude 2004), service learning (Crabtree and Sapp 2005; Youngblood and Mackiewicz 2013), activism (Faber 2002; Jones 2014), and human rights (Dura, Singhal, and Elias 2013; Walton, Price, and Zraly 2013). Although social justice is increasingly relevant to the discipline and pedagogy of technical communication, few resources exist to help teachers explicitly address diversity and social justice in the technical communication classroom. Though prominent in the practice and scholarship of technical communication, social justice and diversity are rarely addressed in our scholarship of teaching, with a few notable exceptions (see Haas 2012; Moore 2013). Without targeted teaching resources, educators will continue to struggle—or, worse, fail altogether—to equip the next generation of technical communication scholars and practitioners for the complex work of recognizing, acting within, and shaping issues of social justice and diversity.

In this chapter we argue that narrative is a useful tool for fostering critical thinking about social justice in the technical communication classroom. Narrative has been examined and employed in technical communication research and pedagogy by a number of scholars (Faber 2002; Perkins and Blyler 1999; Smart 1999). For example, Bridgeford (2002) investigated narratives as "ways of knowing," while Kitalong et al. (2003) used technology autobiographies as a writing assignment in the composition and technical communication classroom. Haas, Tulley, and Blair (2002) used narrative as a method and pedagogy to interrogate technological literacy and mentorship in women, and Moore (2013) employed narrative to explore the intersection of academia and work in public policy, social justice, and inclusiveness.

Building upon this narrative work, this chapter presents four distinct but interrelated capacities of narrative that can be useful for facilitating critical engagement with an eye toward social justice. We present a heuristic for designing teaching resources that use narrative to facilitate this engagement.

CAPACITIES OF NARRATIVE

In this section we extend previous research, connecting narrative as a mechanism for social change with narrative as an instructional tool to help students understand the impact of their writing on diversity and social justice. We ascribe to Arnett's (2002) definition of narrative, with its emphasis on participation and context: "A story has main characters, a plot, and a storyteller. A narrative has all the ingredients of a story, [plus] agreed-on participation, and openness to the needs of a given historical moment" (501). This definition indicates that stories are controlled by the teller, whereas narratives are enacted through participation by the people (501). This distinction implies that for students (or anyone, for that matter) to take action to support social justice, they must do so within the context of the stories they live and tell (Rouse 1990, 181).

Narrative is a promising tool for engaging explicitly with issues of diversity and social justice because of its capacities for fostering identification, facilitating reflexivity, interrogating historicity, and understanding context. These four capacities (identification, reflexivity, historicity, and context) are distinct and useful as singular considerations, but, importantly, they can be connected and are interrelated. Figure 10.1 illustrates one way of conceiving the relationships among the four capacities. **Identification** is key to how we perceive the world, looking through the lens of **historicity**. Those perceptions necessarily occur within a **context**, the boundaries of which are fluid and dynamic, influenced by **reflexivity**.

These capacities can be useful to instructors who focus on one capacity individually and to instructors who examine the interrelations and connections among capacities. In the sample exercises and activities at the end of this chapter, we note a few example connections among capacities. Instructors are encouraged, however, to find their own connections and to investigate other ways they can highlight the value of each capacity.

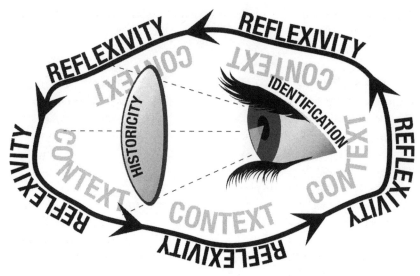

Figure 10.1. Capacities of Narrative Suited to Facilitating Social Justice Pedagogy

(1) FOSTERING IDENTIFICATION

Perkins and Blyler (1999) noted narrative is inextricably tied to culture and therefore to identity: people produce and reproduce collective identity through stories. However, for a collective identity to be developed, individuals must first engage in the rhetorical process of **identification**. From a Burkean perspective, the concept of identification is more complex than just how people identify with individuals, communities, or cultures. Foundationally, identification is about cohering around concepts, values, or beliefs. Burke's rhetorical concept of identification aligns with persuasion and rests on the idea that, in order to persuade an individual (or an audience), a rhetor must connect with the audience through "language by speech, gesture, tonality, order, image, attitude, idea, *identifying* your ways with his" (Burke 1969, 55).

Identification occurs along a continuum and is a process rather than a phenomenon. The ultimate goal of identification is consubstantiality, defined by Jones (2012) as "the recognition of commonality in communion with diversity." Complete identification, or consubstantiality, can be achieved when individuals cohere around common goals and values (202). In other words, a collective identity is developed. Collective identity is important for persuasion in contexts such as social movements and activist organizations (Stewart, Smith, and Denton 2007). Identification is a capacity of narratives that is relevant to social justice and diversity because it (1) presents characters (agents) to encourage consubstantiality and (2) invites co-created assent through climax and use of climactic language.

Encouraging Identification by Presenting Characters (Agents)

Many narratives focus on a character, or agent, who embodies certain values, ideals, and beliefs. The character acts within the confines of the narrative to enact change such as a personal transformation or transformation of a situation or other characters. The character is inextricably connected to the narrative and is thus an important consideration for whether the audience can identify with not only the character, but also with the story that is being told. According to Burke (1969), "an act of persuasion is affected by the character of the scene in which it takes place and of the agents to whom it is addressed. The same rhetorical act could vary in its effectiveness, according to shifts in the situation or in the attitude of audience" (62).

In relation to social justice, central characters often are symbolic representations of a cause. Stewart, Smith, and Denton (2007) noted that who is telling the story matters tremendously: "The narrator's image and audience appeal are so important to the narrative that personal identification overpowers logical rigor" (204). The authors asserted that "storytelling engages people in a communicative relationship defined by the narrator-audience relationship. The narrator and listener create a 'we' through their identification; 'my story' becomes 'our story' through co-creation" (204).

Identification on this level pushes the audience and narrator toward consubstantiality, not only rhetorically, but symbolically and actionally by creating "rhetorical visions." Stewart, Smith, and Denton (2007) quoted Bormann (1972) in defining rhetorical visions as "the composite dramas which catch up large groups of people in a symbolic reality" (205). They argued that these rhetorical visions "arise through communication and provide the themes, heroes, villains, values, and motivations that are invoked in later communication" (205). Extending Bormann's work in a piece on human rights in technical communication, Walton, Price, and Zraly (2013) showed that tools such as symbolic convergence theory and fantasy theme analysis use narratives to trigger an emotional response by drawing upon themes that reflect people's values, priorities, and beliefs. Ultimately, the characters of a narrative (whether narrator, hero, or villain) encourage identification with an audience by tapping into commonly held ideals. This identification, in turn, provides a justification, exigence, and impetus for social action.

Inviting Assent through Climactic Language

The manner in which climax is used to promote rhetorical identification is similar to how climax is used in narratives and stories. In narratives, the climax is the pivotal point for a character, the point at which the character begins to resolve tensions and move toward resolution. Likewise, climax as a rhetorical device (as conceptualized by Burke) can move the audience and the rhetor toward resolution (i.e., consubstantiality). Climax is a narrative device that allows a story to build to an anticipated resolution. Burke addressed climax from a foundational level, examining the use of climactic language as a tool for persuasion and positing that climactic language "invite[s] assent to [the] proposition as doctrine" (Burke 1969, 59). In other words, as the author uses language to build a sense of expectancy in the audience, the audience then becomes more willing to accept the proposition that the author is making. Burke asserted that climax can encourage an audience to "feel how it [a proposition] is destined to develop—and on the level of purely formal assent you would collaborate to round out its symmetry by spontaneously willing its completion and perfection as an utterance" (59).

The climax is co-created by the author and the audience with both parties seeking the same resolution and cohering around an end-goal. As a pedagogical tool, instructors can encourage students to implement this rhetorical and stylistic device when communicating about social justice issues. Burke in fact, encouraged the use of "stylistic identifications" to "establish rapport" with an audience (Burke 1969, 46). These stylistic identifications help an audience to know what to expect and provide textual and rhetorical cues (e.g., consider how alliteration and repetition is used in a speech). Making students aware of how climax can be used to build rapport with an audience can help students become more critical of how climax is manifested in communication about social justice issues, and it can equip students to be more effective in their own communication.

Character and climax can help students understand how the rhetorical process of identification is accomplished through narrative. Further, understanding how and why rhetorical identification is fostered through narratives allows students to better analyze and engage in social action.

(2) FACILITATING REFLEXIVITY

For students to engage critically with issues of social justice, they must be reflexive, and narrative is particularly well suited to facilitating **reflexivity**: "The genre [of narrative] also evokes reflexiveness: critical thought crucial for change and transformation" (Beynon and Dossa 2003, 252). As a pedagogical tool, narratives can enable critical insights through reflexivity. These insights can shape students' understandings of themselves as people and as professionals, as well as their ability to perceive relations of power that structure and operate in social contexts (Blyler 1995). Further, the reflexivity enabled by narrative is useful for considering what actions to take and the ethical merit of those actions, particularly when the production of narratives is interwoven with discussion. As Blyler claimed, "Narrative can provide a basis for reflection, critique, and dialogue" (1995, 295). In exploring narrative's capacity for facilitating reflexivity, we focus on two considerations relevant to social justice pedagogy: (1) supporting ethical reasoning and (2) exploring positioning and revealing privilege.

Supporting Ethical Reasoning

Technical communication ethics must be understood within a social context: "In order to talk about ethics in technical communication, we must understand the place of technical communication in societal and political experience" (Sullivan and Martin 2001, 252). No communication is neutral or apolitical, including—perhaps *especially*—technical communication (Katz 1992), because it powerfully bounds and shapes people's lives in its many instantiations: for example, civic communication (Ornatowski and Bekins 2004), humanitarian communication (Haselkorn and Walton 2009), activist communication (Jones 2014), and business communication (Faber 2002). To adapt a popular saying, with technical communication's great power for affecting lives comes great responsibility for ethical practice. Thus, ethical reasoning is a critical component of technical communication pedagogy (Dragga 1997). However, facilitating ethical consideration of technical communicators' roles, approaches, and communication products is challenging.

One challenge to facilitating ethical reasoning is that technical communication is a field of practice (Hughes and Hayhoe 2008, 4), guided more by communication ethics than by philosophical ethics. As Arnett (2002) explained, philosophical ethics is precise because it evaluates the level of congruence between an ideal conceptual standard and a particular action. Communication ethics is less precise because both the conceptual standard and the action compared against it are considered situational and rhetorical. Communication ethics treats the conceptual standard for ethical behavior as a "rhetorical construct" comprised of the historical moment, topic, and ideal communication form (Arnett

2002, 498). Evaluation informed by communication ethics considers that rhetorical construct in relation to the communication at hand and the contextual factors surrounding that communication: "Communication ethics navigates among a conceptual standard, the players in a story, the historical moment, and the topic. . . . Communication ethics are applied; they do not rest in abstract theory but in the give and take of persons living together" (Arnett 2002, 498). Thus, communication ethics is complex because it is communal and relational, as well as situational and applied. This relational aspect is particularly appropriate because it enables ethical reasoning to get at the heart of why technical communication ethics matters: what we do and write affects people.

Another factor that can make ethical reasoning about technical communication difficult is that technical communication has been historically considered as a means to an end. Situating technical communication within an Aristotelian view of ethics, Sullivan and Martin (2001) said that although technical communication unavoidably has ethical and political dimensions, it can be difficult to investigate technical communication in the context of the ends it promotes because it is doubly removed from these ends: technical communication is a means to enabling use of technology and/or information that is itself a means to accomplishing a particular end (252). Current conceptions of technical communication, its instantiations, and its purposes are much broader than merely enabling the use of technologies, but we believe that Sullivan and Martin's premise holds true: the distance between technical communication and its political and social effects can mask the connection, especially if technical communicators lack reflexivity to prompt ethical reasoning about their work. To cultivate a generation of technical communicators well equipped to engage in ethical reasoning, instructors can use narrative to facilitate reflexivity. Because narratives provide a moral context for actions, they can serve as a tool for complex ethical reasoning, particularly for writers (Jameson 2000; Sullivan and Martin 2001).

One way that narratives can support ethical reasoning is by enabling people to project forward (envisioning possible actions) and reflect back (considering possible ramifications of those actions). Gee (2005) described narratives as simulations that provide low-stakes ways to explore potential courses of action: "We can act in the simulation and test out what consequences follow, before we act in the real world" (75). Narrative reasoning highlights connections between the actor and the range of people potentially affected by his or her actions: "We believe that narrative reason is better than syllogistic reason in apprehending the ethical situation because stories contextualize the proposed action in the unity of an individual's life and in the lives of others" (Sullivan and Martin 2001, 266). This relational aspect of narratives makes them well suited to engaging in communication ethics, a relational and rhetorical approach to ethical reasoning. Sullivan and Martin (2001) proposed envisioning potential outcomes of an action and then considering an explanatory narrative to convey to the full range of affected people the reasoning underlying that action. This narrative should not be considered a way to excuse unethical actions but rather as an envisioning exercise useful for identifying and ruling out unethical actions—that is, actions that would be impossible to explain and support in a narrative directed to the affected parties. Especially relevant to ethical dilemmas with social justice implications, the authors emphasized explicitly considering marginalized people within the audience of affected parties: "The audience we have in mind consists not only of moral experts and those with power, but also those who have been traditionally silenced, those without voice" (Sullivan and Martin 2001, 260).

In addition to helping students to envision and ethically evaluate courses of action, narratives can also aid students in drawing upon values and moral examples embodied by people they admire and to envision the type of person and professional they want to become. As Dragga (1997) explained, ethics addresses character as well as behavior: "Ethics is not just the study of what we do as writers and readers . . . but of who we are as human beings—reason-giving, story-telling human beings" (175). This holistic view of ethics is embraced by McCoy (1992), who advocated for the pedagogical use of narratives like scenarios and case studies because they convey the complexity of human experience and facilitate students in *communally* exploring that complexity. Although behavior-focused approaches to ethics historically dominated technical communication pedagogy (and, we argue, still do), Dragga (1997) instead focused on day-to-day moral integrity as part of one's identity:

> This perspective emphasizes the ethical power of narrative: through stories, a community identifies its heroic individuals and the ideals and behaviors that characterize the heroic. Individuals decipher virtuous and vicious practices and develop their moral vision by interpreting the stories of their lives according to the heroic narratives of their community, by fitting the stories of their lives within the ongoing stories of their community. (164)

Narrative is an important tool for ethical reasoning because other ethical guides such as heuristics, policies, and principles are necessarily vague and can be difficult to apply to specific ethical dilemmas. In fact, industry practitioners indicated that they do not consult ethics policies and heuristics when wrestling with an ethical dilemma in the workplace but instead reflect upon who they want to be, inspired by those they admire (Dragga 1997). That said, we do believe that heuristics can be useful for facilitating engagement with ethically complex issues such as social justice—obviously so, since we introduce a heuristic in this chapter. But we do not expect practitioners to directly draw upon the heuristic when conducting their work. Rather, this heuristic can guide educators in using narrative to foster the kind of complex, visionary, and character-focused engagement advocated by scholars such as Dragga, Blyler, Gee, Sullivan, and Martin.

Exploring Positioning and Revealing Privilege

Narrative's capacity for facilitating reflexivity can also help students to explore positioning—their own and that of others—and to recognize the ways in which certain positions are privileged. This recognition is central to engaging critically with issues of diversity and social justice, as communication is a powerful tool for preserving or disrupting power structures. Summarizing Herndl (1993) and Wells (1986), Blyler (1995) proposed that educators in our field help students to develop " . . . a stronger sense of the way in which communicative practices reproduce certain privileged structures and modes of behavior at the expense of others" (Blyler 1995, 292). In developing this sense of communication's relationship with the existing social order, students who engage in a truly critical way would neither blindly accept nor reject out of hand the social order but would mindfully navigate the structures at play and their own positioning within them (Herndl 1993).

As students navigate these structures, they must consider their *relative* positioning—that is, avoid overemphasizing individualism and rather see themselves and their work as relational. In a move related to historicity, Arnett (2002) described the dangers of taking an extremely individualistic view of positioning and the importance of situating oneself within a historical moment and community:

> Individualism disembodies the person from the historical situation. This type of humanistic assumption about the importance of the individual misses the embedded nature of our lives together in a story that is *a priori* to us, dictating to, and not guiding, communicative partners. Persons living within a narrative structure situate themselves within the story of a community. Humanism focused on the individual conceptualizes communicative meaning within the person. Such a philosophy is long on agency and short on embeddedness. (499)

To facilitate embeddedness, instructors can characterize professional communication as a dialogic process that supports relational work. Students can engage in reflexive exercises on their own positioning within the context of the relational work of our field (Moore 2013). For example, Moore (2013) used storytelling as a teaching tool to help students uncover and understand relationships involved in policy work—including students' own positions as players within the network of stakeholders. Having students read or produce stories conveying multiple perspectives on the same topic can help students develop complex understandings of position and privilege in terms of, for example, race, ethnicity, sexuality, gender identity, nationality, (dis)ability, religion, legal status, culture, education, and socioeconomic level. If students stifle each other's perspectives on sensitive aspects of identity and privilege, it can be helpful to assign readings by authors with a range of perspectives and require students to engage privately with these issues through reflective writing before engaging in class-wide discussion (Beynon and Dossa 2003, 259). Equally important, teachers can scaffold students in critically reflecting upon how perspectives are characterized so as to recognize and reject characterizations of one perspective as central and another perspective as secondary, alternative, or "other" (Haas 2012; Powell 2004). In conclusion, narrative is well suited to this complex work of exploring positioning and revealing privilege because of the dialogic space it opens up: "This genre [of narrative] provides a dialogical space for the articulation of concerns routinely silenced in mainstream discourse" (Beynon and Dossa 2003, 252; citing Ong 1995).

3 INTERROGATING HISTORICITY

Another important capacity of narrative is **historicity**, which posits that ideals, values, and beliefs are developed, articulated, and rearticulated through perceptions of historical and contextual occurrences. Historicity is a complex term with nuanced meanings for scholars in a variety of fields. Ohnuki-Tierney (1990) defined historicity as "the culturally patterned way or ways of experiencing and understanding [and] constructing and representing history" (4). Gramsci (1971) asserted that "reality does not exist on its own, in and for itself, but only in an historical relationship with men who modify" (660). In this sense, historicity relies on cultural context and representation.

We employ historicity to interrogate how persistent historical (mis)representations impact social justice and diversity. Historicity engages the subjective nature of narrative, even narratives that are positioned as objective and value-neutral. Focusing on historicity in narrative is useful for two reasons: (1) it allows for a closer examination of the process of knowledge legitimization and (2) it encourages critical interrogation of dominant ideologies.

Examining the Process of Knowledge Legitimization

Technical communication scholars have long been interested in how knowledge is created and legitimized. For example, Slack et al.'s (1993) pivotal text argued that technical communicators were to be considered authors and knowledge-makers, producers of knowledge rather than transmitters of already-existing knowledge. The authors employed articulation theory to explore how power and identity function in communicative processes of technical communicators, stating that "articulation thus points to the fact that any identity is culturally agreed on or, more accurately, struggled over in an ongoing process of disarticulation and rearticulation" (27). History is similarly articulated and rearticulated. Examining how understandings and presentations of historical concepts and events are then "struggled over" necessarily calls into question which knowledge(s) and ways of knowing are considered legitimate and which are not.

Historicity provides a way of understanding that knowledge and knowledge legitimization are connected to cultural-historical representations and thus enables us to examine knowledge legitimization from a cultural-historical perspective. Ohnuki-Tierney (1990) asserted that even historical records, often seen as sites of legitimate knowledge, "result from the subjective and cultural perspectives, insights, and 'facts'" (4). In this sense, even facts and records undergo a process of legitimization (articulation and rearticulation) that is inextricably connected to "culturally patterned . . . ways of experiencing and understanding history" (4). In this way, the processes of knowledge legitimization and historicity are mutually created. According to Burke (1937), how individuals "construct" an understanding of self, environment, values, and beliefs—the knowledge that they develop—depends on subjective occurrences: "in the face of anguish, injustice, disease, and death one adopts policies. One constructs his notion of the universe or history and shapes attitudes in keeping" (3). This assertion points to the value of asking our students to not only be aware of the subjective nature of knowledge legitimization but also to attempt to unpack why certain ways of thinking and knowing are legitimized while others are not.

Interrogating the Perpetuation of Dominant Ideologies
Historicity provides a rich perspective for interrogating dominant ideologies. Knowledge that is legitimized and presented as "objective fact" often represents dominant ideologies. These ideologies are the narratives that are most frequently heard, most frequently repeated, and least questioned. They are what Giroux (1991) called "master narratives." These narratives help to create and maintain dominant ideologies, ways of knowing, and ways of learning. Historicity can help deconstruct these narratives, acknowledging the subjective nature of the ideologies imbued in the narratives. Scott, Longo, and Wills (2006) called for "research and teaching practices that historicize technical communication's roles in hegemonic power relations" (1). Historicity can help to bring to the fore the social, political, economic, and material contexts in which ideologies are produced and perpetuated.

So, what is the connection between narrative's capacity for examining historicity and the classroom? Unfortunately, classrooms can often function to reinforce dominant ideologies and perpetuate oppressive and marginalizing contexts (see Haas 2012; Selfe and Selfe 1994). In fact, the student-teacher relationship itself can function as an example of a dominant and dominating ideology. Freire (1970) addressed the "narrative character" of traditional student-teacher interactions, claiming that this type of juxtaposition "involves a narrating Subject (the teacher) and patient, listening objects (the students)" (53). The teacher produces the master narrative and fills up the silenced students with knowledge. Freire called this the "banking model" of education—a model in which "the scope of action allowed to the students extends only as far as receiving, filing, and storing the deposits" (53). In contrast, exploring historicity through narrative promotes a genuine exchange of information between student and teacher. Students can engage with subjects, concepts, and topics from their own perspective and worldview. Student narratives are valued and listened to. Students' historical consciousness is then an important part of the educational process: "Education must begin with the solution of the teacher-student contradiction, by reconciling the poles of the contradiction so that both are simultaneously teachers *and* students" (Freire 1970, 53).

Educators can incorporate into their classrooms activities and assignments that help students to examine historicity more closely. Asking our students to consider how knowledge is legitimized and how dominant ideologies are maintained requires that they consider cultural and historical ways of understanding and how they participate in either supporting or dismantling hegemonic practices. In addition, careful examinations of this sort can help instructors to be acutely aware of knowledges and ideologies that they privilege in the classroom.

4 UNDERSTANDING CONTEXT

Narrative also has the capacity to facilitate an understanding of **context**. This is a vital capacity for a pedagogical tool because, as many scholars have argued (Dragga 1997; Kienzler 2001; Sims 1993), our work as technical communication educators centers on preparing students to take ethical, responsible communicative action in situated contexts that include workplaces and communities. It is no small feat to prepare students to reflect upon and act within "the tremendous technological, scientific, and institutional (bureaucratic) complexity that characterizes contemporary life" (Simmons and Grabill 2007, 423). These technological, scientific, and institutional factors create unique contexts that must be critically understood and engaged. Narrative offers a way of constructing, understanding, and acting within varied contexts of contemporary life by helping students (1) to understand how context situates action and (2) to recognize context as encompassing and constructed.

Understanding How Context Situates Action

The concept of context has long been central to technical communication practice, scholarship, and pedagogy. Context creates an "environment" in which agents can act, interact, and identify. As instructors, we tell our students to consider the audience, context, purpose, and use of communication. Context inherently encourages typified responses to social situations—in other words, context is central to the creation of genres (Miller 1984) and genres, in turn, encourage specific social action. Further, context positions actors within an encompassing social, political, economic, and ontological framework. Context sets the scene for a story to be told.

Scholarship in our field has focused on narrative contexts including management in organizations (Zachry 1999), banks (Smart 1999), museums (David 1999), and academic disciplines (Dulek 2008; Powell 2004). Narratives make sense only as far as they are positioned in an "appropriate" context. This position fosters a sense of identification between the addressed (the audience) and the addressee (the speaker or writer). It helps the audience to understand what is "normal," typical, appropriate, and it aids the audience in deciding upon an appropriate social action. Drawing upon Suchman (1987), Kuutti (1996) asserted that "actions are always situated into a context, and they are impossible to understand without that context" (Kuutti 1996, 27). Further, we understand that identification depends on contextual elements (Burke 1969, 62).

Narrative allows exploration of context as it relates to social movements and social justice issues. We see a key role for tools such as critical race theory to provide structure for these narratives, which describe particular rhetorical situations and position both the storytellers and their audiences to recognize injustice and take action (see Edwards, chapter 11 in this collection). The political, economic, and social contexts in which social inequities occur precipitate the development of groups that form to address the inequities. Rhetorically, if the audience understands and identifies with the context of a specific situation, then the audience may be more easily persuaded to respond to the situation in a particular manner. As Burke (1969) posited, "The rhetorician, as such, need operate only on this principle. If, in the opinion of a given audience, a certain kind of conduct is admirable, then a speaker might persuade the audience by using ideas and images that identify his cause with that kind of conduct" (55).

Narratives are a powerful tool for creating communicative spaces that can help students to understand context and reason through the types of action that can be taken within that context.

Recognizing Context as Encompassing and Constructed

Narratives are essential for meaning making and knowledge creation partly because they convey multiple connections and contexts: "we interpret stories as we make knowledge from others and for others—configuring connections and layering contexts" (Perkins and Blyler 1999, 3). Narratives frame events and sequences of events within a context that clarifies the meaning of elements *and* relationships among those elements (Kerby 1991, 3). In other words, stories are relational, taking into account a number of actors, perspectives, time periods, and ethical views. The relational aspect of narratives make them well suited to not only conveying complex contexts but to understanding those contexts (Moore 2013).

The notion of context as encompassing *and* constructed is key to social justice pedagogy, as it illuminates both (1) the power structures enabling and constraining particular actors in particular ways and (2) the creative possibilities for action not only within the existing context but to shape that context (see Edwards, chapter 11 in this collection). Arnett (2002) used the metaphors of story and narrative to frame

Freire's social justice pedagogy, claiming that we are embedded within a context that limits us but that can be creatively shaped by us (500). Context is important for understanding the actors embedded within it, Arnett claimed, that "to miss the context is to miss the person" (500). In other words, context is social, and this social aspect of context is vital to narrative's meaning-making ability. Mumby (1993) said that narrative "takes on meaning only in a social context and . . . plays a role in the construction of that social context" (5).

However, narrative's relation to context and meaning making is not neat and tidy, especially from postmodern perspectives. Defining postmodern as "incredulity toward master narratives" (Lyotard 1984, xxiv), Lyotard introduced terms that have been translated as master narrative or metanarrative (*maître-réci*) and local narratives or little stories (*les petites histoires*). Lyotard's terms allow us to extend the discussion of master narratives from the earlier section on historicity with the additional concept of local narratives. Master narratives legitimate dominant ideologies, while local narratives are shorter, difficult to situate within master narratives, often represent the experiences of marginalized people, and are sometimes characterized by artistry (Klein 1995, 280–281).

Similar concepts, such as dominant cultural narratives and counter-narratives, give us ways to talk about the contexts created by narrative and the ways those contexts enable, constrain, and even define us. Whereas dominant narratives attempt to create a context that designates the boundaries of normative experience, counter-narratives express a wide range of perspectives and experiences that resist dominant narratives (Bamberg and Andrews 2004). These counter-narratives are not diametrically opposed to dominant narratives, nor are they completely separate from them but rather operate in tension with dominant narratives (Torre et al. 2001).

Andrews (2004) highlighted the value of counter-narratives for enabling meaning making outside of dominant narratives:

> Wittingly or unwittingly, we become the stories we know, and the master narrative is reproduced. But of course it is not so simple. When, for whatever reason, our own experiences do not match the master narratives with which we are familiar, or we come to question the foundations of those dominant tales, we are confronted with a challenge. How can we make sense of ourselves, and our lives, if the shape of our life story looks deviant compared to the regular lines of the dominant stories? (11)

Boje's (2011) "antenarratives" are in some ways similar to counter-narratives: disruptive stories that threaten the coherence and authority of (dominant) narratives. Unlike the definition of narrative that informs this chapter, Boje saw narrative as fixed, focused on the past, and monological, whereas he defined antenarrative as "a bridging of past narratives stuck in place with emergent living stories" (3). Antenarratives are "before" stories that threaten to dismantle the neat, coherent, sanctioned version of the story (Boje 2007, 223). Attempting to make sense of possible futures, antenarratives support "prospective sensemaking" (Boje 2011, 3) that can equip people to take action. Because antenarratives have a reciprocal relationship with materiality (i.e., antenarratives inform reality and are shaped by it), they are agential (1). As bridges between the fossilized, dominant version of a story and multiple emergent living stories, antenarratives may break apart contextual boundaries of the sanctioned narrative and offer different, opposing perspectives that are vital for meaning making—especially when considering issues of diversity and social justice. As such, antenarratives offer promise for helping students to look beyond monological, sanctioned versions of a story. For example, Haas's (2007) "Wampum as Hypertext" serves as a disruptive "before" story useful for prospective sensemaking, widening the view of what is possible (and desirable) through hypertext by breaking apart contextual boundaries of what "counts" as hypertext, what constitutes its history, and who is involved in its creation.

For students to explore the complexities of social justice contexts, they must also understand their own capacities to act through communication. In arguing this point, Herndl (1993) claimed that students need to "see how discourse and the reality it constructs are shaped by the political, economic, and material interests of professions and the institutions they create" (354). This claim links the co-constructed nature of context to discourse: context encompasses our experiences and possibilities, but we contribute to its construction through our discourse—discourse that is essential for action and change. Linking words (stories) and action (change), Faber (2002) made a similar point:

> Stories broker change because they mediate between social structures and individual agency. In other words, stories help us negotiate between those factors that restrict and limit our possibilities and our free ability to pursue our own choices . . . Change will be a product of what can be legitimately said (or written) in a specific context at a specific moment in time. (25)

Narratives can help students to develop a complex understanding of context and the varying constraints and affordances within it, setting the stage for action to promote social justice.

CAPACITIES IN ACTION: SUPPORTING SOCIAL JUSTICE

Considered in light of the above capacities, narrative is well suited to brokering change that supports social justice. As Faber (2002) claimed, "... change itself is a story, and stories are acts of change" (21). Similarly, Sullivan and Martin (2001) said that, "to decide what to do next is to ask what stories you are in" (260). In other words, narrative not only *can* contribute to social justice but is *necessary* for change. We agree with Blyler's (1995) assertion that narrative is a useful tool in technical communication classrooms for clarifying "the nature of socially responsible communicative action" (293). But we caution technical communication educators not to stop at helping students to clarify the nature of actions but to support them in *taking action* informed by a complex understanding of context. The importance of this understanding is underscored by scholars such as Scott (2004) and Jones (2017), who encouraged technical communication educators to reject "hyperpragmatism," which decontextualizes students' work from broader, complex social issues and underlying power structures.

To facilitate students (1) in exploring issues of social justice and diversity and (2) in taking action on those issues, narrative must be multivocal and participatory. As Simmons and Grabill (2007) noted, simply receiving a single narrative does not facilitate participatory action:

> We believe the design of civic information must allow for multiple entry points, multiple types of questions, and multiple angles of investigation to allow citizens to invent usable knowledge from the available information. Providing a single narrative of information does not allow for these explorations. Without the ability to invent and produce usable knowledge from available information, full participation in civic issues becomes unlikely. (434)

Thus, we argue that *producing* narratives, especially multiple narratives drawing from the same data about the same topic, is an excellent way to engage students in exploration and invention, both of which Simmons and Grabill claimed are necessary for people to write to change communities. Indeed, what we suggest is similar to the iterative, collaborative practice that Simmons and Grabill (2007) described. The organization they studied collaboratively reviewed and discussed a large body of information and produced summaries to synthesize what they learned and to share with the broader community in order to generate action. Further, the documents they produced were only *part* of their ongoing and iterative efforts to slowly make headway on the issue of concern; this, too, is congruent with how we recommend using narrative as a tool for social justice pedagogy. Social justice is complex and difficult and takes collaborative work over the long-term. We believe that the work of producing narratives—work that includes first reviewing, selecting, and synthesizing information, and then involves sharing, reflecting upon, and facilitating action through the production of multi-vocal narratives—is centrally relevant to social justice pedagogy.

HEURISTIC

Distilled from the above sections, the heuristic in table 10.1 provides succinct guidelines for developing in-class exercises, assignments, discussion guides, and other narratively driven pedagogical tools. For the purpose of this chapter, we borrow from the field of usability and user-centered design and define a heuristic as "design guidelines that serve as a useful evaluation tool" (Desurvire, Caplan, and Toth 2004, 1509). Though Desurvire, Caplan, and Toth apply this definition to usability and software development, we extend this definition for a pedagogical application. In this sense, our heuristic allows instructors to evaluate how an assignment uses narrative to help students in the technical communication classroom engage explicitly with issues of diversity and social justice. In the table below, the first column presents a heuristic in question format and asks instructors to evaluate whether an assignment addresses one of the four major capacities of narrative. The second column provides brief synopses of sample exercises that can be used in the classroom to engage the capacities of narrative. These synopses are purposely conceptual and brief, leaving room for instructors to adapt the exercises for their own use. In the final column, we describe one connection between the capacities that instructors may find useful to highlight. Again, instructors can find additional connections beyond those suggested here.

CONCLUSION

As technical communication scholars recognize the need to explicitly address social justice and diversity in the classroom, the exigence for pedagogical resources that facilitate subjective, reflexive, and critical understandings of technical communication becomes increasingly evident. Because few of these pedagogical resources are available to instructors in our field, we produced a heuristic that can be used to encourage students to consider issues of inclusion and ways that the communication they create can empower and disempower.

Undoubtedly, using the lens of narrative provides a viable entry point for beginning dialogues about inequity and injustice from a

Table 10.1. Heuristic for Using Narrative to Explore Social Justice and Diversity in the Technical Communication Classroom

Heuristic	Sample Exercise	Connections
Does the assignment demonstrate how narrative fosters identification:	Climax Analysis & Character Sketch	Identification and Historicity
- by presenting characters?	Have students analyze professional communication produced by advocacy groups such as environmental organizations or human rights organizations, identifying climactic language and how it functions. Based on this analysis, ask students to create a character sketch of the speaker or the agent in the speech.	This exercise can easily address historicity as well by pairing the analysis and sketch with discussions of (1) how marginalized perspectives, knowledges, and ways of knowing are legitimized in these speeches and (2) how these speeches challenge dominant ideologies.
- by inviting assent through climactic language?		
Does the assignment demonstrate how narrative can facilitate reflexivity:	Response to an Ethical Dilemma	Reflexivity and Context
- by supporting ethical reasoning?	Have students reflect upon an ethical dilemma (e.g., from a scenario you provide, one they select from the news, or one brought in by a guest speaker). Students should select a potential course of action and write a narrative explaining why they selected that course of action and likely outcomes/effects. The audience for their explanatory narrative is those affected by their course of action, with a particular emphasis on those who have little to no power and whose voices have been silenced.	This exercise can easily address context as well when paired with discussions of the range of possible actions (i.e., contextual constraints) and ways that specific actions may shape/construct context moving forward.
- by exploring position and revealing privilege?		
Does the assignment demonstrate how narrative can interrogate historicity:	Stories of Privilege	Historicity and Reflexivity

continued on next page

critical-cultural perspective. The four capacities (identification, reflexivity, historicity, and context) highlight narrative as a tool for social change and as a pedagogical tool. Narratives can bring individuals together and help people think about the impact they have on people

Table 10.1—continued

Heuristic	Sample Exercise	Connections
- by examining how knowledge is legitimized?	Have students write a story about privilege: for example, (1) a moment or experience when they became aware of being/not being privileged in some way; (2) a time when they saw certain ways of knowing legitimized to a greater degree than or at the expense of other ways; (3) a story of using their privilege to challenge dominant ideologies.	In exploring issues of position and privilege, this exercise already addresses reflexivity as well as historicity.
- by challenging how dominant ideologies are perpetuated?		
Does the assignment demonstrate how narrative can enable understanding of context:	Multiple Narratives	Context and Identification
-by demonstrating how context situates action?	Have students compile a shared collection of information (e.g., newspaper articles, facts/statistics, interviews, videos, images) about a social justice issue with local relevance. Ask students to write the local story of this social justice issue individually or in small groups. Compare narratives: what contextual constraints are in play? How could those constraints be lifted, mitigated, or revealed? What got included, and what got left out? What sources did they draw upon most heavily? What makes some voices, messages, information seem credible and to whom? How could other voices and forms of knowledge be legitimized?	This exercise can easily address identification as well by having students intentionally (1) employ climactic language to persuade readers to cohere around a common goal and (2) feature a central character as a symbolic representation of the cause.
- by recognizing context as encompassing and constructed?		

and situations. Narratives allow people to conceptualize and (re)create past experiences and to justify actions in recurring or unique situations. Narratives are inherently subjective and reflexive. Moreover, narratives provide points of reference for how we, as individuals and communities, experience all facets of life—the good, the bad, the exciting, the mundane—and how we can help our students to seek ways to promote more just and equitable experiences for all.

ADDITIONAL RESOURCES

Here we share some additional resources that instructors may find useful in their efforts to explicitly engage with issues of diversity and social justice in the technical communication classroom.

READING LISTS

- The Social Justice Training Institute has an excellent recommended reading list: http://www.sjti.org/suggested_reading.html
- The White Privilege Conference website lists readings and resources, some that are directly downloadable: https://www.whiteprivilegeconference.com/resources.html

GLOSSARIES AND DEFINITIONS

- Linked from the Teaching Social Justice blog by Jennifer Jackson Whitley, this article presents definitions, drawing from Iris Young's "Five faces of oppression" and Paulo Freire's "Historical conditioning and levels of consciousness": https://mrdevin.files.wordpress.com/2009/06/five-faces-of-oppression.pdf
- The Detroit Community-Academic Urban Research Center has an excellent glossary of terms relevant to community-based action research: http://www.detroiturc.org/about-cbpr/cbpr-glossary.html

THOUGHT-PROVOKING COMICS

- This accessible comic about managing privilege is available under a Creative Commons, Attribution, Noncommercial, No Derivatives license: http://www.robot-hugs.com/?attachment_id=894
- This comic presents a counter-narrative to historical accounts of Christopher Columbus's heroism: http://theoatmeal.com/comics/columbus_day (This comic is copyright protected and has The Oatmeal's characteristically blunt tone.)

SPEECHES AND INTERVIEWS

- This video is a Brown Tugaloo Partners' reading of Martin Luther King Jr.'s *Letter from Birmingham Jail*: https://www.youtube.com

/watch?v=xpOx6dwsgHU (Instructors should be aware that the speech includes language that may be unsettling for some students.)

This video presents an abbreviated version of Sojourner Truth's speech on race and feminism *Ain't I A Woman?* performed by Nkechi: https://www.youtube.com/watch?v=eUdxsQ0Qsrc

This video presents an interview of Gloria Steinem about men, women, and power (especially as related to the US economy): https://www.youtube.com/watch?v=TmFxFmrcngk

In this short video, Judith Butler discusses gender as performative: https://www.youtube.com/watch?v=WRw4H8YWoDA

This video presents an interview with Angela Davis in 1972 in which she speaks about revolution and violence: https://www.youtube.com/watch?v=HuBqyBE1Ppw (Instructors should be aware that the interview includes language that may be unsettling for some students.)

MORE SAMPLE EXERCISES

- Ask students to write a narrative envisioning whom they want to become as a professional: tell the story of what kind of technical communicator you are or want to become.
- Assign students a short paragraph about where they stand on a social justice issue and why. Have students exchange with a peer and write a counter-argument. Discuss positions and paragraphs: how difficult was it to write a counter-argument that may or may not have aligned with your own views?
- Encourage students to identify a communication hero and tell the story of someone who supports social justice through communicative acts.

References

Agboka, Godwin Y. 2013. "Participatory Localization: A Social Justice Approach to Navigating Unenfranchised/Disenfranchised Cultural Sites." *Technical Communication Quarterly* 22 (1): 28–49. https://doi.org/10.1080/10572252.2013.730966.

Andrews, Molly. 2004. "Opening to the Original Contributions: Counter-Narratives and the Power to Oppose." In *Considering Counter-Narratives: Narrating, Resisting, Making Sense*, edited by Michael Bamberg and Molly Andrews, 1–6. Amsterdam, NLD: John Benjamins Publishing Company. https://doi.org/10.1075/sin.4.02and.

Arnett, Ronald C. 2002. "Paulo Freire's Revolutionary Pedagogy: From a Story-Centered to a Narrative-Centered Communication Ethic." *Qualitative Inquiry* 8 (4): 489–510. https://doi.org/10.1177/10778004008004006.

Bamberg, Michael, and Molly Andrews. 2004. *Considering Counter-Narratives: Narrating, Resisting, Making Sense*. Amsterdam, NLD: John Benjamins Publishing Company. https://doi.org/10.1075/sin.4.

Beynon, June, and Parin Dossa. 2003. "Mapping Inclusive and Equitable Pedagogy: Narratives of University Educators." *Teaching Education* 14 (3): 249–264. https://doi.org/10.1080/1047621032000135168.

Blyler, Nancy Roundy. 1995. "Pedagogy and Social Action a Role for Narrative in Professional Communication." *Journal of Business and Technical Communication* 9 (3): 289–320. https://doi.org/10.1177/1050651995009003002.

Boje, David M. 2007. "The Antenarrative Turn in Narrative Studies." In *Communicative Practices in Workplaces and the Professions: Cultural Perspectives on the Regulation of Discourse and Organizations*, ed. Mark Zachry and Charlotte Thralls, 219–37. Amityville, NY: Baywood.

Boje, David M., ed. 2011. *Storytelling and the Future of Organizations: An Antenarrative Handbook*. New York: Routledge.

Bormann, Ernest G. 1972. "Fantasy and Rhetorical Vision: The Rhetorical Criticism of Social Reality." *The Quarterly Journal of Speech*, 396–407.

Bowdon, Melody. 2004. "Technical Communication and the Role of the Public Intellectual: A Community HIV-Prevention Case Study." *Technical Communication Quarterly* 13 (3): 325–340. https://doi.org/10.1207/s15427625tcq1303_6.

Bridgeford, Tracy. 2002. "Narrative Ways of Knowing: Re-imagining Technical Communication Instruction." PhD diss., Michigan Technological University.

Burke, Kenneth. 1937. *Attitudes toward History*. Berkeley: University of California Press.

Burke, Kenneth. 1969. *A Rhetoric of Motives*. Berkeley: University of California Press.

Crabtree, Robbin D., and David Alan Sapp. 2005. "'Technical Communication, Participatory Action Research, and Global Civic Engagement: A Teaching, Research, and Social Action Collaboration in Kenya.' *REFLECTIONS: A Journal of Rhetoric*." *Civic Writing and Service Learning* 4 (2): 9–33.

David, Carol. 1999. "Elitism in the Stories of US Art Museums: The Power of a Master Narrative." *Journal of Business and Technical Communication* 13 (3): 318–335. https://doi.org/10.1177/105065199901300305.

Desurvire, Heather, Martin Caplan, and Jozsef A. Toth. 2004. "Using Heuristics to Evaluate the Playability of Games." *ACM CHI 2004 Extended Abstracts on Human Factors in Computing Systems*: 1509–1512. https://doi.org/10.1145/985921.986102.

Dragga, Sam. 1997. "A Question of Ethics: Lessons from Technical Communicators on the Job." *Technical Communication Quarterly* 6 (2): 161–178. https://doi.org/10.1207/s15427625tcq0602_3.

Dulek, Ronald. 2008. "Academic Research: Two Things That Get My Goat—and Three That Offer Meaning." *Journal of Business Communication* 45 (3): 333–348. https://doi.org/10.1177/0021943608317761.

Dura, Lucia, Arvind Singhal, and Eliana Elias. 2013. "Minga Perú's Strategy for Social Change in the Peruvian Amazon: A Rhetorical Model for Participatory, Intercultural Practice to Advance Human Rights." *Journal of Rhetoric, Professional Communication, and Globalization* 4 (1): 33–54.

Faber, Brenton D. 2002. *Community Action and Organizational Change: Image, Narrative, Identity*. Carbondale: Southern Illinois University Press.

Freire, Paulo. 1970. *Pedagogy of the Oppressed*. Trans. Myra Bergman Ramos. New York: Continuum.

Frey, Lawrence R., W. Barnett Pearce, Mark A. Pollock, Lee Artz, and Bren AO Murphy. 1996. "Looking for Justice in All the Wrong Places: On a Communication Approach to Social Justice." *Communication Studies* 47 (1–2): 110–127. https://doi.org/10.1080/10510979609368467.

Gee, James P. 2005. *An Introduction to Discourse Analysis*. New York: Routledge.

Giroux, Henry A. 1991. "Border Pedagogy and the Politics of Postmodernism." *Social Text* 28 (28): 51–67. https://doi.org/10.2307/466376.

Grabill, Jeffrey T., and W. Michele Simmons. 1998. "Toward a Critical Rhetoric of Risk Communication: Producing Citizens and the Role of Technical Communicators." *Technical Communication Quarterly* 7 (4): 415–441. https://doi.org/10.1080/10572259809364640.

Gramsci, Antonio. 1971. *Selections from the Prison Notebooks of Antonio Gramsci: Ed. and Transl. by Quintin Hoare and Geoffrey Nowell Smith*. Edited by Geoffrey Nowell-Smith and Quintin Hoare. International Publishers.

Haas, Angela, Christine Tulley, and Kristine Blair. 2002. "Mentors Versus Masters: Women's and Girls' Narratives of (Re)negotiation in Web-Based Writing Spaces." *Computers and Composition* 19 (3): 231–249. https://doi.org/10.1016/S8755-4615(02)00128-7.

Haas, Angela M. 2007. "Wampum as Hypertext: An American Indian Intellectual Tradition of Multimedia Theory and Practice." *Studies in American Indian Literatures* 19 (4): 77–100. https://doi.org/10.1353/ail.2008.0005.

Haas, Angela M. 2012. "Race, Rhetoric, and Technology: A Case Study of Decolonial Technical Communication Theory, Methodology, and Pedagogy." *Journal of Business and Technical Communication* 26 (3): 277–310. https://doi.org/10.1177/1050651912439539.

Haselkorn, Mark, and Rebecca Walton. 2009. "The Role of Information and Communication in the Context of Humanitarian Service." *Professional Communication, IEEE Transactions on* 52 (4): 325–328. https://doi.org/10.1109/TPC.2009.2032379.

Herndl, Carl G. 1993. "Teaching Discourse and Reproducing Culture: A Critique of Research and Pedagogy in Professional and Non-Academic Writing." *College Composition and Communication* 44 (3): 349–363. https://doi.org/10.2307/358988.

Hughes, Michael A., and George F. Hayhoe. 2008. *A Research Primer for Technical Communication: Methods, Exemplars, and Analyses*. London: Routledge.

Hunsinger, R. Peter. 2006. "Culture and Cultural Identity in Intercultural Technical Communication." *Technical Communication Quarterly* 15 (1): 31–48. https://doi.org/10.1207/s15427625tcq1501_4.

Jameson, Daphne A. 2000. "Telling the Investment Story: A Narrative Analysis of Shareholder Reports." *Journal of Business Communication* 37 (1): 7–38. https://doi.org/10.1177/002194360003700101.

Jones, Natasha N. 2012. "Mediation, Motives, and Goals: Identifying the Networked Nature of Contemporary Activism." PhD diss., University of Washington.

Jones, Natasha N. 2014. "Methods and Meanings: Reflections on Reflexivity and Flexibility in an Intercultural Ethnographic Study of an Activist Organization." *Rhetoric, Professional Communication, and Globalization* 5 (1): 14–43.

Jones, Natasha N. 2017. "Modified Immersive Situated Service Learning (MISSL): A Social Justice Approach to Technical Communication Pedagogy." *Business and Professional Communication Quarterly* 80 (1): 1–23.

Katz, Steven B. 1992. "The Ethic of Expediency: Classical Rhetoric, Technology, and the Holocaust." *College English* 54 (3): 255–275. https://doi.org/10.2307/378062.

Kerby, Anthony Paul. 1991. *Narrative and the Self*. Bloomington: Indiana University Press.

Kienzler, Donna. 2001. "Ethics, Critical Thinking, and Professional Communication Pedagogy." *Technical Communication Quarterly* 10 (3): 319–339. https://doi.org/10.1207/s15427625tcq1003_5.

Kitalong, Karla, Tracy Bridgeford, Michael Moore, and Dickie Selfe. 2003. "Variations on a Theme: The Technology Autobiography as a Versatile Writing Assignment." In *Teaching Writing with Computers: An Introduction*, ed. Pamela Takayoshi and Brian Huot, 29–42. Boston: Houghton Mifflin.

Klein, Kerwin Lee. 1995. "In Search of Narrative Mastery: Postmodernism and the People without History." *History and Theory* 34 (4): 275–298. https://doi.org/10.2307/2505403.

Kuutti, Kari. 1996. "Activity Theory as a Potential Framework for Human-Computer Interaction Research." In *Context and Consciousness: Activity Theory and Human-Computer Interaction*, ed. Bonnie A. Nardi, 17–44. Cambridge, MA: MIT Press.

Longo, Bernadette. 1998. "An Approach for Applying Cultural Study Theory to Technical Writing Research." *Technical Communication Quarterly* 7 (1): 53–73. https://doi.org/10.1080/10572259809364617.

Lyotard, J. F. (1979) 1984. *The Postmodern Condition: A Report on Knowledge.* Trans. Geoff Bennington and Brian Massumi. Minneapolis: University of Minnesota Press.

McCoy, Charles S. 1992. "Narrative Theology and Business Ethics: Story-Based Management of Values." In *A Virtuous Life in Business: Stories of Courage and Integrity in the Corporate World,* ed. Olivier F. Williams and John W. Houck, 51–72. Lanham, MD: Rowman and Littlefield.

Miller, Carolyn R. 1984. "Genre as Social Action." *Quarterly Journal of Speech* 70 (2): 151–167. https://doi.org/10.1080/00335638409383686.

Moore, Kristen. 2013. "Exposing Hidden Relations: Storytelling, Pedagogy, and the Study of Policy." *Journal of Technical Writing and Communication* 43 (1): 63–78. https://doi.org/10.2190/TW.43.1.d.

Mumby, Dennis K. 1993. *Narrative and Social Control: Critical Perspectives.* Newbury Park, CA: Sage. https://doi.org/10.4135/9781483345277.

Ohnuki-Tierney, Emiko, ed. 1990. *Culture through Time: Anthropological Approaches.* Stanford, CA: Stanford University Press.

Ong, Aihwa. 1995. "Women out of China: Traveling Tales and Traveling Theories in Postcolonial Feminism." In *Women Writing Culture,* ed. Ruth Behar and Deborah A. Gordon, 350–72. Berkeley: University of California Press.

Ornatowski, Cezar M., and Linn K. Bekins. 2004. "What's Civic about Technical Communication? Technical Communication and the Rhetoric of 'Community.'" *Technical Communication Quarterly* 13 (3): 251–269. https://doi.org/10.1207/s15427625tcq1303_2.

Palmeri, Jason. 2006. "Disability Studies, Cultural Analysis, and the Critical Practice of Technical Communication Pedagogy." *Technical Communication Quarterly* 15 (1): 49–65. https://doi.org/10.1207/s15427625tcq1501_5.

Perkins, Jane M., and Nancy Roundy Blyler, eds. 1999. *Narrative and Professional Communication.* Stamford, CT: Ablex.

Powell, Malea D. 2004. "Down by the River, or How Susan La Flesche Picotte Can Teach Us about Alliance as a Practice of Survivance." *College English* 67 (1): 38–60. https://doi.org/10.2307/4140724.

Rouse, Joseph. 1990. "The Narrative Reconstruction of Science 1." *Inquiry* 33 (2): 179–196. https://doi.org/10.1080/00201749008602217.

Rude, Carolyn D. 2004. "Toward an Expanded Concept of Rhetorical Delivery: The Uses of Reports in Public Policy Debates." *Technical Communication Quarterly* 13 (3): 271–288. https://doi.org/10.1207/s15427625tcq1303_3.

Savage, Gerald, and Kyle Mattson. 2011. "Perceptions of Racial and Ethnic Diversity in Technical Communication Programs." *Programmatic Perspectives* 3: 5–57.

Saukko, Paula. 2005. "Methodologies for Cultural Studies: An Integrative Approach." In *The Sage Handbook for Qualitative Research,* 3rd ed., ed. Norman K. Denzin and Yvonna S. Lincoln, 343–56. Thousand Oaks, CA: Sage.

Scott, J. Blake. 2004. "Rearticulating Civic Engagement through Cultural Studies and Service-Learning." *Technical Communication Quarterly* 13 (3): 289–306. https://doi.org/10.1207/s15427625tcq1303_4.

Scott, J. Blake, Bernadette Longo, and Katherine V. Wills, eds. 2006. *Critical Power Tools: Technical Communication and Cultural Studies.* Albany: State University of New York Press.

Selfe, Cynthia L., and Richard J. Selfe. 1994. "The Politics of the Interface: Power and Its Exercise in Electronic Contact Zones." *College Composition and Communication* 45 (4): 480–504. https://doi.org/10.2307/358761.

Simmons, W. Michele, and Jeffrey T. Grabill. 2007. "Toward a Civic Rhetoric for Technologically and Scientifically Complex Places: Invention, Performance, and Participation." *College Composition and Communication* 85 (3): 419–48.

Sims, Brenda R. 1993. "Linking Ethics and Language in the Technical Communication Classroom." *Technical Communication Quarterly* 2 (3): 285–299. https://doi.org/10.1080/10572259309364542.

Slack, Jennifer Daryl, David James Miller, and Jeffrey Doak. 1993. "The Technical Communicator as Author Meaning, Power, Authority." *Journal of Business and Technical Communication* 7 (1): 12–36. https://doi.org/10.1177/1050651993007001002.

Smart, Graham. 1999. "Storytelling in a Central Bank: The Role of Narrative in the Creation and Use of Specialized Economic Knowledge." *Journal of Business and Technical Communication* 13 (3): 249–273. https://doi.org/10.1177/105065199901300302.

Stewart, Charles J., Craig Allen Smith, and Robert E. Denton, Jr. 2007. *Persuasion and Social Movements*. Long Grove, IL: Waveland Press.

Suchman, Lucy. 1987. *Human-Machine Reconfigurations: Plans and Situated Actions*. Cambridge University Press.

Sullivan, Dale L., and Michael S. Martin. 2001. "Habit Formation and Story Telling: A Theory for Guiding Ethical Action." *Technical Communication Quarterly* 10 (3): 251–272. https://doi.org/10.1207/s15427625tcq1003_2.

Torre, Maria Elena, Michelle Fine, Kathy Boudin, Iris Bowen, Judith Clark, Donna Hylton, Migdalia Martinez, et al. 2001. "A Space for Co-constructing Counter Stories under Surveillance." *Critical Psychology* 4: 149–166.

Walton, Rebecca, Ryan Price, and Maggie Zraly. 2013. "Rhetorically Navigating Rwandan Research Review: A Fantasy Theme Analysis." *Rhetoric, Professional Communication, and Globalization* 4 (1): 78–102.

Wells, Susan. 1986. "Jurgen Habermas, Communicative Competence, and the Teaching of Technical Discourse." In *Theory in the Classroom*, ed. Cary Nelson, 245–69. Urbana: University of Illinois.

Youngblood, Susan A., and Jo Mackiewicz. 2013. "Lessons in Service Learning: Developing the Service Learning Opportunities in Technical Communication (SLOT-C) Database." *Technical Communication Quarterly* 22 (3): 260–283. https://doi.org/10.1080/10572252.2013.775542.

Zachry, Mark. 1999. "Management Discourse and Popular Narratives: The Myriad Plots of Total Quality Management." In *Narrative and Professional Communication*, ed. Jane M. Perkins and Nancy R. Blyler, 107–20. Stamford, CT: Ablex.

11
RACE AND THE WORKPLACE
Toward a Critically Conscious Pedagogy

Jessica Edwards

Language shapes us. Action connects us. Both language and action give us insight into what it means to be social and what it takes to communicate effectively. I am interested in encouraging students to view writing as a way to make sense of the world, as a tool to help them understand the possibilities and purposes of language. In exploring the possibilities of language, however, I find it important to also consider the ways in which language has been and continues to be used as a tool to oppress and to reify structures related to racism. Recently, a poignant article about language and race appeared in the *Huffington Post* titled "When the Media Treats White Suspects and Killers Better Than Black Victims." The writer, Nick Wing, argues that the media's use of language to report about black victims is often degrading compared to the use of language to describe white suspects. One headline example that the writer juxtaposed is as follows: "*White Suspect*: Son in Staten Island murder brilliant, but social misfit. *Black Victim*: Montgomery's latest homicide victim had history of narcotics abuse, tangles with the law" (Wing 2014, 1). Wing brings up the need for the responsible use of language and a type of awareness about how language becomes a part of how we think and categorize people and how language allows for connections. Critical lenses, particularly related to language, are important to engage and explore in order to foster awareness of the link that should exist between language and action as they have bearing on how meaning is made and understood. As the introduction to this collection puts forth, social justice frameworks help us imagine more socially-just practices. Critical engagements with language and intentionality in the professional writing classroom in order to adopt more conscious and responsible practices, particularly as it relates to conversations about race and racism, would help to inform socially aware communicators and global participants. In this chapter I offer Critical Race Theory as a way to engage students in these conversations and find solutions.

In our field, however, there is apprehension about teaching matters of race as some simply have a difficult time perceiving race as relevant to either theorizing or practicing professional writing. In a 2012 special issue for the *Journal of Business and Technical Communication* titled "Race, Ethnicity, and Technical Communication," editors Miriam F. Williams and Octavio Pimentel note that "many, inside and outside of our field, believe that race is not a relevant concept in our society or field. Some argue that we live in a nonracist society, and thus the need to acknowledge color no longer exists" (Williams and Pimentel 2012, 2). The special edition, comprised of four articles, calls for awareness on the topic and invites discussion and action. For instance, Angela Haas's case study, "Race, Rhetoric, and Technology: A Case Study of Decolonial Technical Communication Theory, Methodology, and Pedagogy," examines the place of race, ethnicity, rhetoric, and technology in a graduate-level technical communication classroom. In addition, Carlos Evia and Ashley Patriarca's article, "Beyond Compliance: Participatory Translation of Safety Communication for Latino Construction Workers," presents data collected that shows how Latino construction workers design safe tools that are culturally appropriate (Evia and Patriarca 2012). Williams and Pimentel's special issue, then, brings awareness about issues of race and ethnicity in professional writing and includes four articles to help situate the context for the larger problem in relation to the myth of a non-racist society. Similarly, Adam Banks, in his book on race and technological access titled *Race, Rhetoric, and Technology: Searching for Higher Ground*, writes that there is an overall rhetorical problem in English departments with recognizing race, let alone its connection to language and action. Banks notes that when "cultural issues are raised, the subject is [often] broached in the service of global capitalism," which moves away from a specific focus on race studies (Banks 2006, 14). The rhetorical problem, then, lies in an inability to address silences about race and the ways in which race, ethnicity, and multicultural issues connect to our overall mission of helping students to think more critically and engage more responsibly as they go out to become participants in a global world as professional writers.

While there has historically been limited information about the study of race or racism and language in our field, there is emerging scholarship on the topic. Using a cultural studies approach to examine Texas regulations and discourse markers that helped further reservations surrounding relationships between black Texans and state/local governments, Miriam F. Williams notes this about language in her book *From Black Codes*: "Disenfranchisement of freed blacks in this country was an intentional breakdown in communication and governance" (Williams 2010, 33). Williams's study about regulatory writing underscores how, historically, language has been used to serve miscommunication and a lack of transparency related to laws that impact black and Latino people in Texas. More recently, Williams and Pimentel's (2014) edited collection, *Communicating Race, Ethnicity, and Identity in Technical Communication*, provides important scholarship, from considerations of race and ethnicity in social media to social justice issues in technical communication. One scholar in the collection, Flourice Richardson, analyzes reports produced during the eugenics movement to interrogate medical and science writing used to disenfranchise and subsequently sterilize women of color in America. Another contributor, Cruz Medina, examines how race has been left out of professional writing conversations and how language practices via social media, namely, Twitter, could offer useful conversations about language diversity, particularly for Latinos.

In addition, Gerald Savage and Natalia Matveeva, in "Toward Racial and Ethnic Diversity in Technical Communication Programs," look at HBCUs (Historically Black Colleges/Universities) and TCUs (Tribal Colleges/Universities) to gauge how to increase diversity in technical communication organizations, such as the Council for Programs in Technical and Scientific Communication (CPTSC) (Savage and Matveeva 2011). Savage and Matveeva make some important points about the history of HBCUs and TCUs and the complexities of programmatic diversity. Also, Gerald Savage and Kyle Mattson, in "Perceptions of Racial and Ethnic Diversity in Technical Communication Programs," surveyed technical communication program directors about diversity (Savage and Mattson 2011). The pair's survey results prompted several useful recommendations for enhancing diverse practices in technical communication programs. Taken together, this emerging body of scholarship offers promise for critical engagement with the connections that exist between language and action. However, there are still few examples that offer specific strategies or practices for teaching professional writing in a way that encompasses practical teaching tools. To fill this gap, I suggest here that the field of professional writing needs to engage Critical Race Theory and attend to race and racism in our pedagogies and language practices.

LANGUAGE, RACISM, AND PROFESSIONAL WRITING

I argue here that if we ignore race and racism as key concerns in our scholarship and practice, we do harm to our field. In other words, if we do not consider race and racism in our field, we fall short in helping students to connect with the details associated with communicative processes that are realities in American society. In 2015, several racialized incidents occurred on college campuses and in workplace environments across the nation. For instance, CVS, a major cooperation, recently made headlines as employees were trained to racially profile Latino/as and customers of African descent as part of their workplace practices. There were several times during the year when nooses were found hanging on different colleges campuses around the country. The noose incidents symbolize a time in American history when black and brown people were treated horribly, tortured and often killed by white mobs; the victims would usually be hanging from trees. Even more recently, several black and brown young men and women have been killed or badly injured across the country through state directed violence situations by professionals. With silence about the history of race and racism in this country, we only perpetuate the problem. Engagement with issues related to race and racism allow for a race-consciousness that will, I argue, equip students with knowledge to help them navigate writing and communication situations. One useful example of the need for consciousness about race is Geneva Smitherman's work on black English. Her 1973 article "God Don't Never Change: Black English from a Black Perspective" insists that scholars teach about racialized realities so that students can understand the "totality and complexity of the communication process" (Smitherman 1973, 829). Some forty years after Smitherman's plea, Natasha Jones and Rebecca Walton's work in this edited collection, "Using Narratives to Foster Critical Thinking about Diversity and Social Justice," provides a heuristic that allows for more intentional engagement with language practices; the writers offer narrative as a way to help move toward social justice to provide agency for those who have historically been underrepresented.

Understanding the nuances associated with language use across races and cultures comes with responsibility to address these inequities. In other words, the problem has to not only be acknowledged but understood in order for equity to happen. Using student work to highlight the connections between professional writing scholarship, workplace perceptions, and race in this chapter, I contend that it is necessary for students to consider the ethical and social responsibilities that undergird their language use and, moreover, that professional communication classrooms are a vital site for promoting students' cultural competence and attention to issues of race.

I suggest that Critical Race Theory (CRT) is one way to rethink and reconsider the professional writing classroom as one that acknowledges issues of race, power, and racism through critical engagement with language. Through CRT, I am beginning to interrupt the silences that exist with discussing, acknowledging, and dealing with the connections between race, racism, and power. Since silence about race and racism in our field often thwarts discussion, CRT offers a launching pad to begin conversation in a way that does not point fingers or criminalize a person or culture. CRT also provides an opportunity to highlight the ways in which systemic problems impact communities of color. In addition, CRT makes it possible to have language for discussion of issues as it creates an entry into recognizing the connections that exist between race, racism, and power. CRT has the potential to help writing teachers engage rhetorically with both communicating and presenting critical instances to students as well as tools for which to analyze materials. CRT also allows for interdisciplinarity as it connects our field to studies in history, sociology, criminal justice, and the sciences. Our field benefits from the insights that CRT provides as we consider ways to make the most of language practices and how they can be used as tools of power for progressive social change. Connecting issues of race and racism with rhetoric and professional writing, then, stands as a way to negotiate how students may use and represent knowledge to be more aware and conscious about systemic racism to, thus, move away from practices that reify old structures.

Ultimately, my mission is to show the promise of framing writing assignments and professional discussions using CRT as a lens. In the next sections I explain CRT and then move into how I implemented CRT into a professional writing course.

WHAT IS CRITICAL RACE THEORY?

Critical Race Theory has been around for some time, but the name was crystallized during the 1970s when law professors and scholars Derrick Bell, Alan Freeman, and Patricia Williams began to critically engage the ways in which law impacted people of color in a different, more demoralizing way than white counterparts. Critical Race Theory thus grew out of the ways in which laws were enacted to oppress people of color, and it argues that "the American legal system, from the very beginning, encoded racism into its workings, and that discursive conventions in both jurisprudence and legal scholarship have ensured the maintenance of that initial code" (Banks 2004, 87). The CRT movement is comprised of scholars and activists "interested in studying and transforming the relationship between race, racism, and power" (Delgado and Stefancic

2012, 3). The movement, which takes on issues similar to those born out of the struggle for civil rights, considers economics, context, and history as well as group/self interests. More specifically, CRT builds on two previous movements, critical legal studies and radical feminism. Scholars and activists in the movement also align themselves with theorists and philosophers like Antonio Gramsci and historical figures like Sojouner Truth and Cesar Chavez. From critical legal studies, the group of scholars articulated the concept of legal indeterminacy—the idea that not every legal case has one correct outcome (Delgado and Stefancic 2012, 5). The idea of legal indeterminacy is important for CRT because it is a concept that speaks to not judging a person based on his or her background, economic position, or racial background. The idea, then, makes apparent the importance of recognizing nuances associated with cases to help make more informed choices. From feminism, CRT scholars were able to gain more insight into the connections that exist between "power and the construction of social roles, as well as the unseen, largely invisible collection of patterns and habits that make up patriarchy and other forms of domination" (Delgado and Stefancic 2012, 5). Feminism and legal indeterminacy, together, make for a rich theoretical tool that has structural racism issues at the center and encourages interrogation of other forms of oppression in the process.

As pioneers of the sub-field, Bell, Freeman, and Williams created CRT to highlight systemic problems. By connecting feminism and legal indeterminacy, the scholars were able to craft five basic tenets to characterize CRT as an accessible theory in law. Since the 1970s, the five tenets have been adopted by many fields and disciplines, including sociology, education, history, math, and some English departments across the country. I use all five tenets as they are useful in understanding just how deep structural problems are in our world and provide language for engaging ideas.

The first tenet is *ordinariness*, which speaks to the colorblind narrative that is often expressed by people who prefer not to see color. We may see a colorblind narrative being used to justify or connect people using phrases like "we are all the same." What ordinariness does, however, is creates a false sense of connectedness that does not acknowledge unique perspectives.

The second tenet is *interest convergence* (or material determinism), which acknowledges that because "racism advances the interests of both white elites (materially) and working class Caucasians (psychically), large segments of society have little incentive to eradicate it" (Delgado and Stefancic 2012, 8).

The third tenet, *social construction*, makes known that "race and races are products of social thought and relations" (Delgado and Stefancic 2012, 8). Social construction acknowledges that language is the primary way a community makes meaning. With CRT, however, social construction means that those in power in a society invent thoughts and structures when convenient. Social construction allows for those in power to pick and choose what is important and ignore facts when necessary.

The fourth tenet, *differential racialization*, speaks to "the idea that each race has its own origins and ever-evolving history" (Delgado and Stefancic 2012, 9). Differential racialization speaks to how dominant society racializes different minority groups and uses stereotypes to categorize groups by difference and race, making one represent all. For example, a group of young boys of color may appear as simple-minded boys who are uninterested in education or life in one era. In another era, the same group may be portrayed as out of control and depressed. In essence, the time changes but the narrative remains the same as the boys are not associated with success or in a positive light.

The final tenet, *legal storytelling* "urges Black and Brown people to recount their own unique experiences with racism and the legal system to apply their own unique perspectives" (Delgado and Stefancic 2012, 10). This tenet speaks to the importance of black and brown people telling their own stories to help break the narratives that have traditionally been taught, used, or explained to create a more diverse knowledge base as well as ownership of one's life.

These five basic tenets of CRT recognize the marriage between language and action; they encourage critical thinking and engagement with language and social structures. The use of CRT, then, allows for a look at systemic structures and offers ways to understand and subvert them in order to reduce inequalities and to provide more opportunity and clarity for diverse practices.

UNDERSTANDING THE LINK BETWEEN CRT AND PROFESSIONAL WRITING

Adopting the perspective of CRT in a professional writing classroom will ultimately produce more awareness and consciousness among students about language and systemic racism. It is helpful to understand race as a social construction, as "a concept that signifies and symbolizes sociopolitical conflicts and interests in reference to different types of human bodies" (Winant 2001, 317). In other words, race is a term that speaks to and represents how difference in color and culture are marked. Racism, a social action, is the result of marking difference as it inscribes itself in practices (discrimination, violence, language, systems, exploitation) that always work to create hierarchal and imbalanced relationships. The inherent connections that exist between language and action in relation to race and racism, then, are important to consider when thinking about diverse practices. Sociologist Eduardo Bonilla Silva's discussion in *Racism without Racists* shows how ideology and the naturalization of race are often understood, noting that "the word 'natural' or the phrase 'that's the way it is' is often interjected to normalize events or actions that could otherwise be interpreted as racially motivated (residential segregation) or racist (preference for whites as friends and partners). But . . . few things happen in the social world are 'natural,' particularly things pertaining to racial matters" (Bonilla-Silva 2003, 37). This type of myth-making and normalization related to language plays into the preservation of power structures and dominant ideologies. Critical Race Theory debunks the myths by acknowledging that there are social, judicial, and cultural ideas needed to engage a dialectical interplay. These moves are needed for professional writing teachers to consider because language use is never objective, and understanding the ways that particular language practices influence, draw upon, and engage different orientations to race, racism, and diversity can equip students with useful tools to better participate in rhetorical situations that they will encounter in the workplace. CRT, then, provides an umbrella that provides language to question an oppressive system. The language provided using CRT offers a way to not only define problems that exist but offer ways to engage complexities to hopefully address problems.

In order to understand and better position CRT in a writing classroom, I look to the sophistic idea of kairos. Aristotle defined rhetoric as "the faculty of observing in any given case the available means of persuasion" (Aristotle 2001, 26). Rhetoric, then, is about using knowledge and transmitting that knowledge through language to solve disagreements about political, religious, and social issues. Kairos takes on a significant part in solving these problems because it requires a type of awareness of the occasion, whether it is writing a novel or writing a prime time public speech. Kinneavy notes that "kairos plays an important role in Aristotle's definition of rhetoric . . . it is also included in his view of such related terms as virtue, equity, fitness, and occasion" (Kinneavy 2002, 66). An understanding of the occasion, the kairotic, for a situation, then, requires respect for persons (virtue). It also requires fairness (equity) in treatment of those involved. The kairotic also makes need for aptitude (fitness) or the ability to be able to make just decisions. And lastly, kairos is careful to take heed to the particular instance (occasion). So, use of rhetoric is nuanced and considers several ideas in order to be effective. In many ways, the modern definition of kairos speaks to situational contexts that hinge on understanding the connections that exist between language and situations, between language and action. Better put, kairos helps to make the relationship between language and situations more transparent; kairos gives an awareness of the occasion, a wherewithal that considers audience, purpose, and context. CRT provides the specific rhetorical situation and looks more closely at how language connects to action. Taken together, both CRT and kairos make for a rich teaching tool that takes language that describes particular rhetorical situations (CRT) and connects it with an occasion for writing (kairos) to try to present more opportunity for ethically grounded conversations about equity and social justice through writing and professional communication.

For writing teachers, CRT offers a lens to craft syllabi and assignments for instruction. CRT can be used as a framework to disrupt trained ways of looking at race and encourage more thinking about the link between language and action using specific rhetorical situations. In many ways, CRT provides a language of intentionality and purpose that demands use of specific rhetorical situations in order to engage concepts. Writing teachers can assign a particular occasion (kairos) for writing like pamphlets, reports, or proposals and use a rhetorical situation (say, like a discussion of social media's role in Ferguson, Missouri, or Baltimore, Maryland, during crisis) to engage a tenet of CRT. More specifically, a teacher may consider developing a class unit devoted to letter writing (kairos). When considering a specific rhetorical situation for the kairotic moment, a teacher may assign students to write a persuasive letter about the function of brown fields in underprivileged communities. Such a task requires both considerations of the kairotic as well as rhetorical appeals.

USING CRT IN THE PROFESSIONAL WRITING CLASSROOM

In what follows, I show how one approach to using CRT in the professional writing classroom enables students to offer more ethically-grounded responses to rhetorical situations that are part of everyday life for professional writers and teachers. In 2012, my task was to teach a junior level professional writing class made up of engineering, finance, and hospitality management majors. Over the first four weeks of instruction, students were asked to read Victor Villanueva's (2001) "On Rhetoric and the Precedents of Racism" as well as Steven Katz's (1992) "An Ethic of Expediency." I chose these two readings because they set a tone for discussion about ethics and race. It was important for me to ground our discussion in the language of the field and to provide context about each reading. When engaging these two pieces, students were able to talk about Katz's notion of ethics and how the Holocaust was a horrifying part of history. Villanueva's piece raised some important questions for students about rhetoric and the concept of race as a social construction. I paired these readings with a then-current scenario about race in the workplace involving former Georgia State Director of Rural Development, Sherri Sherrod, who was cast as a racist after parts of a speech she gave were taken out of context and widely distributed. More specifically, shortly after being appointed as the Georgia State Director of Rural Development, Shirley Sherrod was forced from her position. The late conservative blogger, Andrew Breitbart, produced an edited clip of Sherrod's speech at an NAACP event to show her saying she was not in favor of helping white farmers keep their land. The speech in its entirety, however, showed that Sherrod, who is a black woman, was against unequal practices and racism in general. In the Sherrod case, there was miscommunication as well as a lack of transparency that impeded media interpretations and, as a result, national audience's understanding of Sherrod's speech. In other words, the connection between language and action was not clear as portions of the speech lacked context; the result was Sherrod's firing. I showed the Sherrod case to get students to begin a conversation about the importance of understanding the rhetorical situation before jumping to conclusions. The viewing of the Sherrod case, coupled with the readings, brought up students' thoughts related to affirmative action as well as disparities related to women in the workplace. My professional writing students, however, were not really able to speak directly about race or acknowledge racism in this case beyond what Peggy McIntosh (1988) calls as "individual acts of meanness." Students referenced instances like the 1993 Rodney King beating, but no attention to or connection with systems of oppression that may have led to that incident. Students also could not verbally identify specific current events that they recognized in popular culture or in the news during in-class discussion. In other words, students were aware of the occasion and wrongdoing associated with Mr. Rodney King, for instance. They were not as clear, however, about the connection between language and action within the occasion. Thus, I designed a writing assignment with CRT in mind that I hoped would enable students to make a connection that many refused to acknowledge verbally. My purpose was to highlight structural inequalities by creating a specific rhetorical situation as well as a specific occasion for writing about the situation.

To engage Critical Race Theory in the classroom, students need to explore not only their beliefs about race in the workplace, but also the ways that they understand their own subjectivity with regard to issues of race. So, after our in-class discussion about the readings and the Sherrod case, I asked students to engage in a mock interview situation. While in the interview, students had to pretend that they were applying for their first job and needed to write answers, in at least three pages, to engage two questions about race and diversity.[1] On the assignment sheet, students were asked to engage this particular line from an article in the *Chronicle of Higher Education*:

> from 1980 to 1999 the number of African Americans enrolled in college in the U.S. rose by half, the number of Native Americans almost doubled, the number of international students almost doubled, the number of Hispanics almost tripled, and the number of Asian Americans more than tripled. In contrast, the number of whites enrolled in college barely changed. (Gomstyn 2003, 1)

The first question asked: "What are your thoughts about the projected demographic changes in higher education as well as in most work environments? Do you see the changes causing problems? If so, how would you handle these problems? If not, explain your answer." I created this first interview question with CRT's tenet of social construction in mind and to invite students to challenge the assumptions made in the prompt and to write about how they viewed and understood race in the workplace. It was meant to help them move beyond the situation provided in the prompt to a more varied and diverse thinking based on what they had read and viewed in class.

The second question asked: "How does your racial identity and comfort level influence the people you work with, especially the people of races other than your own whom this job will require you to supervise?" I created the second question, particularly in relation to CRT's tenet of interest convergence, to get students to think about their own subjectivity in relation to the people they will work with; I was also interested in students making connections between their own ideas and the language in the prompt.

ANALYSIS OF ANSWERS FOR QUESTION 1

I will focus my analysis initially on the first question. Of the forty-eight student writers, twenty-three gave specific examples using our class conversations. A few students were able to make some connections between demographic changes, race, and specific connections that show some parallels using the prompt as well as class examples. More specifically, student answers provide us with three points about the benefits of critical race consciousness, and they are the ability to make connections, thoughtfulness about a diversity of voices and experiences, and the importance of critical listening.

Making Connections

In addition to use of the Sherrod case and communicating broader patterns of racism in the United States in professional contexts, students were able to focus on demographic changes to point out that racism remains a problem in the United States despite initiatives related to equality and inclusion. SM notes that, "Dr. Villanueva states in his article that, 'Racism seems to have the greatest depth of trouble, cuts across most other bigotries, is imbricated with most other bigotries, and also stands alone, has the greatest number of layers.' Dr. Villanueva is right that racism is very deep and has many layers." SM's recognition of the complexity of racism as well as connections to class reading highlights a type of consciousness that would be useful in navigating the realities of the world. SM's response acknowledges race and begins to make parallels between language and action, connections that are important for engagement and application of ideas.

Moreover, HC brings in some historical truths to elucidate his or her claims about demographic changes in the workplace. HC notes that, "demographics in higher education have changed in the past half century and continue to change. This should come as no surprise; the Civil Rights Act of 1964 is an example of why demographics have changed." HC goes on to state that as a young person, "sometimes I forget that it was not long ago that people could legally discriminate based on demographics. With the changing atmosphere of the country it would be completely logical that the numbers of 'minorities' in colleges has changed. The change in demographics could represent that people are getting hired for their credentials instead of their demographic." HC's response is a bit curious in that he or she acknowledges the Civil Rights Act of 1964 and even the fact that it should not be jarring that demographics are changing. HC even mentions that such a move is "logical." However, HC moves on to write about the need for folks to be judged by their credentials. HC's response, in many ways, echoes Dr. King's dream for people to be judged by the content of their character and not their skin color. What HC leaves out, however, is some information about why the discussion is important to have in the first place. The connections made here by HC, then, are tangential, but promising.

Respect for Speaker Voices

MD writes about the need for respect in order to be successful with demographic changes in the workplace. MD's personal experience and response about the need for respect says much about how he or she valued his time with different people. The participant notes,

> One group I was in once comprised of myself, a student from Kenya, a veteran who had served two tours in Iraq, and a 32 year-old black man who had been a high school music teacher. I personally learned a lot by simply spending a huge amount of time with these people whom I might not normally, learning about different experiences and perspectives. We all came from very different backgrounds but all knew that we had to work together to accomplish a task and we respected one another. The key is respect.

MD's writing gives off a hopeful tone that speaks to the possibilities of teaching critical race consciousness in the professional and technical writing classroom. The positive personal connection allows MD to talk about respect as a way to engage with and actually work on projects. MD, in many ways, positions respect for diversity as a way to move beyond difference to actually learn. MD's point about a diverse group of workers says something about the need for physical difference in classrooms.

Listening and Attention to Detail

NJ's response is a bit different from others that I have analyzed thus far as he or she speaks about communication in the workplace in a very specialized and specific way. NJ notes that, "the best way to break a communication barrier is to eliminate one-on-one communication that can cause frustration and promote participation of individuals in team discussions." NJ goes on to say that "Team discussions allow employees to practice their communication skills as well as their listening skills without the pressure of one-on-one conversations. Achieving effective communication between team members will create a sense of comfort, and is the most important step in ensuring that a diverse team is functioning at full potential." NJ's response highlights the importance of listening skills. NJ's method of adding more team building is interesting as it suggests a type of dissonance as a result of one-on-one communication in company culture. Without talking about race or difference in a specific way, NJ highlights the importance of creating a safe environment, one that privileges a type of diversity. Although NJ does not use sources in his or her work, what we can take away is some connection to the kairotic as NJ begins to consider audience and context in a fluid way.

In an effort to listen, TW fails to do contextual work that would add more meaning to his or her point. TW writes that,

> In Victor Villanueva's *On the Rhetoric and Precedents of Racism* he argues that racism is still prevalent in America and that multiculturalism has done nothing to help our country (Villanueva 650). I disagree with Villanueva, not to say that racism does not exist but rather to say that minorities make racism what it is today. He mentions the beatings of Rodney King and Alicia Soltero Vasquez, both minorities. He never mentions the incidents of brutality against whites, such as a cop in Atlanta GA punching a white woman in 2011.

TW's response is compelling as he or she uses the source as a discussion point to engage other inequalities. By the end of TW's conversation, he or she makes a claim about how prejudice is not just about people of color. In many ways, TW's comment is connected to CRT's mission to recognize the parallels that exist between race, racism, and power. Unknowingly, TW actually engages in a conversation with race at the center. He or she highlights other structural inequalities that exist related to women as well as those who may have differences in religion or sexual orientation. TW, however, has not yet made the connections that exist between race, class, and gender in what is provided here.

ANALYSIS OF ANSWERS FOR QUESTION 2

Responses to question 2 were equally as direct and focused as those previously analyzed. Of the forty-eight student writers who responded to the second question, only fifteen responded using specific sources. A few students were able to talk about how race may impact their future careers. Others, however, were not able to. More specifically, student answers provide us with two points about race in the workplace: hopeful attitudes about race in their own future workplaces and indifference about race as a factor for the future.

Hopeful about Race in the Workplace

One student was able to bring in specifics related to question two about racial identity using sources from class. MS writes how he or she sees demographic changes as important to a potential job at Boeing by connecting the readiness to personal experience. The writer notes that:

I believe that if we can pull together a diverse group of individuals [as a potential member of] The Boeing Company and lead them to success, racism will not be an issue. They will learn how to grow, communicate, work effectively, trust, and love one another. I can vouch for this through first hand experience with athletics and as the president of my fraternity.

MS's ability to connect personal experiences to the conversation about diversity is important and I think needed in order to engage a type of consciousness that should exist between language and action. At the same time, MS's inability to note the structural problems with race and racism is clear as the writer notes that, through personal experience, he or she believes that "simply pulling together a diverse group" could lead to success when working with the Boeing company.

Moreover, SS laments on how understanding self as well as listening and respecting cultures will help to create and continue a healthy workplace environment. SS notes that,

> developing an awareness of one's own cultural habits and how they are interpreted by others is an integral part of developing respect for and understanding of the cultural behaviors and expectations of others. For example, my background has taught me to express commands and expectations indirectly (for instance, saying "Maybe you should think about doing this" when what I mean is "Do this"). This habit can cause frustrations for me when my expectations are not met, as well as for those I supervise, who may not understand the source of my dissatisfaction. Therefore, I make every effort to give clear directions so that my expectations are understood.

SS has made an important connection between communication and cultural practices. SS's ability to adjust interaction with people in order to communicate in a more effective manner speaks to SS's awareness of critical thinking and engagement.

Indifference about Race as a Factor for the Future

KK's indifference about race as a factor for future work correlates an acknowledgment of race with negativity. KK expresses that, "My racial identity does not negatively affect those that I work with. I am who I am and my skin is the color that it is but that doesn't change the way that I view people or the way that I let them view me. I am very open to working with different demographics as it is refreshing, enlightening, and it molds me into a better person." KK went on to share that, "my most valuable experience working with those of other races was when I was a Crew Trainer at McDonald's. I had the pleasure of working with difference [sic] races and every type of person imaginable. What I learned very quickly is that we are all the same. We all have our problems, but we are all also very smart and we often have similar interests." KK's initial response hinges on the idea of a post-racial society. Then, KK moves into a specific experience that does not provide much evidence to support the claims made. The writer's misunderstanding of race as a concept allows for the ability to be indifferent about its effects and affects.

These students provided their thoughts about changes and they used time, specific circumstances, evidence, and critical thinking to highlight some important points about race, racism, power, and demographic changes in the workplace. In other words, these students were better able to see the ways in which situations are rhetorical and to recognize that rhetoric plays some part in how people react to a situation. They were able to make some connections between language and action. What is interesting about responses to both questions to me, however, is that there is no real explicit connection or conversation about systemic racism, which is discussed at length in class, particularly through our conversations of Villanueva's article as he calls for readers to "break precedent" as well as our discussions of the Sherrod case. Despite that fact, I believe including specific readings, developing particular assignments, and facilitating discussion in a semester long class about the readings and assignments help students to begin to make the connections. Students in my classes were able to show a type of engagement in four weeks that I think is rich and important for teaching race and racism, a type of teaching that helps students to create meaning and to see the rhetorical nature of situations. It is my hope that their experiences with readings and writing in this way translate to thoughtful and responsible practices in the workplace.

IMPLICATIONS

We allow opportunities for a more dynamic consciousness about race and racism when we create particular rhetorical situations to do so. CRT helps to recognize the relationships that exist between race, racism, and power to help students develop more awareness about the function of race in society. It is important to note, however, that some students may not be as comfortable with the material presented. One of my students wrote a response about how he or she believed the entire assignment to be racist and if asked questions posed in an interview setting, they would leave. My student wrote: "I would immediately question the merit of that individual and reflect it upon their company. We live in a globalized world that has changed so dramatically in the last half century that if you question the fact of race any more you are an outcast. What is the difference between a person of color and a white person? Whoever works harder deserves the job." Another student wrote that he or she is all for diversity but does not want "race to become the reason why some people are admitted to Universities and some are not." So, there will be times when students will simply refuse to or will be unable to engage the rhetorical situation. The two student responses are part of the color-blind narrative that is part of the CRT framework, one that scholars are charged to work against. It is clear that this work needs to be done to at least spark critical thought.

Even more, race studies scholar Howard Winant's work on race highlights how concepts are often repeated over time, in different ways, to keep structures and unproductive divisions in our society. He notes that "The rearticulation of (in)equality in an ostensibly color-blind framework emphasizing individualism and meritocracy . . . preserves the legacy of racial hierarchy far more effectively than its explicit defense" (Winant 2001, 35). Winant's work highlights the need for constant critical engagement with the concept of race and awareness of practices that have and continue to feed structural racism to foster transparency about how race still matters. Connecting issues of race and racism with rhetoric and professional writing, then, serves as a way to talk about, understand, and negotiate how students may approach, deal with, and use knowledge to be more aware and conscious about race and systemic racism.

CONCLUSION

CRT challenges professional writing scholars to account for connections that are often ignored or justified by arbitrary notions of truth. CRT gives us moments; CRT provides us with five different windows into what it means to engage matters of race and racism, which ultimately impacts what we do as writers, communicators, and humanists. Just a few months ago in Ferguson, Missouri, an African American, unarmed teenager was fatally shot, six times, by a white police officer. The day of the shooting and the days to follow included many instances when language was not used responsibly by professional writers or oral communicators. More recently, in McKinney, Texas, a white police officer used excessive force on a fifteen-year old African American girl after a pool party. These instances, and many others that happen in communities and workplaces, deserve our critical attention as they represent large structural problems in our society. It may be easier to write off race and racism or the fact that multiple people work in different communities. What cannot be denied, however, is that communication with many people from different backgrounds and races will have to happen in our world, no matter the field. As a result of a critical race conscious pedagogy in professional writing, a social justice framework that moves toward more equitable practices and approaches, students can be encouraged to learn the connections between race, racism, and power as well as to develop responsible possibilities for language and action. CRT provides a healthy start for our discipline to add to practices using a critically conscious pedagogy to help prepare students for work inside and outside of the university setting.

Note

1. I received IRB approval to conduct this study of students in professional writing classes.

References

Aristotle. 2001. *Rhetoric*. In *The Rhetorical Tradition: Reading from Classical Times to the Present*, ed. Patricia Bizzell and Bruce Hertzberg, 26. New York: Bedford St. Martins.
Banks, Adam. 2004. *Race, Rhetoric, and Technology: Searching for Higher Ground*. Mahwah, NJ: Lawrence Erlbaum Associates.
Bonilla-Silva, Edwardo. 2003. *Racism without Racists: Color-Blind Racism and the Persistence of Racial Inequality in the United States*. Oxford: Rowman and Littlefield.
Delgado, Richard, and Jean Stefancic. 2012. *Critical Race Theory: An Introduction*, 3–9. New York: NYU Press.
Gomstyn, Alice. 2003. "Minority Enrollment in Colleges More Than Doubled in Past 20 Years, Study Finds." *Chronicle of Higher Education*, October 17, 2003. Accessed August 10, 2012. http://chronicle.com/article/Minority-Enrollment-in/20000/.
Katz, Steven. 1992. "The Ethic of Expediency." *College English* 54 (3): 255–275. https://doi.org/10.2307/378062.
Kinneavy, James. 2002. "Kairos in Classical and Modern Rhetoric." In *Rhetoric and Kairos: Essays in History, Theory, and Praxis*, ed. Phillip Sipiora and James S. Baumlin, 66. Albany: State University of New York Press.
McIntosh, Peggy. 1988. "White Privilege: Unpacking the Invisible Knapsack." Working Paper 189. Wellesley, MA: Wellesley College Center for Research on Women.
Savage, Gerald, and Kyle Mattson. 2011. "Perceptions of Racial and Ethnic Diversity in Technical Communication Programs." *Programmatic Perspectives* 3:5–57.
Savage, Gerald, and Natalia Matveeva. 2011. "Toward Racial and Ethnic Diversity in Technical Communication Programs." *Programmatic Perspectives* 3:55–85.
Smitherman, Geneva. 1973. "God Don't Never Change: Black English from a Black Perspective." *College English* 34 (6): 828–833. https://doi.org/10.2307/375044.
Villanueva, Victor. 2001. "On Rhetoric and the Precedents of Racism." In *Cross-Talk in Comp Theory: A Reader*, ed. Victor Villanueva, 645–661. Urbana: NCTE.
Williams, Miriam. 2010. *From Black Codes to Recodification: Removing the Veil from Regulatory Writing*. Amityville, NY: Baywood Publishing Company.

Williams, Miriam, and Octavio Pimentel. 2012. "Introduction: Race, Ethnicity, and Technical Communication." *Journal of Business and Technical Communication* 26 (3): 271–276. https://doi.org/10.1177/1050651912439535.

Williams, Miriam, and Octavio Pimentel. 2014. *Communicating Race, Ethnicity, and Identity in Technical Communication*. Amityville, NY: Baywood Publishing Company.

Winant, Howard. 2001. *The World is a Ghetto: Race and Democracy Since World War II*. New York: Basic Books.

Wing, Nick. 2014. "When the Media Treats White Suspects and Killers Better than Black Victims." *Huffington Post*, August 14, 2014. Accessed August 25, 2014. https://www.huffingtonpost.com/2014/08/14/media-black-victims_n_5673291.html.

12
SHIFTING GROUNDS AS THE NEW STATUS QUO
Examining Queer Theoretical Approaches to Diversity and Taxonomy in the Technical Communication Classroom

Matthew Cox

> *People are different from each other. It is astonishing how few respectable conceptual tools we have for dealing with this self-evident fact.*
> Eve Kosofsky Sedgewick, Epistemology of the Closet

After spending over a decade working in industry as a technical communicator, I came to technical communication as an academic discipline knowing this: talking about race, gender, sexuality, and (dis)ability in the workplace can be tricky and complex at best and risky and dangerous at worst. Difference in the workplace more often seemed quashed and hushed than discussed head-on or harnessed for a greater good. As a gay man, my own workplaces were often sites of uncertainty and fear. Upon becoming a technical communication instructor and researcher, this uncertainty and fear continued as I struggled with the extent to which I should or even could discuss issues of identity and of race, gender, sexuality, and (dis)ability with my students. But as workplaces, the academy, technical communication as a field, and society and culture in general become increasingly differentiated and complex, avoiding discussions of these things seems unfair to students who we will ask to navigate these increasingly complex workplaces. In addition to my own technical communication experience (both in the workplace and in the academy, and points in between), I also began to research, study, and draw from LGBT studies, queer theory, and my own experiences as a queer scholar. As these experiences and approaches have overlapped, meshed, and interacted, a queer rhetorics approach in the technical communication classroom has become vital for me.

DOI: 10.7330/9781607327585.c012

By "queer rhetorics approach," I mean one that accounts for the intersectionality of queer and LGBT persons both internally (with their own race, class, sex, dis/ability, gender, and faith tradition markers) as well as externally (with their own and other cultures, groups, and peoples). At the heart of this approach is the question: how does the work of technical communication impact these intersectional bodies and how is it impacted by them? It is an acknowledgment of existing power structures and a desire to free such structures from monolithic and hegemonic control and ownership that silences the many positionalities and identities that intersect in workplaces. With such an approach in mind, this chapter seeks to answer the questions: (1) how do we accommodate different discourses of diversity, more specifically, what does a queer rhetorics theoretical framework look like and help accomplish this goal, and (2) how do queer theory and cultural rhetorics serve as foundations for such a framework?

CHAPTER STRUCTURE

This chapter asserts that essentialized ideas of "diversity" and the monolithic categorized identity politics of the 1990s are no longer serving us in the academy and in technical communication studies. I illustrate this here through the lens of a hybrid graduate technical communication course (partially online students and partially campus-based students) taught in the fall semester of 2013. Building upon scholars in the field who have already called for a more meaningful engagement of cultural situations and topics in the technical communication field (Grabill 2000; Haas 2012; Katz 1992; Miller 1979; Scott, Longo, and Wills 2006; etc.), my goal for this chapter is to:

- discuss the current state of technical communication in relation to queer rhetorics
- define intersectionality as it relates to the idea of cultural consciousness
- investigate queer theory and cultural rhetorics as sites that can offer foundational ideas and practices in the technical communication classroom
- provide examples from my own graduate course of both the practice and content following these foundations
- explore implications and takeaways both present and evolving for the reader as well as our larger discipline

NOT GOING BACK: CULTURAL CONSCIOUSNESS AND TECHNICAL COMMUNICATION

Concerns about race, gender, sexuality, and (dis)ability, have in corporate and business settings often been dictated first by local and federal laws, courts, shareholder desires, public policy, and, to a lesser extent, by public opinion and common practice. In this sense, the corporate/business world does not often lead or innovate in these areas. Instead, corporate settings have often avoided discussions of identity in the workplace altogether (unless mandated by local, regional, or national laws), or, when "leading," have relegated these issues to human resources settings and occasional corporate training. The practice often seems to be to cover the "bare minimum" on what are otherwise seen as difficult or unpleasant subjects that may just cause trouble or undue discomfort for all in the workplace. In my own workplace and business experiences over the years (outside of academic spaces), and in the experiences of many coworkers and colleagues, there seemed to be a noticeable focus on what these workplaces and corporations believe (falsely) to be "non-raced," "non-gendered," "non-sexed," and "non-embodied" areas such as means of production, corporate content and behavior, and the day-to-day operations of the businesses themselves.

As mentioned above, there have been some calls in technical communication studies to move beyond oversimplified ideas of "diversity" and monolithic categories that do not allow for multiplicity and complexity. In addition, there have been conversations in other fields, such as communications studies (Allen 1995), for decades. But these have not been plentiful or pervasive. In the introduction to their 2007 edited collection, *Critical Power Tools: Technical Communication and Cultural Studies*, Scott, Longo, and Wills write that, as opposed to a "relative abundance of work on cultural studies approaches to composition," they have seen "only a handful of technical communication scholars" using a cultural studies approach (6). Of course, these sparse and general calls and conversations simply show that technical communication has not yet been ready to address, or at a place to address, queer and LGBT concerns directly and in detail. This chapter seeks to continue to connect these by naming such shortcomings, calling our field of technical communication to listen more closely and communicate more openly with queer rhetorics and with the concerns of its LGBT and queer participants and audiences. And, here, I hope to give some idea of approaches that may help to fulfill these goals.

In the twentieth century as technical communications grew as an academic discipline, many began to call for a humanist oriented approach to technical communication and workplace writing (perhaps Miller's "Humanistic Rationale" best represents this humanistic turn). This initial call gave way to other turns within technical communication (the social turn of Blyler and Thralls 1993; and Herndl 1993, the cultural turn of Porter, Sullivan, Blythe, Grabill, and Miles 2000, and the social justice turn of Haas and others). Here, at the turn of the millennium, more and more new voices are entering the field and taking up both the cultural message and social justice message within technical communication. But even given these new and increasingly articulate and frequent voices, many of us in technical communication studies feel that not enough is being done, specifically when it comes to new frameworks and examples to lead us as researchers and teachers to be more fully conscious of social justice concerns and culturally conscious approaches. This apparent lack of resources and examples is what motivated me to write here.

(SHIFTING) THEORETICAL FOUNDATIONS: DEFINING CULTURAL CONSCIOUSNESS AND INTERSECTIONALITY

Rhetoric is foundational to technical communication. My own definition of rhetoric draws on the idea that meaning making practices are both "enabled and constrained by culture and identity . . . " and cannot be defined in one authoritative way. Rather, modern understandings of rhetoric are "bifurcated by conventional notions" (e.g., the destabilization of power structures of knowledge production and consumption) (Powell et al., 2009, 4556). So then, all rhetoric is cultural at its core. And if rhetorics are the myriad ways we make meaning both individually and culturally, then technical communication has immediate and powerful implications for rhetoric.

There are examples of a cultural lens being brought to technical communication studies. Recently, Scott, Longo, and Wills (et al.) examine technical communication through a cultural studies lens. Cultural Studies comes out of the critical theory of the Birmingham School of the 1980s in the United Kingdom (Stuart Hall 1980 and others). While culturally relevant and timely, this lens comes out of a philosophy and critical theory background. The study of Cultural Rhetorics, on the other hand, grows firmly out of rhetoric and writing studies and, specifically, indigenous rhetorics, African and African American rhetorics, Asian rhetorics, Latino/Latina rhetorics, feminist, and queer rhetorics studies. The *Cultural Rhetorics Theory Lab* at Michigan State University explains:

> When we say culture, we mean a system of practices, both physical and discursive, that help build and sustain communities. Additionally, when we say rhetoric, we don't mean the traditional notion of classical Greek & Roman rhetoric we've received via Scholastic and Early Modern era scholars; instead, we're referring to the act of studying systems of discourse through which meaning was, is, and will continue to be made in a given culture. (CR Theory Lab 1)

In keeping with this definition, I believe a queer rhetorics approach is one that takes into continual account and consideration that such systems (of physical and discursive practices) not only exist but are at work at all times. Similarly, Haas (2012) states: " . . . rhetoric is always already cultural. It takes into account that subjectivity and knowledge are interrelated" and that rhetoric can only be "fixed momentarily" in its definition and understanding (287). I, too, see the discipline of rhetoric studies as foundational to technical communication studies.

QUEERING THE CONVERSATION: QUEER THEORY AND QUEER RHETORICS AS LENSES

So why queer theory and queer rhetorics, then, as a key lens for my course if all other identity markers and intersections carry equal and considerable weight? For me, it is because queer theory accounts for these intersections inherently. David Halperin (2012) notes that: "Queer is by definition whatever is at odds with the normal, the legitimate, the dominant. There is nothing in particular to which it necessarily refers. It is an identity without an essence. 'Queer' then, demarcates not a positivity but a positionality vis-à-vis the normative" (47). So queer is whatever is outside dominant cultural spaces. Whatever is "normal." Queer theory is not to be confused with LGBT Studies, a predecessor that is much more concerned with literary criticism and representation of the essentialized categories of lesbian, gay, bisexual, and transgender. Queer theory, on the other hand, is not only concerned with whatever is non-normative but also seeks to constantly disrupt the normative as well. For the queer theorist (drawing from but differing from and building upon the ideas of feminism, women's studies, and LGBT studies), labels and categories should be kept in constant flux and always interrogated and torn down even as new categories are constantly being built up in their place. In this sense there is no "destination" to work toward in terms of righting old wrongs. The fact that we have searched for "right" in the first place is the problem. Chaos, failure, and disruption are not only productive; they are desirable as a permanent state.

Prominent queer theorists argue that such failure may be the way to disrupt hegemony and dominant power structures we've been searching for. In *The Queer Art of Failure,* Judith Halberstam (2011) states that: "From the perspective of feminism, failure has often been a better bet than success. Where feminine success is always measured by male standards, and gender failure often means being relieved of the pressure to measure up to patriarchal ideals, not succeeding at womanhood can offer unexpected pleasures" (4). When old boundaries and expectations around gender and sexuality are swept away and one's failure to ever "measure up" to them is accepted, pleasure can enter the picture. In queer theory, this personal pleasure and happiness are the pinnacle of self acceptance and awareness. Indeed, Sara Ahmed (2010), in *The Promise of Happiness,* writes: "Happiness becomes a more genuine way of measuring progress; happiness, we might say, is the ultimate performance indicator" (4). We live in a culture where happiness and smoothness and efficiency as success are so expected that they seem to keep us from taking risks. But risks may be the very thing that our students need most to truly have transformational experiences.

In "What Does Queer Theory Teach Us about X," Lauren Berlant and Michael Warner state that "queer commentary takes on varied shapes, risks, ambitions, and ambivalences in various contexts. The word queer itself can be a precious source of titillation . . . " (Berlant and Warner n.d., 344). As a queer-identified scholar and instructor, it was important to me to bring in this type of queered commentary throughout the semester.

Other cultural rhetorics driven understandings helped collectively piece together this approach. These included both Indigenous North American rhetorical methods and rhetorical frameworks, critical race theory frameworks, disability studies and rhetorics, and feminist theory and rhetorics. Queer Rhetorics is, like cultural rhetorics, a fairly nascent area of formal study in the academy. Non-heteronormative rhetorical strategies, tools, and practices, however, are not new. As long as queer people have sought community with one another, they have made their own workable rhetorical practices. Examples of these include the rants and political pamphlets of twentieth century groups such as the Daughters of Bilitis, Mattachine Society, ACT-UP, and others as well as academic and theoretical work (including Judith Butler 2004; Michel Foucault 2012; Eve Sedgwick 1990; Gloria Anzaldúa 1998, and others). Queer rhetorics, within rhetoric studies, is interested in studying, explicating, and understanding these queer rhetorical methods and practices. In this way, queer rhetorics is interested and draws from both cultural rhetorics and queer theory and works with these to form my queered rhetorical approach, both in this course and in this piece.

INTERSECTIONALITY AT WORK

I (and our class) use the term intersectionality to refer to the ways that both individual identity and cultural identity is layered, intertwined, complex, and always connected. Discussions of race, for example, are underserved and oversimplified if they cannot take into considerations issues of gender, class, sexuality, and (dis)ability (and so on). So then, keeping in mind that a queer rhetorics and intersectional approach means a recognition of the systematic and interrelated nature of communication, we must take care not to separate out these aspects and layers of identity.

The generic idea of "diversity" based in singular categories/taxonomies (e.g., "first we talk only about gender, and then we talk only about race, and then we move on to sexuality, etc.) have, to be sure, served us (at times very well) as we struggle to account for difference. Spivak's (1988) concept of "strategic essentialism" says that we take on names and labels and essentialized identities to help ourselves and others understand difference and to represent that difference in ways that it might otherwise be ignored or silenced. But we must also not fall prey to always categorizing in this essentialized way. This is where intersectionality comes into play. It asks us to see identity as always and permanently intertwined within itself. As a gay man, for example, I cannot ever separate out my experience of being male from my experience of being gay as well. A differently abled Chicana lesbian cannot separate out the experience of being (dis)abled from the experiences of being Chicana and lesbian.

Gloria Anzaldúa's (1987) journey through her own intersectional identity and fragmented sense of place served also as a strong example for myself and my students. As I write this, I recall Anzaldúa's passage:

> But it is not enough to stand on the opposite river bank, shouting questions, challenging patriarchal, white conventions. A counterstance locks one into a duel of oppressor and oppressed; locked in mortal combat, like the cop and the criminal, both are reduced to a common denominator of violence . . . At some point, on our way to a new consciousness, we will have to leave the opposite bank, the split between the two mortal combatants somehow healed so that we are on both shores at once and, at once, see through serpent and eagles eyes. Or perhaps we will decide to disengage from the dominant culture, write it off altogether as a lost cause, and cross the border into a wholly new and separate territory. Or we might go

another route. The possibilities are numerous once we decide to act and not react." (100–101)

Anzaldúa believes we must step outside of the "rules" that the oppressor and dominant culture have set out for us—in this case convenient, compartmentalized "diversity" categories that force us to speak only from one part of ourselves. Instead, a queer rhetorics approach demands that we are not simply a woman, or a black person, or a gay person, but that we are these things at all times and in unison. This intersectional approach says to the dominant culture "I will not fit into your tidy idea of who or what I am and what my voice should sound like or say."

For my own part, it was important for me to be transparent about my own intersectionality. I am, after all, not simply a gay man, or a queer person, I am also a white male, an able-bodied individual, and a white-collar worker. But I also come from a working-class background and am a first-generation college graduate. I have long-overlooked and ignored (by my family) indigenous ancestry, and I am someone who has risen into middle-class ranks. I was raised in a highly religious, politically conservative family and area and find myself now to be a marginally religious, politically liberal/progressive person. These things all contribute to my own outlook and approaches to teaching.

Because of this, I believe that the best approach, both theoretically and pedagogically, is to keep these ideas of identity mixed in and intertwined upon each other at all times. Discussions of one of these identity markers should never happen without an understanding that others are also present and on the table.

COURSE EXAMPLES: CULTURAL CONSCIOUSNESS AND INTERSECTIONALITY AT WORK

To show examples of queer rhetorics frameworks in a graduate rhetoric and technical communication studies course, I draw from my own experiences a few semesters ago. As I put this course together in the months leading up to the start of the semester, I asked myself the same questions I have started out with in this chapter: how do I bring the ideas of intersectionality to the course and as a matter of both practice/methodology as well as content?

This special topics course focused on cultural issues in technical communication—specifically gender, race, ethnicity, sexuality, (dis)ability, and class. It tried to operationalize a variety of cultural-theoretical frameworks by integrating them throughout the semester rather than the typical graduate course structure of a "week for disability studies, a week for LGBT studies," and so on. Instead, this course tried to weave cultural issues and concepts through the semester and build on them as inherently foundational.

The following examples show that students were encouraged to (and did) queer/disrupt ideas of online discussion board space. Examples included responses that often ranted, felt comfortable showing anger and frustration, and showed emotional vulnerability and discomfort with difficult topics and experiences and that often used video recorded responses between one another. Students also showed a willingness to complicate within topic areas and lived experiences. For example, two female African American students often disagreed on issues of African American experience and rhetorics across generational and sexuality lines. Also, female students (across race and age boundaries) often brought conflicting understandings of feminism and women's scholarship into discussions. In addition, male and female students showed differing reactions at times to the role feminisms should play in the academy. Students also worked toward a final course collage project that encouraged them to mix narrative and academic discourses in new and creative-oriented ways.

EXAMPLE ONE: SYLLABUS DESIGN (READINGS, ASSIGNMENTS, INTERACTIONS)

Before the semester began, I thought hard about how to queer the class itself in structure (both content and in process). The fact that this graduate course consisted of both distance students as well as campus-based face-to-face students meant it was already somewhat queer. I had seen others take this type of course and simply turn it into an "all online" course using course management system (CMS) based discussion boards and online assignments, but I wanted to disrupt that as much as I could.

Of course, when disrupting any course structure, I always believe it's also important to leave enough benchmarks or processes that are familiar so that students do not become completely disoriented by changes. If you do make radical changes and leave less in the course that is familiar, I think you need to account for that steep learning curve in the grading/ evaluation system. And so I set about to create a hybrid (part distance and part face-to-face class) that incorporated some familiar elements and some newer, lesser-used elements. My familiar elements were course readings, online discussion boards (on Blackboard), and weekly group leaders for reading discussions. My newer elements included the use of Google

Hangouts (multi-user video/audio chat rooms along with the recording and archiving of these Hangouts), non-traditional writing assignments/major projects, and the use of guest participants/speakers nearly each week of the semester. I will elaborate on each of these elements here:

- Course Readings: I tried to disrupt expected course readings by complicating understandings of the terms professional writing and communication. I also tried to blur the lines between personal/professional, public/civic, and private/individual. This meant bringing in readings in academic discourse that covered rhetorical studies from indigenous and decolonial rhetorics (Haas 2012; Riley Mukavetz 2014; and Powell 2002) for example, to queer rhetorics (Alexander and Banks 2004; Cox 2012; Dadas 2011; etc.), technical and professional writing studies (Faber 2002; Frost 2013; Grabill 2000; Haas 2012; Miller 1979; etc.), disability studies (Brueggeman 2001; Dolmage 2007; Palmeri 2006), and critical theory (Mignolo 2011; Warner 2002). In addition, I attempted to disrupt genre expectations by assigning not only critical theory and traditionally framed academic discourse (journal articles, books, dissertation segments, etc.) but also including pieces like Thomas King's (2003) *The Truth about Stories: A Native Narrative*, which frames conversations about narrative theories and rhetorical approaches from non-Euro/Western understandings and from an Indigenous American perspective. Students immediately found these unexpected pieces thought-provoking and useful as lenses for the rest of the semester's work.

- Online Discussion Boards: Many courses with a distance/online education component utilize discussion boards. Over the years, many graduate students have told me that they actually prefer to have these available because they are familiar with them as a space to process ideas and concepts from the class over the semester. I did try to disrupt this form a bit, however, by telling students early on that the discussion boards would mainly be a space for them to interact with one another and not for them to interact with me (which I encouraged them to instead do in emails to me and in Google Hangout time—as I will discuss below). I told students that I would check in on the discussion boards to ensure that everyone was participating and in order to read and get a sense of what students were discussing but that I would be less likely to get involved and intervene than other instructors they may have had. Of course, I was willing to step in if I felt that students overall were going off on any unproductive tangents or became too disrespectful with one another. I did allow and sometimes even encouraged a certain level of debate and ranting, though, as productive given the nature of our content that semester.

- Weekly Group Leaders for Reading Discussions: While the concept of designating individual students or teams of students to proctor/lead class discussion across various weeks is in no way a new idea in graduate courses, I tried to disrupt the status quo here as well. Because most weeks of the semester involved an invited guest speaker/participant (see below), I put the students assigned for the particular week's discussion in touch with the speaker/participant who would be visiting class. Different student teams took advantage of this at varying levels of involvement. For example, some made minimal contact with the speaker—only emailing them once to introduce themselves and ask a brief question on the speaker's chosen readings, and other teams were much more involved, contacting the speaker multiple times and inviting the speaker's continuing feedback even after the speaker's official class visit (although I was sure to have a conversation with students about effective rhetorical engagement and being mindful of the speaker's busy schedule and limited time to interact with us as a class).

- Google Hangouts: The use of Google Hangouts (and occasionally Skype for one-on-one office hours with students) for class meetings was met with great approval and seeming pedagogical success. Early in the semester some distance students expressed anxiety and worry that they would not receive the same kind of instruction and guidance that campus students in the course did. But this was quickly assuaged through the use of video conferencing. Distance students immediately commented on the fact that seeing and interacting with the teacher via video (where they could see facial expressions and hear verbal intonation) was so preferable to text-only interaction. They also enjoyed being able to see and interact with each other across the distance.

- Non-Traditional Writing Assignments/Major Projects: One of the most influential practices in my own doctoral education, especially mentoring under Dr. Malea Powell and Dr. Trixie Smith at Michigan State University, had been the use of unorthodox genres and forms in doctoral writing. Both of these mentors encouraged the use of creative nonfiction and collage/patchwork essays that blended both personal narrative and academic explication and discourse. In this course, because we were negotiating ideas of personal identity and professional identity and writing/communication, I had students begin the semester by writing a personal identity "walkabout" in which they wrote about their own sense of identities and stories from their lives. In the mid-term of the semester, students did an annotated bibliography of the readings we had done to that point in the semester and were also encouraged to include a few works (both academic and non-academic) that were related to the discussions we'd had over the semester as well. And then, as a final project, students were asked to put these first two projects into conversation with each other in a collage/patchwork essay that incorporated personal narrative, creative writing, and academic/theoretical pieces from the semester. Students commented frequently that they had never before in a class had an opportunity to reconcile their own stories and personal identities with expectations and ideas of academic discourse and identity.

- Guest Participants/Speakers: Perhaps the most popular and useful (and important) aspect of the semester, according to students in the course, were my efforts to bring guest participants into the course around every-other week. Over the sixteen-week semester, we had eight guest participants. These participants were:
    - Dr. Andrea Riley-Mukavetz (indigenous rhetorics scholar at Bowling Green State University)
    - Dr. Caroline Dadas (digital and queer rhetoric scholar at Montclair State University)
    - Dr. Michael Faris (digital and queer rhetorics scholar at Texas Tech University)
    - Dr. Trixie Smith (queer rhetorics and Women and Gender Studies scholar at Michigan State University)
    - Dr. Michelle Eble (scholar of technical communication, gender studies, and mentoring at East Carolina University)
    - Dr. Erin Clark Frost (feminist rhetorics and technical communication scholar at East Carolina University)
    - Dr. Tracy Morse (scholar in (dis)ability Studies and rhetorics at East Carolina University)
    - Beth Keller, MA (PhD Candidate at Michigan State University studying women and mentorship in the technical communication workplace)

These participants were mostly off campus/out-of-town and were brought in via Google Hangouts to speak with the class. Also, most participants were faculty at various universities around the country. One participant was an advanced doctoral candidate at another university. These participants were invited because of their own research and knowledge around issues of race, class, (dis)ability, gender, sexuality, and difference in our field. These participants were also encouraged to "cross boundaries"—that is, even though one week's speaker may have had the most knowledge in the area of (dis)ability studies, they were still encouraged to lead the discussion across lines of gender, class, and race as well. Students were also encouraged all semester to complicate notions of "this week is a week to talk about gender and professional writing." Students were very excited to be introduced to names in our field that they had often only read and never made a personal connection to. So, in addition to the generative discussion in our class, students were also able to make a new networked connection in the field. Many students even emailed and kept in touch with these guest faculty (and student) well after the class had ended.

EXAMPLE TWO: STUDENT REACTIONS AND SHARING

Previously in this collection, Jones and Walton describe the ways narrative can be an engagement of diversity and social justice. To this end, I want to share here a few of these graduate students' end-of-semester narratives/reflections on the course's content and operation. At the time the course was being taught, I was just beginning to envision writing about the experiences of planning and teaching this course. I asked several of my students if they would be interested in sharing pieces of their own experiences during our class's semester together. The narratives I share here were shared with me first by them and with their permission. I share these to demonstrate how implementing a queer rhetorics and intersectional course design and content into practice was helpful for students (in their own words).

One African American female student talked about the course's continual emphasis on discussions of race in technical and professional communication. She wrote:

> The discussion of race, both historical and contemporary, is a conversation that is sometimes to be avoided at all costs. I've often been told myself by well-intentioned people that I should just get over that Black thing or that slavery thing, and deal with the world that they think exists now "post-racially" in this country and around the world. That really is such a joke, "post-racial," you've only to turn on the nightly news or read a newspaper to know that we live far from a post-racial country. In fact in many ways we have gone backward in time when it comes to the "isms," you know racism, sexism, ageism, etc., things like voter rights, which were fought and won (or so we thought) ... One of the things that I have enjoyed the semester in both my TPC classes is the discussion of how social cultural studies should be incorporated into our field of study. I've often thought that it should be incorporated into many other academic fields of study, often in the context that is given by the evolution of academia itself, which often lacks diverse lenses.

For this student, open and honest discussions of difference were not only critical to dealing with questions of identity as they related to technical and professional communication (TPC) but also something she wished were present in her courses in other fields and disciplines.

A white female student wrote of her own experiences growing up and the way that our course's utilization of feminist theory and pedagogy was a revelation for her. She wrote that:

> Erin Frost's "Theorizing an Apparent Feminism in Technical Communication," initiated an epiphany within my previous consideration of medical rhetoric(s) and the way feminism interacts therein. I'd read *Our Bodies, Ourselves* as a teenager; I remember thinking, even then, this particular

agency given to women, through this work was so radical, so different from anything I'd ever read in the past. Because the notion of women not having a right to a voice was for me (mostly instigated by my Dad), I thought the book was somehow very risqué, and somehow its ideas had to be too radical for consumption by "ordinary" people. Were women really supposed to consider their bodies in such a manner as the text made plain? Surely, we weren't supposed to assign "agency" to our private parts, among so many other aspects of ourselves, were we? The idea of women having power—power over their thoughts and actions, their sexuality, was one that was not exactly encouraged in the household I grew up in.

For this student, the idea that women might find a voice and take control of her own academic voice and identity (even personal voice and identity) was not something she expected in a technical and professional communication course.

Finally, a white male student writes about his own personal and professional journey as highly intersectional (especially where sexuality was concerned), and he references Malea Powell's "Listening to Ghosts" (which we read that semester):

> I could not have seen that our (class's) talk about freedom, unity, and equality was what I needed when I signed up for it last spring. We have all been able to tell our own ghost stories in an academic setting or framework, yet we have accomplished this without following the way Powell describes modern academia: dissecting, examining, and categorizing every minute detail in order to "figure out" our lives. The soul-searching sojourn of the last year has only been enhanced by the coursework, which was indisputably moving and cathartic for me, even though some of the subject matter most definitely haunts me still.

To some in the academy, a technical and professional communication course that emphasizes both personal and professional identity might seem "off topic" (when only professional discourse is expected to be covered). And yet, time and time again, these three students and others mentioned the ways that being able to write about and discuss their personal journeys and intellectual musings helped them bridge those intersectional spaces that led back to their professional sense of self.

PRESENT AND SHIFTING IMPLICATIONS: WHAT CAN BE DONE NOW AND WHERE OUR FIELD MUST MOVE

I've laid out here one example across one semester, of how implementing intersectional methods and content (what Edwards, in the previous chapter, might call a "critically conscious pedagogy") can help turn theory into practice that changes students' fundamental learning experiences. Are there other ways and examples of how to do this intersectional work? Yes, many. But, I would venture that we must first (as Halberstam says) let go of some of the most strongly embedded ideas of classroom "successes" and fears of classroom "failures."

Many of my students expressed a fear from the first day of class that writing about and discussing openly conversations of personal and identity issues would bring on arguments, clashes, emotional outbursts, and cognitive dissonance. Students were so used to certain types of sanctioned classroom discourses that they feared change. We, as instructors, I think also fear this. We worry that students must always feel successful, that the flow of the classroom and semester themselves must also feel successful, and that students and instructors must always come away feeling happy and content with the course. This seemed even more important as this particular class had fully one-third of students who self-identified as LGBTQ (often the spaces of discussion actually allowed some of these students to come out both to the entire class and to me privately).

In addition to feminist and critical race theory focused and postcolonial lenses, our focus on queer rhetorics became another important foundation, especially in disrupting otherwise often fairly stagnant/conservative online course structures within most college level programs (that often rely overly on Blackboard and/or common course management systems). In technical and professional communication, we have too long decided to take the easy way out by accepting and not questioning the wider societal and cultural trope of "professional" that is both whitewashed and simplified (for instance, the acceptance and portrayal of a standard office work environment as full of mostly straight white males in suits and ties—perhaps with women or people of color sprinkled in here-and-there for representativeness only). When, and if, we have allowed conversations about "diversity," we have fit these into neat packages that don't complicate one another. What would our discipline's conversations and our classroom conversations look like if we not only disrupted traditional and expected ideas of professional but also pushed beyond "bare minimum" ideas of representation and integration of the skills, leadership, and abilities of non-normative workers? For example, instead of asserting that workplaces are too white and too straight and too male and that we should include more people of color, LGBT persons, and women, what would it look like to say "how could we benefit from a majority-minority workplace?" or "how would an LGBT manager or CEO benefit a corporate workplace?" These are conversations that can seem messy or uncomfortable, and so, I believe, we often avoid them past a certain "safe zone" of conversation.

And sure enough, as our class progressed through the semester, it was interesting to see how issues in queer theory and discussions of how sexuality might relate to the study of technical communication made students confused and uncomfortable. Queer theory and, by association, queer rhetorics seeks to revel in and disrupt by its very sexual and non-dominant nature. In this sense, a specifically queer rhetorics approach within cultural rhetorics is perhaps one of the most radical and unaddressed methodological approaches to technical communication studies. Queer rhetorics does not seek to "play nicely" within the constraints of what is professional or successful or efficient. In this way, its dissonance offers technical communication some of the most useful, and risky, frameworks and tools for thinking about and working within an increasingly global and disparate set of workplaces and situations. Until we take such risks in our classrooms and put our theoretical approaches into practice, we will not see the change we continue to state must happen.

References

Ahmed, Sara. 2010. *The Promise of Happiness*. Durham, NC: Duke University Press. https://doi.org/10.1215/9780822392781.

Alexander, Jonathan, and William P. Banks. 2004. "Sexualities, Technologies, and the Teaching of Writing: A Critical Overview." *Computers and Composition* 21 (3): 273–293. https://doi.org/10.1016/j.compcom.2004.05.005.

Allen, Brenda J. 1995. "'Diversity'; and Organizational Communication." *Journal of Applied Communication Research* 23 (2): 143–155. https://doi.org/10.1080/00909889509365420.

Anzaldúa, Gloria. 1987. *Borderlands/La Frontera: The New Mestiza*. San Francisco: Aunt Lute Books.

Anzaldúa, Gloria. 1998. "Too Queer the Writer." In *Living Chicana Theory*, ed. Carla Trujillo, 263–276. Berkeley, CA: Third Woman Press.

Berlant, Lauren, and Michael Warner. n.d. "What Does Queer Theory Teach Us about X."

Blyler, Nancy, and Charlotte Thralls. 1993. *Professional Communication: The Social Perspective*. Newbury Park: Sage.

Brueggeman, Brenda Jo. 2001. "An Enabling Pedagogy: Meditations on Writing and Disability." *Journal of Advanced Composition* 21 (4): 791–820.

Butler, J. 2004. *Undoing Gender*. New York: Psychology Press.

Cox, Matthew B. 2012. "Through Working Closets: Examining Rhetorical and Narrative Approaches to Building LGBTQ and Professional Identity Inside a Corporate Workplace." Diss., Michigan State University. Ann Arbor: UMI.

Dadas, Caroline. 2011. "Writing Civic Spaces: A Theory of Civic Rhetorics in a Digital Age." Diss., Miami University. Ann Arbor: UMI.

Dolmage, Jay. 2007. "Mapping Composition: Inviting Disability in the Front Door." In *Disability and the Teaching of Writing: A Critical Sourcebook*, ed. Brenda Jo Brueggeman and Cynthia Lewiecki-Wilson, 14–27. New York: Bedford/St. Martin's.

Faber, Brenton D. 2002. *Community Action and Organizational Change: Image, Narrative, Identity*. Carbondale: Southern Illinois University Press.

Foucault, M. 2012. *The History of Sexuality, vol. 2: The Use of Pleasure*. New York: Vintage.

Frost, Erin. 2013. "Theorizing an Apparent Feminism in Technical Communication." Diss., Illinois State University. Ann Arbor: UMI.

Grabill, Jeffrey T. 2000. "Shaping Local HIV/AIDS Services Policy through Activist Research: The Problem of Client Involvement." *Technical Communication Quarterly* 9 (1): 29–50. https://doi.org/10.1080/10572250009364684.

Haas, Angela. 2012. "Race, Rhetoric, and Technology: A Case Study of Decolonial Technical Communication Theory, Methodology, and Pedagogy." *Journal of Business and Technical Communication* 26 (3): 277–310. https://doi.org/10.1177/1050651912439539.

Halberstam, Judith. 2011. *The Queer Art of Failure*. Durham, NC: Duke University Press. https://doi.org/10.1215/9780822394358.

Hall, S. 1980. "Cultural Studies: Two Paradigms." *Media, Culture and Society* 2 (1): 57–72.

Halperin, D. M. 2012. *One Hundred Years of Homosexuality: And Other Essays on Greek Love*. London: Routledge.

Herndl, Carl G. 1993. "Teaching Discourse and Reproducing Culture: A Critique of Research and Pedagogy in Professional and Non-academic Writing." *College Composition and Communication* 44 (3): 349–363. https://doi.org/10.2307/358988.

Katz, S. B. 1992. "The Ethic of Expediency: Classical Rhetoric, Technology, and the Holocaust." *College English* 54 (3): 255–275.

King, Thomas. 2003. *The Truth about Stories: A Native Narrative*. Minneapolis: University of Minnesota Press.

Mignolo, Walter. 2011. *The Darker Side of Western Modernity: Global Futures, Decolonial Options*. Durham, NC: Duke University Press. https://doi.org/10.1215/9780822394501.

Miller, Carolyn. 1979. "A Humanistic Rationale for Technical Writing." *College English* 40 (6): 610–617. https://doi.org/10.2307/375964.

Mukavetz, A. M. R. 2014. "Towards a Cultural Rhetorics Methodology: Making Research Matter with Multi-Generational Women from the Little Traverse Bay Band." *Rhetoric, Professional Communication, and Globalization* 5 (1): 108–125.

Palmeri, Jason. 2006. "Disability Studies, Cultural Analysis, and the Critical Practice of Technical Communication Pedagogy." *Technical Communication Quarterly* 15 (1): 49–65. https://doi.org/10.1207/s15427625tcq1501_5.

Porter, James E., Patricia Sullivan, Stuart Blythe, Jeffrey T. Grabill, and Libby Miles. 2000. "Institutional Critique: A Rhetorical Methodology for Change." *College Composition and Communication* 51 (4): 610–642. https://doi.org/10.2307/358914.

Powell, Malea. 2002. "Listening to Ghosts: An Alternative (non)Argument." In *ALT DIS: Alternative Discourses and the Academy*, ed. C. L. Schroeder, H. Fox, and P. Bizzell, 11–22. Portsmouth, NH: Heinemann.

Powell, Malea, Stacey Pigg, Kendall Leon, and Angela Haas. 2009. "Rhetoric." In *Encyclopedia of Library and Information Sciences*, 3rd ed., 4548–56. New York: Taylor and Francis. Published online December 9, 2009. https://doi.org/10.1081/E-ELIS3-120043867.

Scott, J. Blake, Bernadette Longo, and Katherine V. Wills, eds. 2006. *Critical Power Tools: Technical Communication and Cultural Studies*. Albany, NY: SUNY Press.

Sedgwick, Eve Kosofsky. 1990. *Epistemology of the Closet*. Berkeley: University of California Press.

Spivak, G. C. 1988. *In Other Worlds: Essays in Cultural Politics*. London: Routledge.

Warner, Michael. 2002. *Publics and Counterpublics*. New York: Zone Books.

Afterword
FROM ACCOMMODATION TO TRANSFORMATION

J. Blake Scott

Edited by two of our field's most generous and creative scholars, teachers, and mentors, *Key Theoretical Frameworks for Teaching Technical Communication in the Twenty-First Century* does more than extend technical/professional communication's cultural turn; it re-orients and re-energizes this turn by mobilizing us "toward a collective disciplinary redressing of social injustice," especially through our curricula and pedagogy (Haas and Eble, this collection). In the introduction to our *Critical Power Tools* collection, Bernadette Longo, Katherine V. Wills, and I overview several moves central to a critical-cultural orientation, including accounting for technical communication's sociohistorical and cultural relations, reflexively critiquing its functions in power-laden struggles over knowledge and agency, and laying the groundwork for intervening in disempowering practices (Scott, Longo, and Wills 2006, 16). Powered by a range of intersectional social justice frameworks, this collection shifts these moves into hyper-drive. As part of its critique of technical communication's hyper-pragmatism, *Critical Power Tools* argued for replacing the goal of accommodation with that of transformation. Developed by scholar-teacher-activists whose passion for social justice is matched by their carefully enacted theoretical and pedagogical frameworks, this collection can ignite the field's social justice transformation. As the editors suggest, transforming a field requires a collective commitment but also a diverse range of approaches and frameworks, and this collection provides both. Taken together, its chapters raise and respond to exigencies and questions about social justice that span the *stases*, including questions about how to understand the causes, functions, and effects of technical communication practices, how to define social justice, how to evaluate technical communication practices and pedagogies, and how to, as the editors so deftly state, "intervene in global and local technical communication problems at the macro and micro levels

DOI: 10.7330/9781607327585.c013

in the face of asymmetrical power relations and limited agency—and teach current and future practitioners to do the same" (Haas and Eble, this collection).

This collection is *kairotic* for several reasons. First, our home institutions, corporate and civic partners, and students' employers have increasingly recognized civic and professional literacies as complementary, creating renewed exigency for their pedagogical cross-infusion but also, given the danger of hegemonic co-option, the need to define these literacies and their goals. Second, the continued growth of undergraduate and graduate programs that focus on or incorporate technical and professional communication affords us expanded opportunities to engage future technical communicators with the knowledge of our field. While some programs are putting technical and professional communication in conversation with other areas of study (Giberson and Moriarty 2010), our field is also considering new ways to align our curricular goals, such as through a collective outcomes statement (Bridgeford and Ilyasova 2014). Third, thanks to teacher-mentors like Angela Haas and Michelle Eble, a new generation of scholars is (cross) trained and positioned to lead us in advancing social justice frameworks in more intersectional directions. Finally, we continue to witness vivid reminders of racial, gendered, sexed, environmental, and other intersecting and systemic forms of injustice and structural violence—from police killings of unarmed black men to the deportation of neighbors in our communities to unregulated environmental violence—that implicate technical communication and the institutions it serves, and that demand justice-oriented responses.

So what can, and what must, a social justice framework for teaching and enacting technical communication entail? Haas and Eble (this volume) clarify that such a framework must "explicitly seek to redistribute and reassemble—or otherwise redress—power imbalances that systematically and systemically disenfranchise some stakeholders while privileging others" (4). Most directly indebted to recent technical communication scholarship by Haas (2012), Godwin Agboka (2013), Miriam Williams and Octavio Pimentel, and others, the contributors to this collection offer an array of reinforcing answers to the question above, answers that are illuminated by their self-reflexive discussions of their civic and scholarly engagement, teaching practices, and student responses (Williams and Pimentel 2014). Because much of my own research and teaching focuses on health and medical rhetoric, my understanding of a social justice framework has been informed by the work of Paul Farmer—the physician, medical anthropologist, and health activist who founded the global NGO Partners in Health. Drawing on liberal theology and contrasting social justice to charity and developmental approaches to global health, Farmer (2004) explains that social justice seeks to redistribute resources and transform the structural conditions that enable inequities in the first place. On its webpage about a World Day of Social Justice, Partners in Health (2013) invokes Farmer's work to posit that social justice "requires immediate, pragmatic action paired with a larger critical analysis of, and fight against, structural violence." The Partners in Health webpage explains that social justice should unwaveringly prioritize the needs of "the poor in health care," and here we could extrapolate to consider students, users, publics, and other technical communication stakeholders who are least powerful and most in need. Beyond being guided by and giving preference to those who are most in need, social justice efforts must be "carried out in pragmatic solidarity" with them, the website qualifies. For Farmer, this means that those experiencing or facing injustice must participate in the planning, decision-making, implementation, and assessment of social justice efforts (155–156). In this collection, Natasha Jones and Rebecca Walton echo most of these characteristics in calling for social justice research and pedagogy that "investigates how communication, broadly defined, can amplify the agency of oppressed people—those who are materially, socially, politically, and/or economically under-resourced. Key to this definition," they emphasize, "is a collaborative, respectful approach that moves past description and exploration of social justice issues to taking action to redress inequities" (242). In another chapter in this collection, Donnie Johnson Sackey similarly highlights the agential imperative of participative justice, stating that those needing redress must have the "right to participate in decision-making within and receive the benefits of the economic institutions that will directly and indirectly affect the quality of their lives" (141). As these examples make clear, the social justice approaches in this collection, even more so than other critical-cultural approaches to technical communication, call for a radical revision of David Dobrin's (1983) now classic definition of "writing that accommodates technology to the user" (227). Rather than asking how we can accommodate users, the approaches in this collection call on us to ask how we can responsively redress those users who are marginalized by technical communication through the redistribution of resources (technological, rhetorical, and otherwise). This movement from responsibility to responsiveness entails teaching more than critical attunement, as Jones and Walton point out—it requires teaching methodologies of preferential action and empowerment.

So we come to the question of how? How can we teach toward social justice and the responsiveness it requires? This collection's chapters develop and illustrate a range of ways to answer this question with critical definitions, evaluative analyses, and action-oriented pedagogies. Because the collection's introduction so thoroughly overviews the contributions of these chapters by topic, in what follows I call out other types of contributions extended by this volume and the larger social justice turn.

First, the chapters in this collection insist on *linking, through concrete examples of practice and pedagogy, the epistemological and ideological dimensions* of technical communication, dimensions that have often been treated as separate in the field's scholarship. Extending Slack, Miller, and Doak's (1993) understanding of technical communication as articulation, these chapters make explicit the power relations that sponsor and are served by technical communication practices, arguing that it is not enough to value what we do and teach as producing meaning or even improving knowledge-making practices; we must also work to create more just practices. In his chapter, for example, Marcos Del Hierro explains how hiphop's knowledge-making *techne* of digital booklets and improvisational ciphers can create for marginalized technical communication students more validating spaces of intellectual engagement and rhetorical experimentation. Jones and Walton turn to narrative as embodying and teaching moral decision-making, helping students "envision and ethically evaluate courses of action" and identify "moral examples" for their own professional development (249). In her chapter on how technical communication teachers and students can adapt tenets of Collins's black feminist epistemology, Kristen Moore discusses these tenets (e.g., "lived experience as a criterion of meaning") as ways of understanding the world, substantiating knowledge, and valuing the "experiences and wisdom" of subjugated citizens (192).

Second, the chapters explain and illustrate how social justice frameworks can help us *understand and respond to interdependent contexts of power on multiple scales at once*. As Haas and Eble explain, technical communication practices are embedded in and shaped by globalizing and localizing forces across a number of contextual levels and dimensions, including those of global markets, cultural histories, community dynamics, institutional and organizational structures, and embodied experiences. In this collection's frameworks, the contexts of technical communication practices are more than backdrops, and they entail immediate and local dynamics but also broader histories and ecologies of exigencies, operations, and effects. For example, Sackey's teaching of environmental social justice entails a multilayered mapping of intersectional subjectivity that "engages local and global concerns by continuously asking how issues of gender, race, class, sexuality, and ability shape the way environments look and relate to other spaces and even how spaces often are designed in ways to produce specific subject positions" (153). Some of the frameworks in the collection draw on cultural geography to help teachers and students to look beyond typical classroom and vocational contexts in an examination of, as Elise Verzosa Hurley puts it, broader histories and ecologies of technical communication "in relation to practices of domination, subjugation, resource depletion, and human rights—all of which have been brought to bear on cultures, peoples, and bodies through space and spatialization" (106; see also Agboka, this volume). Other frameworks—such as Cruz Medina and Kenny Walker's approach to grading contracts and Barbi Smyser-Fauble's critique of accommodation statements—focus on the ways our teaching practices are implicated in broader institutional, educational, and cultural relations of power, offering more familiar and concrete models for redistributing pedagogical and institutional agency.

Third, the teaching frameworks developed in this collection *understand and approach social injustice and justice as intersectional*. As Farmer (2004) and some of the collection's contributors argue, social justice approaches require the "simultaneous consideration of various social 'axes' of oppression" in order to discern economies of power and avoid forms of cultural essentialism (43). In foregrounding a recognition of intersectional oppression developed by Crenshaw and others (see Moore), the collection's chapters leverage combinations of insights from different, intersecting bodies of thought and practice—including critical race, feminist/womanist, disability, queer, environmental, decolonial/postcolonial, and cultural geography traditions. For example, Marie Moeller uses a feminist disability theoretical framework to help students explore ways to "de-stabilize harmful, multi-layered, normative medical narratives about bodies and health" (212). As Jessica Edwards and Matthew Cox explain, the critical race theory and queer theory on which they respectively draw are at their core intersectional. After explaining how critical race studies came out of critical legal studies and radical feminism (273), Edwards discusses how this movement's understanding of structural racism as embodied in language can be used to help students expand the ways they recognize linguistic markers of racism in the workplace contexts they will face. Cox explains his approach as accounting "for the intersectionality of queer and LGBT persons both internally (with their own race, class, sex, dis/ability, gender, and faith tradition markers) as well as externally (with their own and other cultures, groups, and peoples) (288). In its critique and action, Erin Frost's framework of "apparent feminism" is also intersectional, in part by inviting "participation from allies who do not identify as feminist but do complementary work" (27). These and other contributors provide us with frameworks and tools for investigating and ethically transforming the multi-faceted identities and embodied experiences teachers, students, users, and other subjects disciplined by technical communication's intersecting power dynamics.

Fourth, the contributors' discussions provide *phronesis-driven strategies for implementing and adapting social justice frameworks* across specific technical communication contexts, and *practice-level examples* of how they have done this themselves as teachers. Moore (this volume) asserts that, "More cases, theoretical discussions, field reports, and research projects are needed to effectively equip practitioners and scholars in technical communication with the expanded methodologies necessary to respond to the diverse contexts in which they work, learn, and research" (207). This collection offers a resource guide of such tools, methods, and practices for transforming technical communication through practical wisdom. Each chapter provides a theoretical-ideological framework, pedagogical heuristic, and examples of specific assignments and students' responses to them, from sample grading contracts (Medina and Walker) to critical analysis strategies (Moeller) to collaborative knowledge-building spaces (Del Hierro) to heuristics for analyzing and producing narratives (Jones and Walton). Each chapter additionally offers a self-reflexive, behind-the-scenes look at how the author(s) has navigated the challenges of teaching toward social justice in a field of practice and study that continues to be dominated by corporate values and hyper-pragmatism. Technical communication teachers and students undertrained in social justice approaches (myself included) will benefit from the nuanced yet concrete discussions of "how injustice is not just a problem in technical communication but also one that we can solve *with* technical communication" (Haas and Eble, this volume, 8).

Along with providing a toolbox of intersectional teaching approaches for understanding and responding to unjust technical communication practices, the chapters in this collection raise a number of accompanying challenges and ethical questions with which the field must continue to grapple. The contributors' self-reflexive discussions call attention to the need to examine the ways our well-intended teaching practices can still be implicated in forms of social injustice. Teaching about and for social justice requires us to be vulnerable, self-critical, and open to a collaborative shaping of our work and its specific agendas. As several of the teaching stories told in this collection attest, this vulnerability and openness carries risk and takes substantial time and energy. Given the importance of career preparation to students and universities, what do we risk by foregrounding social justice concerns, and what do we risk and potentially gain by connecting these concerns to user experience, accessibility, information design, and other widely taught and highly valued technical communication *techne*? How can we meet our obligations to prepare students for success in the workplace while teaching them to critically analyze and begin to redress some of the ways technical communication works within larger power assemblages to enact forms of injustice? This latter question begs a number of other related questions around how to adequately *prepare* our students and ourselves for social justice work. How do we decide which technical communication practices to critique, which forms of injustice on which to focus, and which methodologies on which to draw? To what extent and how should students learn about the histories, socio-cultural contexts, and power dynamics of specific forms of technical communication-enabled injustice? Some of the contributors to this collection have addressed such questions by having students begin with the ways institutional power disciplines them and their fellow students through familiar teaching and learning practices. Others contributors have engaged their students around community-based social justice work. To prepare for such work and help students better understand and critique forms of injustice, how can we ensure that they respectively and responsively engage others who have lived these forms? As Moore (this volume) asks, how can we "expose and value (rather than subjugate or suppress) local knowledge" through "humbling ourselves to community members' wisdom, experience, and dialogues" (198)? Of course, some of the social justice methodologies described in this collection offer ways to negotiate these challenges, but we might further ask: how much should students study and practice such methodologies before enacting them in higher-stakes contexts? and how can we respectfully adapt specific social justice methodologies to account for forms of power (involving technical communication) that they were not originally designed to counter, and what might get lost in doing so?

The contributors to this collective bravely negotiate these questions and others, articulating pedagogical commitments to social justice and reflexively assessing these commitments according to what we might call their "consequential validity," to adapt an assessment concept discussed by Cruz Medina and Kenny Walker (this volume). In this vein, the consequences and, more specifically, the preferential benefits of social justice approaches will determine their worth. From my perspective, this collection is more than worthy of the field's attention, as it promises to edify us all.

References

Agboka, Godwin Y. 2013. "Participatory Localization: A Social Justice Approach to Navigating Unenfranchised/Disenfranchised Cultural Sites." *Technical Communication Quarterly* 22 (1): 28–49. https://doi.org/10.1080/10572252.2013.730966.

Bridgeford, Tracy, and K. Alex Ilyasova. 2014. "Establishing an Outcomes Statement for Technical Communication." In *Sharing our Intellectual Traces: Narrative Reflections from Administrators of Professional, Technical, and Scientific Communication Programs*, ed. Tracy Bridgeford, Karla Saari Kitalong, and Bill Williamson, 53–80. Amityville, NY: Baywood Publishing.

Dobrin, David N. 1983. "What's Technical about Technical Writing?" In *New Essays in Technical and Scientific Communication: Research, Theory, Practice*, ed. Paul V. Anderson, John R. Brockman, and Carolyn R. Miller, 227–50. Amityville, NY: Baywood Publishing.

Farmer, Paul. 2004. *Pathologies of Power: Health, Human Rights, and the New War on the Poor*. Berkeley, CA: University of California Press.

Giberson, Greg A., and Thomas A. Moriarty, eds. 2010. *What We Are Becoming: Developments in Undergraduate Writing Majors*. Logan: Utah State University Press. https://doi.org/10.2307/j.ctt4cgppw.

Haas, Angela M. 2012. "Race, Rhetoric, and Technology: A Case Study of Decolonial Technical Communication Theory, Methodology, and Pedagogy." *Journal of Business and Technical Communication* 26 (3): 277–310. https://doi.org/10.1177/1050651912439539.

Partners in Health. 2013. "World Day of Social Justice." February 20, 2013. Accessed December 10, 2014. http://www.pih.org/article/world-day-of-social-justice-what-it-means-to-pih-and-how-you-can-help.

Scott, J. Blake, Bernadette Longo, and Katherine V. Wills, eds. 2006. *Critical Power Tools: Technical Communication and Cultural Studies*. Albany: State University of New York.

Slack, Jennifer Daryl, David James Miller, and Jeffrey Doak. 1993. "The Technical Communicator as Author: Meaning, Power, Authority." *Journal of Business and Technical Communication* 7 (1): 12–36. https://doi.org/10.1177/1050651993007001002.

Williams, Miriam, and Octavio Pimentel, eds. 2014. *Communicating Race, Ethnicity, and Identity in Technical Communication*. Amityville, NY: Baywood Publishing.

ABOUT THE AUTHORS

GODWIN Y. AGBOKA is an associate professor of professional and technical communication at University of Houston-Downtown, where he has taught courses in his area of expertise, including intercultural communication, technical communication, and medical and science writing. His research interests include intercultural technical communication, social justice, and human rights perspectives in technical communication, decolonial methodologies, and the rhetoric of science and medicine. He's the current graduate director of the MS in Technical Communication at the University of Houston-Downtown. Agboka's publications have appeared in forums such as *Technical Communication Quarterly*, *Journal of Technical Writing and Communication*, and *Connexions • International Professional Communication Journal*. He's a co-guest editor of the special issue on "Professional Communication, Social Justice, and the Global South" (*Connexions • International Professional Communication Journal*).

MATTHEW COX is an associate professor of English, specializing in technical and professional communication, at East Carolina University. He holds a doctoral degree in rhetoric and writing studies with an emphasis on cultural rhetorics from Michigan State University. Cox also has over eleven years of industry experience in technical and professional communication with information technology and software companies as well as publishing in information technology. His research areas of interest are LGBT and queer rhetorics, specifically as they relate to workplace identity and communication. Cox's scholarship has also been published in the *Journal of Technical Writing* and *Communication and Present Tense: A Journal of Rhetoric and Society* and is also forthcoming in the *Journal of Business and Technical Communication* and in the edited collection *Re/Orienting Writing Studies: Queer Methods/Queer Projects*.

MARCOS DEL HIERRO is an assistant professor of English at the University of New Hampshire. His research and teaching interests include technical communication, hiphop studies, social justice, American Indian rhetorics, and critical race studies. He is particularly interested in how black, Latinx, and indigenous cultural traditions influence the development and use of hiphop rhetorics and technologies. His essay, "Fighting the Academy One Nopal at a Time," appeared in *El Mundo Zurdo: Selected Works from the Meetings of the Society for the Study of Gloria Anzaldúa* in 2012.

MICHELLE F. EBLE is an associate professor at East Carolina University. She specializes in rhetoric, writing, and technical communication, and her research and teaching interests include technical writing theory and practice, especially as it relates to social justice, scientific, health, and medical contexts, gender studies, and technology, as well as feminist mentoring networks. She has published in *Computers and Composition*, *Technical Communication*, and *Technical Communication Quarterly*. She is the co-editor of *Stories on Mentoring: Theory and Praxis* (Parlor Press, 2008), co-author of *Primary Research & Writing: People, Places, Spaces* (Routledge, 2016), and co-editor of *Reclaiming Accountability: Using the Work of Re/Accreditation and Large-Scale Assessment to Improve Writing Instruction and Writing Programs* (Utah State UP, 2016). Currently, she serves as Chair of East Carolina University's Behavioral and Social Sciences Institutional Review Board and Past President of the Association of Teachers of Technical Writing (ATTW).

ABOUT THE AUTHORS

JESSICA EDWARDS is an assistant professor of English at the University of Delaware in Newark, DE. Dr. Edwards has developed and taught courses in technical writing, critical race studies, diversity studies, composition studies, and African American literature. Her scholarship considers ways to engage critical race theory, the intersections of race, racism, and power, in writing classrooms. Dr. Edwards was a Faculty Scholar with the Center for Teaching, Learning, and Assessment at UD, and her scholarship has appeared in *Computers and Composition Online* as well as the edited collection *Citizenship and Advocacy in Technical Communication: Scholarly and Pedagogical Perspectives*. She also served as a consultant editor for the first-year composition reader, *Black in America: A Broadview Topics Reader*, published in 2018 by Broadview Press.

ERIN A. FROST is an associate professor at East Carolina University who specializes in technical communication, rhetoric, and composition. She has an employment history as an investigative journalist, and she uses that experience to inform her work as a teacher and researcher. Recently, Frost has taught courses focused on scientific writing, public interest writing, risk communication, and digital writing. Her scholarly interests center on issues of gender and feminism in technical communication, most often as they manifest in healthcare policy and risk communication. She was awarded the 2015 Conference on College Composition and Communication Outstanding Dissertation Award in Technical Communication, and her work has appeared in journals including *Computers and Composition*, *Technical Communication Quarterly*, *Programmatic Perspectives*, and *Peitho*.

ANGELA M. HAAS is an associate professor of rhetoric, technical communication, ethnic, and women's and gender studies and serves as the Graduate Program director for the Department of English at Illinois State University. Her research and teaching interests include technical communication, American Indian rhetorics, cultural rhetorics, decolonial theory and methodology, digital and visual rhetorics, indigenous feminisms, and technofeminisms—and has been recognized with the Conference on College Composition and Communication (CCCC) Technology Innovator Award, the Computers & Composition (C&C) Ellen Nold Best Article Award, and the C&C Hugh Burns Best Dissertation Award. Haas worked as a technical communicator in the automotive industry for over ten years and currently serves on the CCCC Executive Committee, as president of the Association of Teachers of Technical Writing (ATTW), and as Coordinator of the Computers & Writing Graduate Research Network. Her scholarship has been published in *Computers & Composition*, *Computers & Composition Online*, *Journal of Business and Technical Communication*, *Pedagogy*, *Studies in American Indian Literatures*, as well as in other journals and edited collections.

ELISE VERZOSA HURLEY is an assistant professor of rhetoric, composition, and technical communication at Illinois State University, and editor of *Rhetoric Review*. Her research and teaching interests include visual and spatial rhetorics, professional and technical communication theory and pedagogy, multimodal composition, feminist rhetorics, public rhetorics, and civic and community engagement. Her scholarship has been published in *Technical Communication Quarterly*, *Kairos: A Journal of Rhetoric, Technology, and Pedagogy*, *Composition Forum*, *Res Rhetorica*, *Rhetoric Review*, and other edited collections.

NATASHA N. JONES is an associate professor at the University of Central Florida. She is a graduate of the University of Washington's Human Centered Design & Engineering Department (2012). Her research focuses on activism, social justice, and narrative in technical communication, and technical communication pedagogy. Her work has been published in *Technical Communication Quarterly*, the *Journal of Technical Writing and Communication*, and *Rhetoric, Professional Communication, and Globalization*. Natasha serves as the current chair of the Council on Programs in Technical and Scientific Communication's (CPTSC) Diversity Committee and is the co-creator (with Gerald Savage) of CPTSC's Diversity and Social Justice Network Listserv.

About the Authors

CRUZ MEDINA is an assistant professor of rhetoric and composition at Santa Clara University where he is a Bannan Institute Scholar on Racial and Ethnic Justice and teaches courses on writing, social justice, multicultural rhetoric, and digital writing. His book, *Reclaiming Poch@ Pop: Examining the Rhetoric of Cultural Deficiency* (Palgrave 2015), looks at pop culture responding to anti-Latinx legislation in Arizona. He is co-editing a collection on misrepresentations of race in online media, and his current research centers on issues of language, race, and citizenship with Central American immigrants in the Bay Area of northern California. His writing has appeared in the collections *Communicating Race and Ethnicity in Technical Communication* and *Decolonizing Rhetoric and Composition Studies*, as well as the journals *College Composition and Communication*, *Reflections: A Journal of Public Rhetoric, Civic Writing and Service Learning*, and *Present Tense: a Journal of Rhetoric in Society*.

MARIE E. MOELLER is a professor of rhetoric and professional and technical communication at the University of Wisconsin–La Crosse. Currently, she is serving as the Associate Dean of the College of Liberal Studies and Chair of Global Cultures and Languages. Moeller's research and teaching interests include disability rhetorics, gender and fat studies, critical advocacy studies, bioethics, and technical communication theories and pedagogies. Moeller's scholarship has been published in venues such as *Communication Design Quarterly*, *College English*, *Disability Studies Quarterly*, and *Technical Communication Quarterly*.

KRISTEN R. MOORE is an associate professor of technical communication and rhetoric at University at Buffalo–SUNY. Her research interests include institutional rhetoric and change, technical communication, public participation and engagement, cultural rhetorics, and critical methodologies. As a founding member of Women in Technical Communication, Moore also researches feminist mentoring and its potential for advocacy in organizations. Moore teaches graduate and undergraduate courses on technical communication, public rhetoric, contemporary rhetorics and critical field methods, and she is an affiliate of the Cross Cultural Academic Advancement Center and the concentration in Literature, Social Justice, and the Environment. Her scholarship has been published in the *Journal of Technical Writing and Communication*, *Technical Communication Quarterly*, *Journal of Business and Technical Communication*, *IEEE Transactions on Professional Communication*, and a variety of edited collections.

DONNIE JOHNSON SACKEY is an assistant professor of rhetoric and writing at the University of Texas at Austin. His research centers on the dynamics of environmental public policy deliberation, environmental justice, and environmental cultural history. Sackey is currently a co-principle investigator on two grant-funded projects from the National Institutes of Health and the Institute for Population Studies, Health Assessment, Administration, Services and Economics. One project assesses communication issues surrounding lead and legionella contamination in municipal water resources in Flint, Michigan; the other is a community-based approach to study adolescent asthma through personalized air-monitoring. In the past he has been funded by the Center for Urban Responses to Environmental Stressors to investigate the ecological impact of airborne petroleum coke on vulnerable populations in Detroit and Dearborn, MI. His work has appeared in *Computers and Composition* and *Present Tense* and several edited collections.

GERALD SAVAGE is professor emeritus of rhetoric and technical communication, retired from Illinois State University. His research focuses on diversity and social justice issues in international and intercultural contexts of technical communication teaching and practice. He has written numerous articles and book chapters, and co-edited four books and three special issues of professional journals. In addition to teaching, he worked for many years in technical, management, editorial, and writing roles in business, industry, education, and government. ATTW, CPTSC, STC, and Illinois State University have recognized his contributions to technical and professional communication with awards for teaching, research, and service.

ABOUT THE AUTHORS

J. BLAKE SCOTT is professor of writing and rhetoric at the University of Central Florida, where he was founding associate chair and director of Degree Programs for the department. With Lisa Meloncon, he is a founding co-editor of the new journal *Rhetoric of Health & Medicine* (*RHM*). In addition to teaching a range of professional communication and rhetoric courses, he is involved in a number of campus-wide initiatives around undergraduate learning and faculty development. He is at work on a book that rhetorically analyzes transnational risk conflicts around pharmaceutical access and regulation.

BARBI SMYSER-FAUBLE is an instructional assistant professor of technical communication at Illinois State University. Outside of her academic training, she also has eight years of professional experience in the healthcare and pharmaceutical industries. Her research and teaching interests include disability studies, feminist theories, medical rhetorics, cultural rhetorics, digital and visual rhetorics, multimodal composition, new media, and technical communication histories and theories. Her scholarship has been published in *Rhetoric, Professional Communication and Globalization*, *Rhetoric Review*, and is forthcoming in two edited collections.

KENNETH WALKER is an assistant professor of English at the University of Texas, San Antonio, where he specializes in environmental rhetoric, rhetoric of science, technology, and medicine, and technical writing. His research is concerned with ecological approaches to rhetorics of risk and uncertainty. His work has been published in *Technical Communication Quarterly*, *Journal of Business and Technical Communication*, *Rhetoric Review*, *Environmental Humanities*, and *POROI*, as well as in other journals and edited collections.

REBECCA WALTON is an associate professor of technical communication and rhetoric at Utah State University. Her research interests include social justice, human rights, and qualitative methods for cross-cultural research. Primarily a field researcher, she has collaborated with organizations such as the Red Cross, Mercy Corps, and World Vision to conduct research in countries including Uganda, Kyrgyzstan, and Bolivia. Her work has appeared in *Technical Communication Quarterly*, *Journal of Business and Technical Communication*, and other journals and edited collections. Three of her co-authored articles have won national awards: the Nell Ann Pickett Award in 2017 and 2016, as well as the STC Distinguished Article Award in 2017.

INDEX

abuse, 80, 115–118, 121, 124–125, 127, 129, 131, 268
accessibility, 14, 37, 68–69, 72–74, 78, 80, 82, 84–90, 310
accommodation, 13–14, 68–73, 75–87, 227, 233, 304, 308; reasonable, 69–70, 77
accountability, 82, 192, 196, 198, 200–201, 204, 206
advocacy, 4, 6, 15–17, 49, 189, 193, 212–224, 228, 260
African and African-American rhetorics, 123, 125, 167, 172, 180, 290
Africana and African feminism(s)/ist(s), 186, 187, 189, 191, 195, 295
Agboka, Godwin, 9, 12, 14, 48, 49, 107, 114–137, 139, 188, 242, 305, 308, 313
agency, 10–14, 17, 42, 151, 153, 167, 171, 242, 250, 257, 271, 300, 304–306, 308; distributed, 4, 8; limited, 5, 86; student, 47, 53, 55–56, 105
altruism, 15, 212–239
Americans with Disabilities Act (ADA), 13, 68–70, 77, 79, 81, 82, 84, 87
antenarrative, 256–257
Anzaldúa, Gloria, 101, 174–175, 202
apparent feminism(s)/ist(s), 13, 17, 23–45, 48, 87, 299, 309
Association of Teachers of Technical Writing (ATTW), 6, 164, 168, 313, 314

Banks, Adam, 55, 165, 167, 180, 269, 272, 296
barriers, 74, 80, 86, 138, 183, 231, 280; physical, 72, 84; social, 74–76, 85
Black feminism(s)/ist(s), 15, 185–208, 307
Blackmon, Samantha, 191
Blyler, Nancy, 117, 138, 156
bodies, 12–14, 17, 23, 25–30, 35–36, 38–39, 41–42, 49, 69, 98, 145, 152–153, 165–166, 169, 177–179, 182–183, 191, 212–216, 224–225, 227–229, 232–235, 274, 288, 300, 308
body, 24–25, 75, 148, 151–152, 215–216, 223, 227–228, 232–233, 235
Bowdon, Melody, 9, 29, 33, 41, 70, 71, 86, 87, 116, 242
breast cancer, 223–224, 227, 229–232, 234

Christian, Barabara, 189
Cole, Johnetta, 188
collective, 305; ethics, 227; identity, 224; work as social justice, x, 5, 47, 304
colonialism, 108, 121, 132, 165
Combahee River Collective, 188
consequential validity, 47–50, 52–53, 56, 58
contact zones, 96, 179
contract grading, 52. *See also* grading contracts
course design, 69, 95, 100, 101, 103, 105–107, 299. *See also* curriculum
Cox, Matthew, xv, 16, 28, 287–302, 308, 313
Crenshaw, Kimberlé, 186–188, 308
critical race theory (CRT), 16, 255, 268, 270–272, 275, 278, 292, 301, 308, 314
critical thinking, 17, 132, 242, 271, 274, 282, 283
cultural geography, 14, 95, 97–101, 105–108, 116, 146, 156, 308
cultural rhetorics, 9, 16, 288, 290, 292, 293, 302, 313–316
cultural studies, ix, x, 5–9, 12, 24–26, 33, 49, 95, 99, 241, 269, 289–290, 299, 302
curriculum, 6–9, 32, 34, 46–47, 49, 51, 58, 64, 81, 106, 117, 131, 177. *See also* course design

decolonial, 42, 121, 132, 163–164, 166, 167, 198, 207, 308; methodologies/theory, 99, 149, 164, 313–314; rhetorics, 296
Del Hierro, Marcos, xiii, 15, 40, 47, 63, 163–184, 198, 307, 309, 313
digital rhetoric(s), 102, 105, 166–167
disability studies, 13, 16, 68, 69, 72–76, 84–86, 88, 131, 198, 212–216, 226, 228, 292, 295–296, 315, 316
disenfranchised, 4, 86, 88, 108; disenfranchise, 270, 305; disenfranchisement, 131, 269. *See also* unenfranchised
Doak, Jeffrey, 307
Dolmage, Jay, 74, 75, 84, 86, 101, 226, 227, 296

Eble, Michelle F., x, 3–22, 25, 31–32, 42, 46–47, 185, 198, 298, 304–305, 307, 309, 313
ecofeminism(s)/ist(s), 14, 145, 148–149
Edwards, Jessica, 16, 41, 255, 268–286, 300, 308, 314
embodiment, 12, 13, 24, 25, 28–29, 31, 35–36, 38–42, 68, 75, 94, 105
energy, 114–115, 142
environment (ecosystems), xi, 5, 8–9, 11–14, 24, 26, 30, 39, 64, 97, 115–116, 118, 120, 122–125, 132, 138, 138–157, 205, 260, 305, 307–308, 315, 316
ethic of care/caring, 11, 195–196, 199, 201–203, 206
experiential knowledge(s), 36, 74–76, 88, 147, 194, 206

failure, 78, 118, 129, 141, 175, 177, 291, 292, 301
Farber, Jerry, 63, 67, 52–53
feminism(s). *See* African and/or Africana feminism(s)/ist(s); apparent feminism(s)/ist(s); Black feminism(s)/ist(s); ecofeminism(s)/ist(s); feminist disability studies (FDS); feminist materialism; feminist political ecology; feminist rhetorics
feminist disability studies (FDS), 13, 41, 69, 75–76, 85–86, 88, 198, 212–216, 228
feminist materialism, 145
feminist political ecology, 14, 145–147
feminist rhetorics, 298, 314
Freire, Paulo, 46, 65, 177, 253, 256, 262
Frost, Erin A., 9, 12–13, 17, 23–45, 48, 87, 296, 298, 299, 309, 314

Garland-Thomson, Rosemarie, 75, 76, 215, 228
gender, 9, 12, 28, 31, 36–37, 40, 56, 68, 70, 75–76, 97, 99, 117, 121, 131, 138–139, 145, 147–148, 152, 154, 156, 164, 168, 186–188, 214–216, 226, 231, 251, 263, 281, 287–289, 291–294, 298, 305, 308, 313, 314–315
globalization, 3–4, 7, 9, 11, 12, 14, 94, 107, 116, 139, 206; anti-globalization, 144
Grabill, Jeffrey, 9, 24, 29, 35, 41, 132, 138–139, 151, 192–193, 198, 219, 241, 254, 258, 288, 290, 296
grading contracts, 13, 46–49, 51–53, 55–56, 58, 60, 63–65, 87, 308–309. *See also* contract grading
graduate course, 8, 31, 288, 295–296

Haas, Angela M., x, 3–19, 23–25, 42, 46–47, 49–50, 56, 58, 99, 117, 132, 138, 165–167, 185, 188, 207, 241–242, 251, 253, 257, 269, 288, 290–291, 296, 304–305, 307, 309, 314
Heuristic, 14, 16, 85, 97, 139, 145, 151, 153, 157, 168, 243, 249–250, 259, 260, 271, 309
Hill Collins, Patricia, 15, 187–193, 195–196, 198, 202, 206–208
hiphop, 15, 163–184, 307, 313
historicity, 243, 251–253, 256, 260–261
hooks, bell, 101, 178, 186
human rights, 5, 11–12, 14, 107–108, 114–126, 129–133, 242, 245, 260, 308, 313, 316; International Bill of Human Rights (IBHR), 115, 120, 132

identification, 75, 243–246, 254–255, 260–261; self-identification, 37, 40, 42, 83, 208
ideology, ix, x, 130, 139, 154, 166, 235, 253, 275
indigenous, 14, 26, 29, 60, 124, 129, 143, 166, 294, 296, 313–314; indigenous rhetorics, 290, 292, 296, 298
injustice, x, xi, 4–5, 8, 10–11, 17, 46–47, 70, 78, 108, 110, 121, 130, 132, 142, 145, 252, 255, 259, 304–306, 308, 310
intersectionality, 16, 41, 151, 187, 288, 290, 293–294, 299–300, 304–305, 308–309

Jones, Natasha, 16, 25, 99–100, 108, 188, 241–267, 271, 299, 306–307, 309, 314

knowledge(s), 12, 15, 40, 49, 51, 60, 63, 95, 99–100, 139, 171–172, 175–182, 199–201, 206, 217, 252–255, 275; embodied, 13, 36, 75; institutional/structural, 98, 290, 304; legitimization, 98, 192, 260–261; through lived experience, 36, 74–76, 88, 147, 194; making of, 4, 5, 143–144, 190, 193, 197, 203–204, 228, 233, 307, 309; situated/contextual, 150–151, 157, 198

latino/latina/latinx, 53, 167, 196, 217, 269, 270–271, 290, 313, 315
legal writing, 3, 70–71, 272, 274. *See also* regulatory writing
liberation, 121, 131, 177
Longo, Bernadette, 5, 23, 49, 50, 93, 95, 98, 241, 253, 288–290, 304
Lorde, Audre, 186, 202

medical/medicine, 9, 15–17, 36, 38, 41, 70, 72, 78, 79, 84, 115, 125, 212–216, 220, 222–228, 231–236, 270, 299, 305, 308, 313, 316
Medina, Cruz, 13, 35, 46–67, 87, 204, 270, 308–310, 315
Moeller, Marie, 15, 41, 198, 212–238, 308–309, 315
Moore, Kristen R., 15, 24, 41, 154, 185–211, 215, 242–243, 251, 255, 307–310, 315

narrative, 16, 29, 35, 52, 56, 94, 131–132, 165, 179, 189, 191, 194, 212–214, 216, 220–221, 226, 234–236, 241–267, 271, 273–274, 284, 295–297, 299, 307, 308–309, 311, 314
non-governmental organization (NGO), 120, 123, 132
non-profit organization (NPO), 3, 15, 60, 64, 108, 196, 207, 213–214, 216–224, 232
normalcy, 214–216, 220, 223, 226–228

objective/objectivity, 5, 25, 27–28, 31, 39–40, 48, 95, 98, 103, 155, 164, 252–253, 275
Ogoni people, 14, 116–117, 123–129, 133
Ogunyemi, Clenora, 186

Palmeri, Jason, 9, 72, 117, 241, 296
participatory design/action, 193, 258
perception, 16, 25, 38–40, 47, 55, 83, 88, 138, 243, 251, 270–271
Pimentel, Octavio, 9, 23, 49–50, 99, 185–186, 188, 197, 269–270, 305
political ecology, 14, 145–147
politics, ix, x, 23, 27, 95, 96, 105, 139, 147, 152, 154, 164, 214–215, 223, 288
postcolonial, 9, 99, 108, 121, 132, 149, 308
privilege, 5, 11, 17, 35, 46–55, 58–60, 63–64, 74–75, 142, 148, 168–169, 175, 177, 179, 187–188, 191, 215, 247, 250–251, 260–261
public intellectual, 42

queer rhetorics/queer theory, 16, 23, 63, 108, 174, 242, 287–303, 308, 313

reflexivity/reflexive, 5, 76, 191, 243, 246–248, 250–251, 259–261, 304–305, 309–310
regulatory writing, 70, 270. *See also* legal writing
resistance, 46, 99, 171; student resistance, 34–35, 40, 52, 53, 55–57

Richardson, Flourice, xiii, 188, 270
risk/risk communication, ix, 9, 12–13, 23–45, 71, 138, 146, 150, 172, 177, 190–191, 217, 226, 228–232, 292, 302, 310, 314, 316
Rude, Carolyn, 4, 96, 216–217, 219, 242

Sackey, Donnie Johnson, 12, 14, 26, 108, 138–160, 205, 306–307, 315
Savage, Gerald, ix–xi, xiii, 8–9, 12, 25, 99–100, 108, 116–118, 122, 130, 138, 185, 188, 242, 270, 314, 315
science, 5, 7, 24, 76, 143–144, 146–147, 150, 156, 182, 270, 272, 313, 316
Scott, J. Blake, xiii, 5, 9, 23, 26, 29, 41, 49–50, 58, 95, 98, 109, 193, 198, 241, 253, 258, 288–290, 304–311, 16
service learning, 58, 193, 217–218, 242
sexuality, 68, 70, 75–76, 108, 117, 121, 148, 154, 186–187, 215–216, 226, 231, 251, 287, 289, 292, 293–294, 298, 300, 302, 308
Shiva, Vandana, 143, 148–149, 152
Simmons, Michele, 9, 24, 29, 35, 41, 138–139, 151, 192–193, 221–222, 241, 254, 258
Slack, Jennifer Daryl, 133, 252, 307
Smith, Barbara, 186, 188
Smyser-Fauble, Barbi, 9, 13, 41, 68–90, 308, 316
space and place, 12, 14, 95, 96–97, 108, 150–152
spatial turn, 14, 94–97
struggle, 14, 46, 98, 104, 119, 121, 130, 166, 176, 178, 180, 194, 252, 273, 287, 304; cultural, 198, 205; of students, 35, 174
Sullivan, Patricia, 96, 102, 116–117, 132, 187, 191–192, 199, 200, 207, 247–250, 258, 290
syllabus, 28, 35, 68, 71, 79, 81, 85, 87, 295

Taylor, Ula, 188
Truth, Sojourner, 186, 263
Tuhiwai Smith, Linda, 198, 202

unenfranchised, 108, 115, 130. *See also* disenfranchised
universal design, 73–74, 81, 84
usability, 59, 61, 68–69, 72–74, 76, 82, 84–85, 87, 96, 221, 259

Verchick, Robert, 144, 150
Verzosa Hurley, Elise, xiv, 11, 14, 33, 93–113, 155, 308, 314
Visvanathan, Shiv, 144

Walker, Kenneth, 13, 35, 46–67, 87, 308–310, 316
Walters, Shannon, 9, 72–73
Walton, Rebecca, 16, 25, 117, 241–267, 271, 299, 306–307, 309, 316
Williams, Miriam, 6, 9, 23, 50, 70, 99, 104, 185, 188, 193, 197, 201, 269–270, 305

Winner, Langdon, 154
womanism, 41, 186, 191, 207, 208
workplace, 5, 8–9, 13–16, 24, 61–62, 98, 116, 169, 175, 195, 207, 217, 220–221, 250, 254, 287, 288–290, 298, 301–302, 308, 310, 313; race and, 268–286; writing, 50, 218, 219